The Linear Algebra Survival Guide: Illustrated with *Mathematica*

The Linear Algebra Survival Guide

Illustrated with *Mathematica*

Fred E. Szabo, PhD
Concordia University
Montreal, Canada

AMSTERDAM • BOSTON • HEIDELBERG • LONDON • NEW YORK • OXFORD
PARIS • SAN DIEGO • SAN FRANCISCO • SINGAPORE • SYDNEY • TOKYO

Academic Press is an Imprint of Elsevier

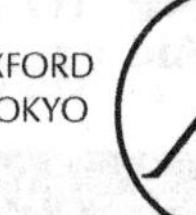

Academic Press is an imprint of Elsevier
125, London Wall, EC2Y 5AS.
525 B Street, Suite 1800, San Diego, CA 92101-4495, USA
225 Wyman Street, Waltham, MA 02451, USA
The Boulevard, Langford Lane, Kidlington, Oxford OX5 1GB, UK

ISBN: 978-0-12-409520-5

British Library Cataloguing in Publication Data
A catalogue record for this book is available from the British Library

Library of Congress Cataloging-in-Publication Data
A catalog record for this book is available from the Library of Congress

For information on all Academic Press
visit our website at **http://store.elsevier.com/**

Printed and bound in the USA

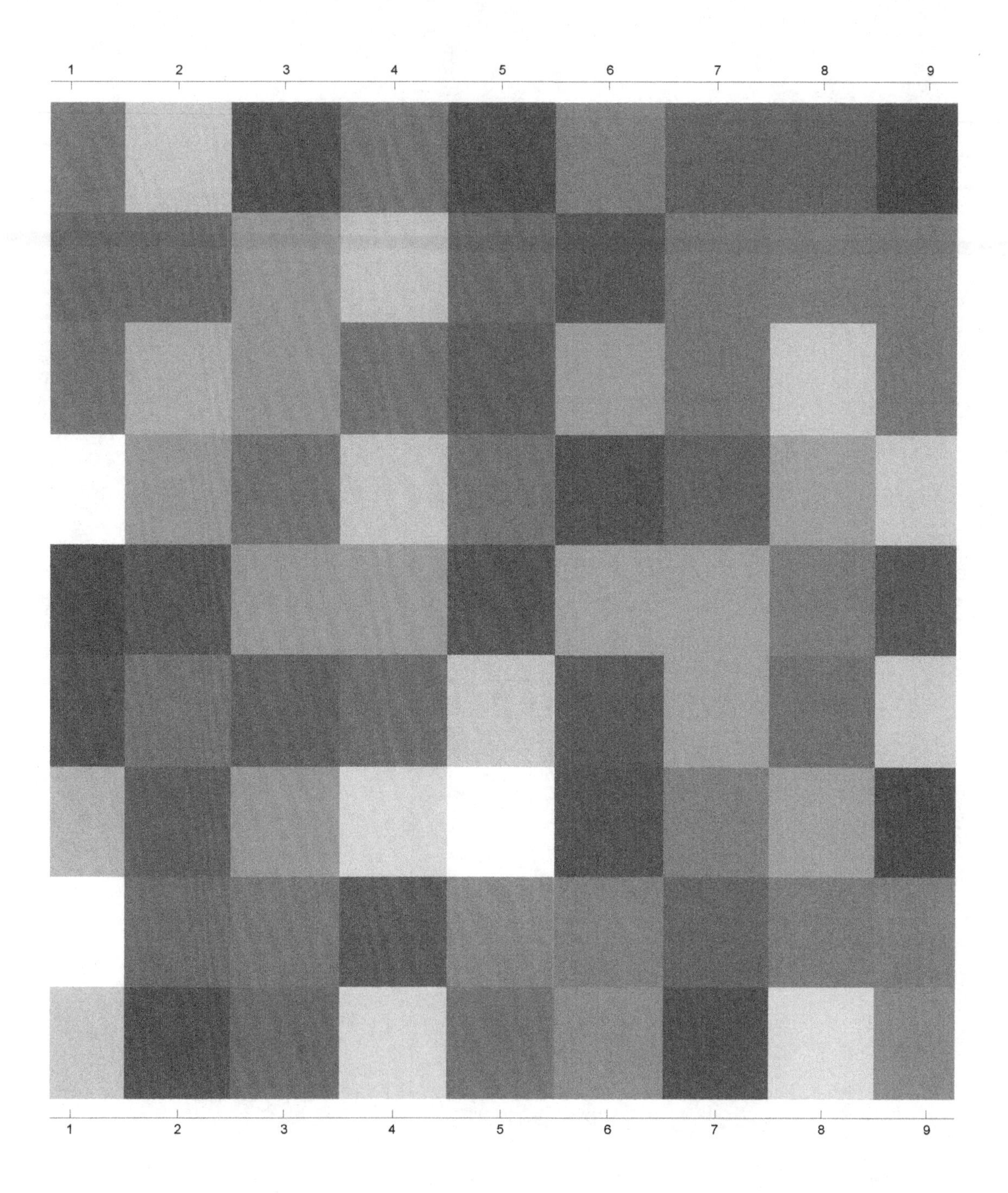
1
2
3
4
5
6
7
8
9
1
2
3
4
5
6
7
8
9

About the Matrix Plot

The image on the previous page is a *Mathematica* matrix plot of a random 9-by-9 matrix with integer elements between -9 and 9.

Random matrices are used throughout the book where matrix forms are required to illustrate concepts, properties, or calculations, but where the numerical content of the illustrations is largely irrelevant.

The presented image shows how matrix forms can be visualized as two-dimensional blocks of color or shades of gray.

```
MatrixForm[A = RandomInteger[{-9, 9}, {9, 9}]]
```

$$\begin{pmatrix} -4 & -9 & 1 & 6 & 8 & 5 & 6 & 7 & -8 \\ 1 & -5 & -5 & 2 & -8 & 8 & 8 & 1 & -8 \\ -7 & -4 & 2 & -1 & 9 & 3 & 1 & 7 & 6 \\ -3 & -1 & 6 & 4 & -9 & 1 & -2 & 0 & 9 \\ 7 & -8 & -4 & -1 & -6 & -8 & 5 & 5 & 1 \\ 7 & -5 & -3 & -3 & -1 & -2 & -9 & 8 & -1 \\ -1 & 6 & 5 & 6 & -5 & 4 & 7 & -9 & 5 \\ -7 & 8 & -7 & 5 & 4 & -1 & 1 & -5 & 4 \\ 8 & -2 & 8 & 7 & -8 & -9 & 4 & -3 & -7 \end{pmatrix}$$

```
MatrixPlot[A]
```

Table of Contents

Preface

The principal goal in the preparation of this guide has been to make the book useful for students, teachers, and researchers using linear algebra in their work, as well as to make the book sufficiently complete to be a valuable reference source for anyone needing to understand the computational aspects of linear algebra or intending to use *Mathematica* to extend their knowledge and understanding of special topics in mathematics.

This book is both a survey of basic concepts and constructions in linear algebra and an introduction to the use of *Mathematica* to represent them and calculate with them. Some familiarity with *Mathematica* is therefore assumed. The topics covered stretch from adjacency matrices to augmented matrices, back substitution to bilinear functionals, Cartesian products of vector spaces to cross products, defective matrices to dual spaces, eigenspaces to exponential forms of complex numbers, finite-dimensional vector spaces to the fundamental theorem of algebra, Gaussian elimination to Gram–Schmidt orthogonalization, Hankel matrices to Householder matrices, identity matrices to isomorphisms of vector spaces, Jacobian determinants to Jordan matrices, kernels of linear transformations to Kronecker products, the law of cosines to *LU* decompositions, Manhattan distances to minimal polynomials, vector and matrix norms to the nullity of matrices, orthogonal complements to overdetermined linear systems, Pauli spin matrices to the Pythagorean theorem, *QR* decompositions to quintic polynomials, random matrices to row vectors, scalars to symmetric matrices, Toeplitz matrices to triangular matrices, underdetermined linear systems to upper-triangular matrices, Vandermonde matrices to volumes of parallelepipeds, well-conditioned matrices to Wronskians, and zero matrices to zero vectors.

All illustrations in the book can be replicated and used to discover the beauty and power of *Mathematica* as a platform for a new kind of learning and understanding. The consistency and predictability of the Wolfram Language on which *Mathematica* is built are making it much easier to concentrate on the mathematics rather than on the computer code and programming features required to produce correct, understandable, and often inspiring mathematical results. In addition, the included *manipulations* of many of the mathematical examples in the book make it easy and instructive to explore mathematical concepts and results from a computational point of view.

The book is based on my lecture notes, written over a number of years for several undergraduate and postgraduate courses taught with various iterations of *Mathematica*. I hereby thank the hundreds of students who have patiently sat through interactive *Mathematica*-based lectures and have enjoyed the speculative explorations of a large variety of mathematical topics which only the teaching and learning with *Mathematica* makes possible. The guide also updates the material in the successful textbook “Linear Algebra: An Introduction Using *Mathematica*,” published by Harcourt/Academic Press over a decade ago.

The idea for the format of this book arose in discussion with Patricia Osborn, my editor at Elsevier/Academic Press at the time. It is based on an analysis of what kind of guide could be written that meets two objectives: to produce a comprehensive reference source for the conceptual side of linear algebra and, at the same time, to provide the reader with the computational illustrations required to learn, teach, and use linear algebra with the help of *Mathematica*. I am grateful to the staff at Elsevier/Academic Press, especially Katey Birtcher, Sarah Watson and Cathleen Sether for seeing this project through to its successful conclusion and providing tangible support for the preparation of the final version of the book. Last but not least I would like to thank Mohanapriyan Rajendran (Project Manager S&T, Elsevier, Chennai) for his delightful and constructive collaboration during the technical stages of the final composition and production.

Many students and colleagues have helped shape the book. Special thanks are due to Carol Beddard and David Pearce, two of my teaching and research assistants. Both have helped me focus on user needs rather than excursions into interesting but esoteric topics. Thank you Carol and David. Working with you was fun and rewarding.

I am especially thankful to Stephen Wolfram for his belief in the accessibility of the computable universe provided that we have the right tools. The evolution and power of the Wolfram Language and *Mathematica* have shown that they are the tools that make it all possible.

Fred E Szabo
Beaconsfield, Quebec
Fall 2014

Dedication

To my family: Isabel, Julie and Stuart, Jahna and Scott, and Jessica, Matthew, Olivia, and Sophie

About the Author

Fred E. Szabo
Department of Mathematics, Concordia University, Montreal, Quebec, Canada

Fred E. Szabo completed his undergraduate studies at Oxford University under the guidance of Sir Michael Dummett, and received a Ph.D. in mathematics from McGill University under the supervision of Joachim Lambek. After postdoctoral studies at Oxford University and visiting professorships at several European universities, he returned to Concordia University as a faculty member and dean of graduate studies. For more than twenty years, he developed methods for the teaching of mathematics with technology. In 2012 he was honored at the annual Wolfram Technology Conference for his work on "A New Kind of Learning" with a Wolfram Innovator Award. He is currently professor and Provost Fellow at Concordia University.

Professor Szabo is the author of five Academic Press publications:
- The Linear Algebra Survival Guide, 1st Edition
- Actuaries' Survival Guide, 1st Edition
- Actuaries' Survival Guide, 2nd Edition
- Linear Algebra: Introduction Using Maple, 1st Edition
- Linear Algebra: Introduction Using Mathematica, 1st Edition

Introduction

How to use this book

This guide is meant as a standard reference to definitions, examples, and *Mathematica* techniques for linear algebra. Complementary material can be found in the Help sections of *Mathematica* and on the *Wolfram Alpha* website. The main purpose of the guide is therefore to collect, in one place, the fundamental concepts of finite-dimensional linear algebra and illustrate them with *Mathematica*.

The guide contains no proofs, and general definitions and examples are usually illustrated in two, three, and four dimensions, if there is no loss of generality. The organization of the material follows both a conceptual and an alphabetic path, whichever is most appropriate for the flow of ideas and the coherence of the presentation.

All linear algebra concepts covered in this book are explained and illustrated with *Mathematica* calculations, examples, and additional *manipulations*. The *Mathematica* code used is complete and can serve as a basis for further exploration and study. Examples of interactive illustrations of linear algebra concepts using the **Manipulate** command of *Mathematica* are included in various sections of the guide to show how the illustrations can be used to explore computational aspects of linear algebra.

Linear algebra

From a computational point of view, linear algebra is the study of algebraic linearity, the representation of linear transformations by matrices, the axiomatization of inner products using bilinear forms, the definition and use of determinants, and the exploration of linear systems, augmented matrices, matrix equations, eigenvalues and eigenvectors, vector and matrix norms, and other kinds of transformations, among them affine transformations and self-adjoint transformations on inner product spaces. In this approach, the building blocks of linear algebra are systems of linear equations, real and complex scalars, and vectors and matrices. Their basic relationships are linear combinations, linear dependence and independence, and orthogonality. *Mathematica* provides comprehensive tools for studying linear algebra from this point of view.

Mathematica

The building blocks of this book are scalars (real and complex numbers), vectors, linear equations, and matrices. Most of the time, the scalars used are integers, playing the notationally simpler role of real numbers. In some places, however, real numbers as decimal expansions are needed. Since real numbers may require infinite decimal expansions, both recurring and nonrecurring, *Mathematica* can represent them either symbolically, such as e and π, or as decimal approximations. By default, *Mathematica* works to 19 places to the right of the decimal point. If greater accuracy is required, default settings can be changed to accommodate specific computational needs. However, questions of computational accuracy play a minor role in this book.

In this guide, we follow the lead of *Mathematica* and avoid the use of ellipses (lists of dots such as "...") to make general statements. In practically all cases, the statements can be illustrated with examples in two, three, and four dimensions. We can therefore also avoid the use of sigmas (Σ) to express sums.

The book is written with and for *Mathematica* 10. However, most illustrations are backward compatible with earlier versions of *Mathematica* or have equivalent representations. In addition, the natural language interface and internal link to Wolfram/Alpha extends the range of topics accessible through this guide.

Mathematica cells

Mathematica documents are called *notebooks* and consist of a column of subdivisions called *cells*. The properties of notebooks and cells are governed by *stylesheets*. These can be modified globally in the *Mathematica* Preferences or cell-by-cell, as needed. The available cell types in a document are revealed by activating the toolbars in the *Window > Show Toolbar* menu. Unless *Mathematica* is used exclusively for input–output calculations, it is advisable to show the toolbar immediately after creating a notebook or to make *Show Toolbar* a default notebook setting.

Mathematica documentation

Mathematica Help is extensive and systematic. To look for help, the *Help > Documentation Center* command will produce access to the search field of the Documentation Center.

Quitting the *Mathematica* kernel

It sometimes happens that we would like to abort a calculation or other Mathematica activity. The command **Quit** aborts the current computation and annuls all computed and assigned values to variables and other objects. Selecting Evaluation > Quit Kernel > Local is equivalent to invoking the **Quit** command.

Clearing assigned and computed values

The commands **Clear** and **ClearAll** can be used to remove previously assigned values to specific variables and symbols without resetting other definitions and assignments to their default. For example, typing **Clear**[*x*,*y*] into an input cell will remove values previously assigned to the variables *x* and *y*.

Generalizing illustrations with *Manipulations*

Many entries of the guide contain interactive manipulation sections. The manipulations can be used to explore the effect of numerical input changes on outputs. In particular, they provide a setting for "what if?" type questions. The ranges of the **Manipulate** parameters are usually arbitrarily chosen. They are easily modified to explore specific numerical questions. The **Manipulate** feature of *Mathematica* is explained and documented in the Wolfram Documentation section of *Mathematica*.

Predictive interface

Starting with *Mathematica* 9, the writing of *Mathematica* commands has become amazingly simple. A *Suggestion Bar* usually appears in any new input cell that is alphabetically organized and tries to anticipate both the built-in and user-defined concepts and definitions the user is about to type. This feature is amazing. It not only facilitates the writing of *Mathematica* code but also avoids having to respect specific grammatical conventions and spellings. The Suggestion Bar contains other features that make working with *Mathematica* a joy and much less code-dependent.

Assumptions about prior knowledge

This guide focuses on the learning, teaching, and review of linear algebra. *Mathematica* is the principal tool for doing so. It is therefore assumed that the reader has a basic knowledge of *Mathematica*. However, the illustrations can be followed, modified, and extended by mimicking the given examples.

The *Wolfram Language*

All *Mathematica* commands begin with capital letters. It is therefore advisable to use lower case letters to name defined objects. When naming matrices, this recommendation is usually not followed in this guide in order to make the presentation conform to the usual notations of linear algebra.

Vectors and matrices are often presented in row form as lists and lists of lists, surrounded by curly brackets ({}). Two-dimensional displays of vectors and matrices can be built by using *Mathematica* palettes. *Command + Enter* and *Command + Comma* in OS X, or *Control + Enter* and *Control + Comma* in Windows add columns and rows to the palettes for larger vectors and matrices. For easier readability or conceptual visualization, two-dimensional outputs of matrices are always forced by embedding the specification of the matrices in the **MatrixForm** commands.

Matrix multiplication and matrix-vector products must always be linked by a period (.). The arguments of functions such as f[x] must be enclosed in square brackets ([]). The metavariables used to define functions must be followed by underscores (f[x_]:=).
The two sides of equations written in *Mathematica* must be separated by double equal signs (==). Single equal signs are used for definitions, the naming of objects, and similar purposes.

Mathematica commands can be written on several lines, written in the same cell and separated by pressing Enter. The command *Shift + Enter* evaluates a cell. *Mathematica* notebooks can be evaluated globally by selecting the *Evaluation > Evaluate Notebook* menu item. Conversely, all *Mathematica* outputs in a notebook can be removed at once by selecting the *Cell > Delete All Output* menu item.

In addition to the material available in the *Help* file, relevant resources can also be found on the Internet by simply typing a topic of interest followed by "with mathematica." However, some of the results found on the Internet no longer apply to recent versions of *Mathematica*.

Only input cells can be used for computations. Depending on the document style, input cells are the default cells of *Mathematica* notebooks and the new cell selection automatically begins a new input cell. All cell types available in a particular notebook are listed in the *Toolbar* associated with a particular notebook style.

Since lists of numbers correspond to row vectors and rows of matrices, the transpose is often required when working with columns. However, vectors cannot be transposed since *Mathematica* is designed to recognize from the context whether a row or column vector is required for a specific calculation. If column vectors are explicitly required, curly brackets must be used to separate the elements of the vectors.

The examples included in this guide are designed to make it easy to understand the *Mathematica* syntax required for linear algebra. *Mathematica* and the Wolfram Research website contain tutorials that facilitate the learning of *Mathematica* basics.

To replicate some of the material in this guide it may be necessary to click on *Enable Dynamics* if it appears at the top of the active *Mathematica* document.

Matrices

Most items in this guide involve matrices. In the Wolfram Language, a matrix is a list of lists. More specifically, a list of rows. The statement,

```
A = {{1, 2, 3}, {4, 5, 6}}
```

for example, displays the "matrix" `{{1,2,3},{4,5,6}}` and names it *A*. *Mathematica* can be forced to display the matrix *A* in the customary two-dimension form in several ways.

- Using **MatrixForm** to display a matrix in two-dimensional form

```
MatrixForm[A = {{1, 2, 3}, {4, 5, 6}}]
```

$$\begin{pmatrix} 1 & 2 & 3 \\ 4 & 5 & 6 \end{pmatrix}$$

However, internally, *Mathematica* still considers *A* to be a list of lists.

```
A
{{1, 2, 3}, {4, 5, 6}}
```

- Using **TraditionalForm** to display a matrix

```
TraditionalForm[A = {{1, 2, 3}, {4, 5, 6}}]
```

$$\begin{pmatrix} 1 & 2 & 3 \\ 4 & 5 & 6 \end{pmatrix}$$

- Using **//MatrixForm** as a suffix to display a matrix in two-dimensional form

```
A = {{1, 2, 3}, {4, 5, 6}} // MatrixForm
```

$$\begin{pmatrix} 1 & 2 & 3 \\ 4 & 5 & 6 \end{pmatrix}$$

```
A
```

$$\begin{pmatrix} 1 & 2 & 3 \\ 4 & 5 & 6 \end{pmatrix}$$

By adjoining **//MatrixForm** to the definition of the matrix *A*, we force *Mathematica* to produce a two-dimensional output. The price we pay is that the output is no longer a computable object.

```
Head[A]
MatrixForm
```

The **Head** command shows that instead of being a list (of lists) the output is no more than a **MatrixForm**.

- Forcing two-dimensional output in a notebook

If we evaluate the command

```
$Post := If[MatrixQ[#], MatrixForm[#], #] &
```
(1)

at the beginning of a notebook, all matrices will be displayed in two-dimensional form. However, this does not change their internal definition as lists of lists.

```
$Post := If[MatrixQ[#], MatrixForm[#], #] &
```

```
A = {{1, 2, 3}, {4, 5, 6}}
```

$$\begin{pmatrix} 1 & 2 & 3 \\ 4 & 5 & 6 \end{pmatrix}$$

```
A
```

$$\begin{pmatrix} 1 & 2 & 3 \\ 4 & 5 & 6 \end{pmatrix}$$

```
A + A
```

$$\begin{pmatrix} 2 & 4 & 6 \\ 8 & 10 & 12 \end{pmatrix}$$

The **Quit** command erases the **Post** command and all other ad hoc definitions, assignments, and computed values. The command **Clear**[*A*] can also be used to remove an assignment without clearing the **Post** command.

```
Quit[]
```

```
A = {{1, 2, 3}, {4, 5, 6}}
```

```
{{1, 2, 3}, {4, 5, 6}}
```

- Using the *Mathematica* Preferences to force **TraditionalOutput** in all notebooks.

The *TraditionalOutput* format of matrices and all other objects for which several output options exist can be reset globally in the *Mathematica* Preferences. The choice of

`Preferences > Evaluation > Format type of new output cells > TraditionalOutput`

forces all outputs to be in **TraditionalForm**.

In OS X, the Preference menu is found in the *Mathematica* drop-down menu, whereas in Windows, it is found in the *Edit* drop-down menu. For matrices this means that they are output in two-dimensional form. The **Quit** command will not change this option back to **StandardForm**.

- Capital letters and the names of matrices

In standard linear algebra books, matrices are named with capital letters. However, *Mathematica* uses capital letters for built-in symbols, commands, and functions. Hence naming matrices with capital letters may be unwise. In particular, the letters C, D, E, I, K, N, and Q are reserved letters and should not be used. For easier readability and since this guide is exclusively about linear algebra, this recommendation is not followed. In most illustrations, matrices are named with the capital letter *A*.

- Matrices as two-dimensional inputs

The built-in palettes can be used to construct matrices as two-dimensional inputs:

`Palettes > Basic Math Assistant > Typesetting >` $\begin{pmatrix} \square & \square \\ \square & \square \end{pmatrix}$

- Special symbols

Special symbols, such as the imaginary number i, can be entered in several ways. The quickest way is to type Esc ii Esc (the Escape key followed by two i's followed by the Escape key). Similar instructions on inputting special symbols occur in various places in the chapters below. The palettes in the *Palettes* menu can also be used to input special symbols.

Palettes > Basic Math Assistant > Typesetting > i

also inputs the symbol i. The required equivalent keyboard entry using Esc can be seen by letting the cursor rest on the symbol i in the menu. The same holds for all other displayed symbols.

Random matrices

Most illustrations and manipulations in this guide use relatively small random matrices with integer elements if the numerical content of the examples is irrelevant. This makes the examples more concrete and simplifies the notation. Two-dimensional notation for vectors and matrices is often used to conform to the standard ways of representation, although internally *Mathematica* uses the one-dimensional form.

- A two-dimensional random matrix with integer elements

```
MatrixForm[A = RandomInteger[{0, 9}, {3, 5}]]
```

$$\begin{pmatrix} 8 & 7 & 3 & 0 & 4 \\ 3 & 8 & 2 & 7 & 9 \\ 3 & 4 & 0 & 6 & 1 \end{pmatrix}$$

produces a random 3-by-5 matrix with integer elements between 0 and 9 in two-dimensional form. However, the name *A* stands for the same matrix as a one-dimensional list of rows. The conversion of the input cell containing the command

```
MatrixForm[A = RandomInteger[{0, 9}, {3, 5}]]
```

and the output cell containing the result of the computation

$$\begin{pmatrix} 8 & 7 & 3 & 0 & 4 \\ 3 & 8 & 2 & 7 & 9 \\ 3 & 4 & 0 & 6 & 1 \end{pmatrix}$$

to *DisplayFormula* cells ensures that the generated matrix is not changed by a second evaluation command (*Shift* + *Enter*). This change of cell types preserves the internal meaning *A* as a computable object.

- A one-dimensional output of the matrix *A*

```
A
```

```
{{8, 7, 3, 0, 4}, {3, 8, 2, 7, 9}, {3, 4, 0, 6, 1}}
```

- A two-dimensional random real matrix with non-integer elements

```
MatrixForm[A = RandomReal[{0, 9}, {2, 3}]]
```

$$\begin{pmatrix} 2.3487 & 6.53058 & 2.2666 \\ 6.60772 & 0.398364 & 5.2241 \end{pmatrix}$$

The resulting matrix output can be preserved by changing the *Output* cell in which the matrix is displayed to a *DisplayFormula* cell.

```
MatrixForm[A]
```

$$\begin{pmatrix} 2.3487 & 6.53058 & 2.2666 \\ 6.60772 & 0.398364 & 5.2241 \end{pmatrix}$$

The usual *Copy* and *Paste* commands display the matrix in the 19-place default format in which *Mathematica* works with the generated "real" numbers.

$$\begin{pmatrix} 2.348698292382359\text{`} & 6.53057615064426\text{`} & 2.266599355944713\text{`} \\ 6.607721548969266\text{`} & 0.3983641332231951\text{`} & 5.224099904694331\text{`} \end{pmatrix}$$

Illustrations

Topics in illustrations are identified by a descriptive title and introduced with a square bullet (■). In most cases, the illustrations are based on random matrices with integer coefficients if this entails no loss of generality. To preserve the specific matrices generated in this way, the matrices are named and displayed in *DisplayFormula* cells.

The matrices used in the illustrations are usually embedded in a **MatrixForm** command. This ensures a two-dimensional output for easy viewability. But it also preserves the one-dimensional nature of *Mathematica* output required internally for specific calculations.

The command

```
MatrixForm[A = {{1, 2, 3}, {4, 5, 6}}]
```

$$\begin{pmatrix} 1 & 2 & 3 \\ 4 & 5 & 6 \end{pmatrix}$$

produces a computable object named *A*:

```
A
```

```
{{1, 2, 3}, {4, 5, 6}}
```

The **Head** function tells us that *A* is a list (hence a computable object):

```
Head[A]
```

```
List
```

The command

```
B = {{1, 2, 3}, {4, 5, 6}} // MatrixForm
```

$$\begin{pmatrix} 1 & 2 & 3 \\ 4 & 5 & 6 \end{pmatrix}$$

produces a two-dimension object named *B*:

```
B
```

$$\begin{pmatrix} 1 & 2 & 3 \\ 4 & 5 & 6 \end{pmatrix}$$

However, the **Head** function confirms that *B* is a non-computable object called a **MatrixForm**.

```
Head[B]
```

```
MatrixForm
```

In all but a few illustrations, we avoid the suffix *//***MatrixForm**.

An illustration can be as simple as

- A 3-by-3 real matrix with three distinct real eigenvalues

```
A = {{2, 3, 1}, {0, 3, 2}, {0, 0, 4}};
```

```
Eigenvalues[A]
```

```
{4, 3, 2}
```

or as complicated as

- An orthogonal projection in $\mathbb{R}^2$

```
projection = Graphics[{Arrow[{{1, 2}, {10, 2}}], Arrow[{{1, 2}, {5, 8}}],
   Arrow[{{5, 8}, {5, 2}}], Arrow[{{1, 1.8}, {5, 1.8}}]}, Axes → True]
```

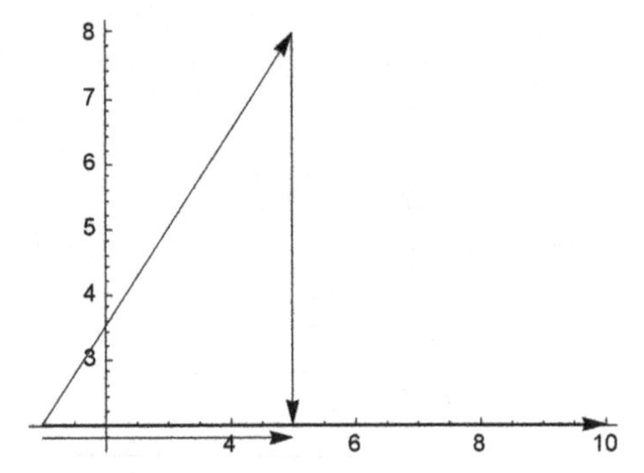

Manipulations

Many *Mathematica* functions and commands can be embedded in **Manipulation** environments controlled by parameters for varying numerical inputs. By changing the values of the parameters, the effect of numerical changes on the properties of the illustrated mathematical objects can be studied.

Here is a simple example to illustrate how **Manipulation** works. How do changes in the elements of the following matrix *A* affect the eigenvalues of the matrix?

```
MatrixForm[A = {{0, 1}, {-1, 0}}]
```

$$\begin{pmatrix} 0 & 1 \\ -1 & 0 \end{pmatrix}$$

```
Eigenvalues[A]
```

```
{i, -i}
```

Let us add parameters to some of the elements of *A* and explore the invertibility of the resulting matrices. For what integer values of $-6 \leq a \leq 6$ and $-6 \leq b \leq 6$ does the matrix $B = \{\{a, 1\}, \{-1, b\}\}$ have real eigenvalues?

```
Manipulate[{B = {{a, 1}, {-1, b}}, Eigenvalues[B]}, {a, -6, 6, 1}, {b, -6, 6, 1}]
```

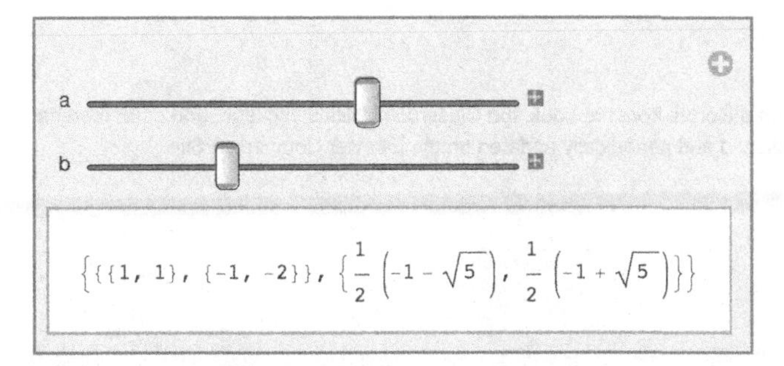

We can combine **Manipulate** and **Eigenvalues** to explore the eigenvalues of 2-by-2 matrices with integer elements. If we let $a = 1$ and $b = -2$, for example, the manipulation produces a matrix that has real eigenvalues. By letting the **Manipulate** parameters range over real numbers, we can force *Mathematica* to produce decimal outputs.

```
Manipulate[{B = {{a, 1}, {-1, b}}, Eigenvalues[B]}, {a, -6, 6}, {b, -6, 6}]
```

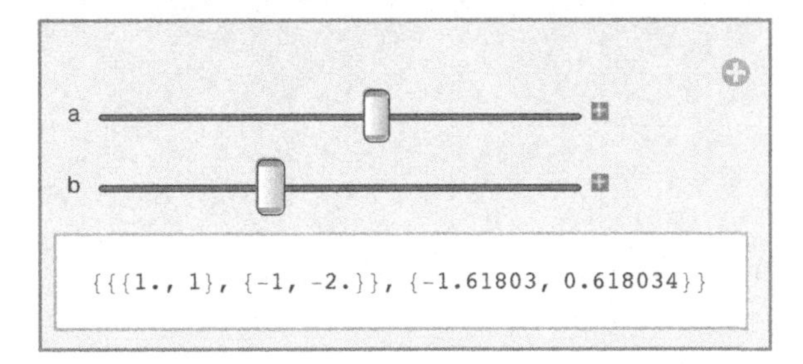

The **N** function confirms that the associated results are equal:

$$N\left[\left\{\frac{1}{2}\left(-1-\sqrt{5}\right), \frac{1}{2}\left(-1+\sqrt{5}\right)\right\}\right]$$

```
{-1.61803, 0.618034}
```

Notation

Unless *Mathematica* terminology requires something else, the following notations are used in the *Text* cells of this guide. *Vectors* are named in regular bold format. The symbol **v** denotes a vector. *Matrices* are usually named in capital italic format. The symbol A denotes a matrix. A matrix with n rows and m columns is called an n-by-m matrix. Built-in *Mathematica* functions and commands are written in regular bold format. The random integer function and the command for solving linear systems, for example, are referred by writing **RandomInteger** and **LinearSolve.** Variables are written in italic format. The letters x, y, z, for example, denote variables. Constants are also written in italic format. The letters a, b, c, for example, denote constants. Defined functions and commands are numbered within the alphabetical items in which they occur. In *Text* cells, numerals are written in regular text font. The symbols 1, 2, 3, for example, denote the numbers 1, 2, and 3.

Duplications

In order to make the individual entries of this guide as readable as possible, minor duplications of illustrations became inevitable. In most cases, they are differentiated as much as possible in the presentations by emphasizing different nuances of overlapping ideas and techniques.

Companion Site

Interactive *Mathematica* **Manipulation** excerpts from the book, the ClassroomUtilities package, and other material complementary to this guide are published and periodically updated on the Elsevier Companion Site.

A

Addition of matrices

Matrices are added element by element. It works provided the matrices to be added have the same dimensions (the same number of rows and columns). In *Mathematica*, a plus sign between two matrices defines addition.

Properties of matrix addition

$$(A + B) + C = A + (B + C) \tag{1}$$

$$A + B = B + A \tag{2}$$

$$A + O = A \tag{3}$$

$$A + (-1)A = O \tag{4}$$

Mathematical systems satisfying these four conditions are known as *Abelian groups*. For any natural number $n > 0$, the set of n-by-n matrices with real elements forms an Abelian group with respect to matrix addition.

The *sum* of an n-by-m matrix A and an n-by-m matrix B is the matrix $(A + B)$ whose ijth element is $(A_{[[i,j]]} + B_{[[i,j]]})$.

Illustration

- Addition of two matrices

```
MatrixForm[A = {{1, 2, 3}, {4, 5, 6}}]
```

$$\begin{pmatrix} 1 & 2 & 3 \\ 4 & 5 & 6 \end{pmatrix}$$

```
Dimensions[A]
```

{2, 3}

```
MatrixForm[B = {{a, b, c}, {d, e, f}}]
```

$$\begin{pmatrix} a & b & c \\ d & e & f \end{pmatrix}$$

```
Dimensions[B]
```

{2, 3}

```
MatrixForm[A + B]
```

$$\begin{pmatrix} 1+a & 2+b & 3+c \\ 4+d & 5+e & 6+f \end{pmatrix}$$

```
Dimensions[A + B]

{2, 3}
```

- Sum of two 3-by-2 matrices

```
MatrixForm[A = {{a, b}, {c, d}, {e, f}}]
```

$$\begin{pmatrix} a & b \\ c & d \\ e & f \end{pmatrix}$$

```
MatrixForm[B = {{1, 2}, {3, 4}, {5, 6}}]
```

$$\begin{pmatrix} 1 & 2 \\ 3 & 4 \\ 5 & 6 \end{pmatrix}$$

```
MatrixForm[A + B]
```

$$\begin{pmatrix} 1+a & 2+b \\ 3+c & 4+d \\ 5+e & 6+f \end{pmatrix}$$

- Two matrices whose sum is not defined

```
MatrixForm[A = RandomInteger[{0, 9}, {3, 4}]]
```

$$A = \begin{pmatrix} 4 & 9 & 8 & 7 \\ 2 & 6 & 0 & 4 \\ 0 & 6 & 3 & 5 \end{pmatrix};$$

```
MatrixForm[B = RandomInteger[{0, 9}, {3, 3}]]
```

$$B = \begin{pmatrix} 0 & 5 & 8 \\ 9 & 9 & 6 \\ 6 & 0 & 1 \end{pmatrix};$$

```
A + B
```

Thread::tdlen : Objects of unequal length in {0, 5, 8} + {4, 9, 8, 7} cannot be combined. ≫

Thread::tdlen : Objects of unequal length in {9, 9, 6} + {2, 6, 0, 4} cannot be combined. ≫

Thread::tdlen : Objects of unequal length in {6, 0, 1} + {0, 6, 3, 5} cannot be combined. ≫

General::stop : Further output of Thread::tdlen will be suppressed during this calculation. ≫

```
{{0, 5, 8} + {4, 9, 8, 7}, {9, 9, 6} + {2, 6, 0, 4}, {6, 0, 1} + {0, 6, 3, 5}}
```

- Sum of a nonzero and zero matrix

```
MatrixForm[A = {{1, 2, 3}, {4, 5, 6}}]
```

$$\begin{pmatrix} 1 & 2 & 3 \\ 4 & 5 & 6 \end{pmatrix}$$

```
MatrixForm[Z = {{0, 0, 0}, {0, 0, 0}}]
```

$$\begin{pmatrix} 0 & 0 & 0 \\ 0 & 0 & 0 \end{pmatrix}$$

```
A + Z == A
```

```
True
```

- Sum of a matrix A and the matrix (-1) A

```
MatrixForm[A = {{1, 2, 3}, {4, 5, 6}}]
MatrixForm[(-1) A]
```

$$\begin{pmatrix} 1 & 2 & 3 \\ 4 & 5 & 6 \end{pmatrix}$$

$$\begin{pmatrix} -1 & -2 & -3 \\ -4 & -5 & -6 \end{pmatrix}$$

```
MatrixForm[A + (-1) A]
```

$$\begin{pmatrix} 0 & 0 & 0 \\ 0 & 0 & 0 \end{pmatrix}$$

- Subtraction of two matrices

```
MatrixForm[A = {{1, 2, 3}, {4, 5, 6}}]
```

$$\begin{pmatrix} 1 & 2 & 3 \\ 4 & 5 & 6 \end{pmatrix}$$

```
MatrixForm[B = {{a, b, c}, {d, e, f}}]
```

$$\begin{pmatrix} a & b & c \\ d & e & f \end{pmatrix}$$

```
MatrixForm[A + (-1) B]
```

$$\begin{pmatrix} 1-a & 2-b & 3-c \\ 4-d & 5-e & 6-f \end{pmatrix}$$

```
A + (-1) B == A - B
```

```
True
```

Manipulation

- Addition of two 3-by-4 matrices

```
Clear[a, b]
```

```
A = {{1, 3 a, 0, 5}, {8, 6, 8, 3}, {3, 3, 6, 2}};
```

```
B = {{6, 2, 9, 3}, {2 b, 4, 7, 3}, {1, 5, 7, 3}};
```

```
Manipulate[Evaluate[A + B], {a, -3, 5, 1}, {b, -2, 4, 1}]
```

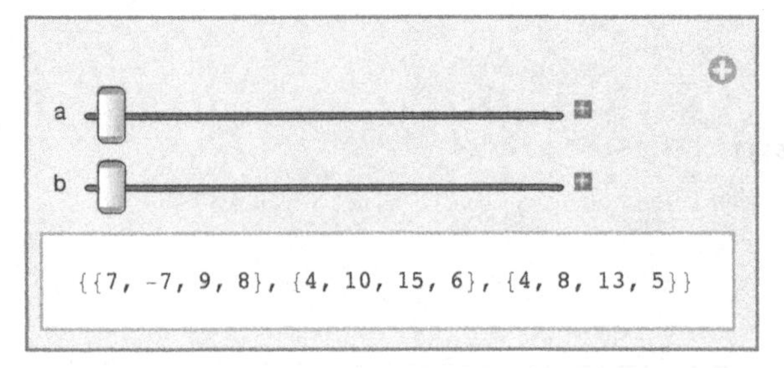

We use **Manipulate** and **Evaluate** to explore the sum of two matrices. The displayed matrix is obtained by letting $a = -3$ and $b = -2$.

Adjacency matrix

The *adjacency matrix* of a simple labeled graph is the matrix A with $A_{[[i,j]]} = 1$ or 0 according to whether the vertex v_i is adjacent to the vertex v_j or not. For simple graphs without self-loops, the adjacency matrix has 0s on the diagonal. For undirected graphs, the adjacency matrix is symmetric.

Illustration

- The adjacency matrix of an undirected graph

```
Graph[{1 ⟷ 2, 2 ⟷ 3, 3 ⟷ 1}]
```

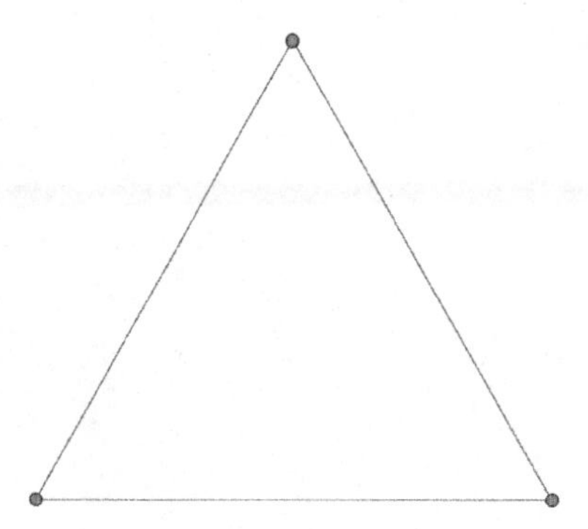

The arrow ⟷ can be created by typing Esc ue Esc.

```
MatrixForm[AdjacencyMatrix[%]]
```

$$\begin{pmatrix} 0 & 1 & 1 \\ 1 & 0 & 1 \\ 1 & 1 & 0 \end{pmatrix}$$

- The adjacency matrix of a directed graph

```
Graph[{1 → 2, 2 → 3, 3 → 1}]
```

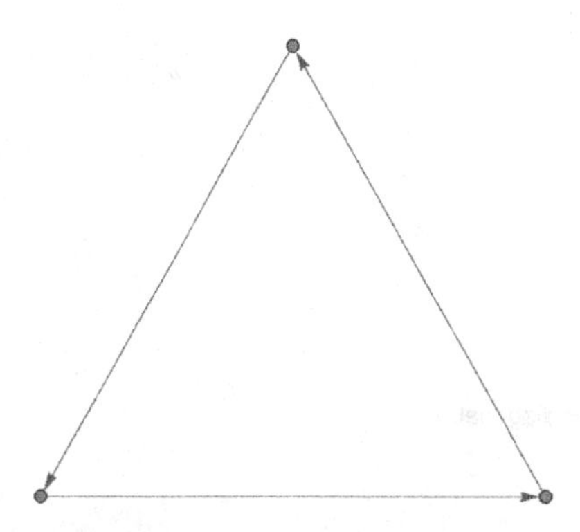

The arrow → can be created by typing Esc de Esc.

```
MatrixForm[AdjacencyMatrix[%]]
```

$$\begin{pmatrix} 0 & 1 & 0 \\ 0 & 0 & 1 \\ 1 & 0 & 0 \end{pmatrix}$$

- The adjacency matrix of an undirected graph is symmetric

```
Graph[{1 ↔ 2, 1 ↔ 3, 2 ↔ 3, 2 ↔ 4, 3 ↔ 4}]
```

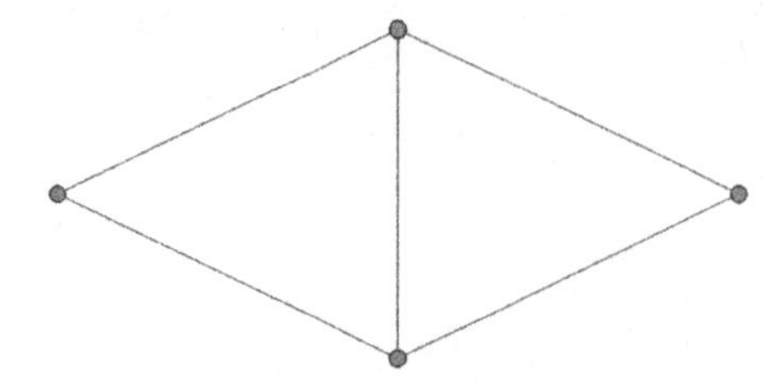

```
MatrixForm[AdjacencyMatrix[%]]
```

$$\begin{pmatrix} 0 & 1 & 1 & 0 \\ 1 & 0 & 1 & 1 \\ 1 & 1 & 0 & 1 \\ 0 & 1 & 1 & 0 \end{pmatrix}$$

- The adjacency matrix of a directed graph can be nonsymmetric

```
Graph[{1 → 2, 2 → 1, 3 → 1, 3 → 2, 4 → 1, 4 → 2}]
```

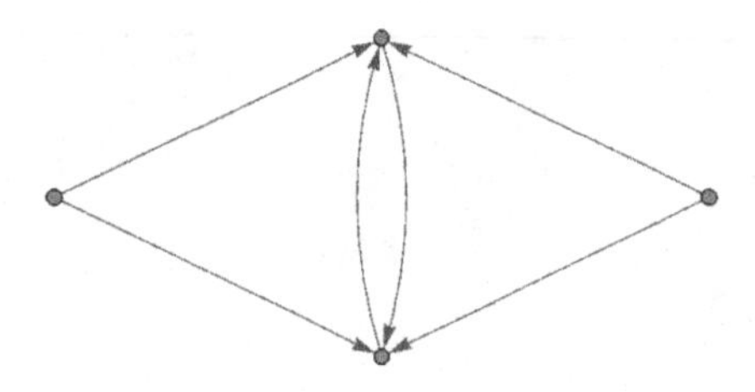

```
MatrixForm[AdjacencyMatrix[%]]
```

$$\begin{pmatrix} 0 & 1 & 0 & 0 \\ 1 & 0 & 0 & 0 \\ 1 & 1 & 0 & 0 \\ 1 & 1 & 0 & 0 \end{pmatrix}$$

- The adjacency matrix of the graph with self-loops has 1s on the diagonal.

```
Graph[{1 ⟷ 2, 2 ⟷ 3, 3 ⟷ 1, 2 ⟷ 2}]
```

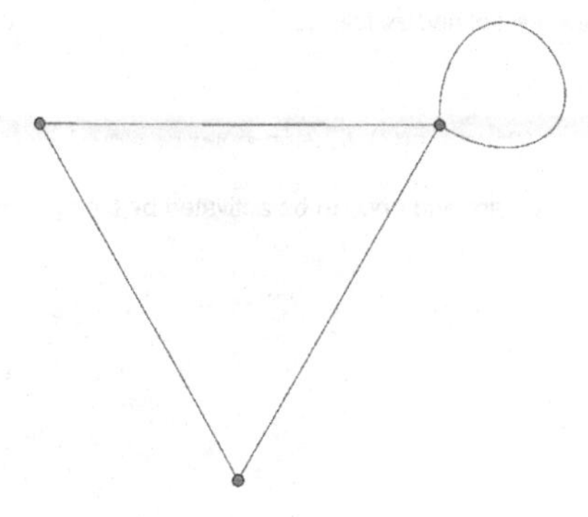

```
MatrixForm[AdjacencyMatrix[%]]
```

$$\begin{pmatrix} 0 & 1 & 1 \\ 1 & 1 & 1 \\ 1 & 1 & 0 \end{pmatrix}$$

- The adjacency matrix of a large graph

```
Graph[Table[i ⟷ Mod[i^2, 10^3], {i, 0, 10^3 - 1}]];

A = AdjacencyMatrix[%]
```

```
MatrixPlot[A]
```

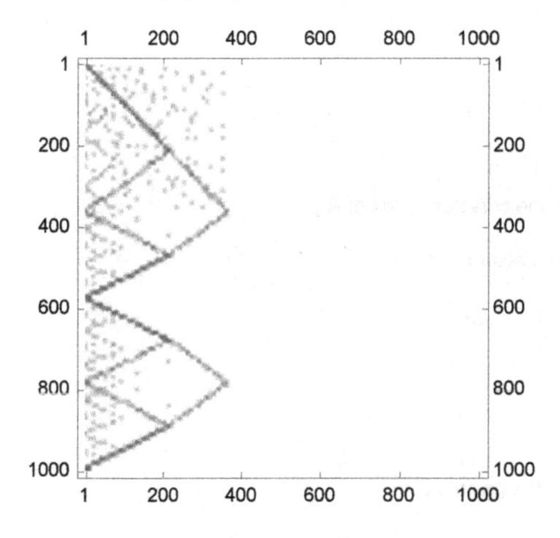

Adjoint matrix

The *adjoint* of a square matrix *A* is the transpose of the cofactor matrix of *A*. In *MathWorld*, at http://mathworld.wolfram.com/-Cofactor.html, the *Mathematica* functions for building adjoint matrices are defined as follows:

```
MinorMatrix[m_List?MatrixQ] := Map[Reverse, Minors[m], {0, 1}]          (1)

CofactorMatrix[m_List?MatrixQ] := MapIndexed[#1 (-1)^(Plus @@ #2) &, MinorMatrix[m], {2}]          (2)
```

These defined functions are not included in *Mathematica*'s function repertoire and need to be activated by typing *Shift + Enter*.

Illustration

- The cofactor matrix of a 3-by-3 matrix with integer entries

```
MatrixForm[A = {{1, 2, 3}, {4, 5, 6}, {7, 8, 10}}]
```

$$\begin{pmatrix} 1 & 2 & 3 \\ 4 & 5 & 6 \\ 7 & 8 & 10 \end{pmatrix}$$

```
MinorMatrix[m_List?MatrixQ] := Map[Reverse, Minors[m], {0, 1}]

CofactorMatrix[m_List?MatrixQ] := MapIndexed[#1 (-1)^(Plus @@ #2) &, MinorMatrix[m], {2}]

MatrixForm[MinorMatrix[A]]
```

$$\begin{pmatrix} 2 & -2 & -3 \\ -4 & -11 & -6 \\ -3 & -6 & -3 \end{pmatrix}$$

```
MatrixForm[cfA = CofactorMatrix[A]]
```

$$\begin{pmatrix} 2 & 2 & -3 \\ 4 & -11 & 6 \\ -3 & 6 & -3 \end{pmatrix}$$

```
Transpose[Det[A] Inverse[A]] == cfA

True
```

The adjoint matrix of the given matrix *A* is the transpose of the cofactor matrix of *A*:

- The adjoint matrix of a 3-by-3 matrix computed from a cofactor matrix

```
MatrixForm[adjA = Transpose[cfA]]
```

$$\begin{pmatrix} 2 & 4 & -3 \\ 2 & -11 & 6 \\ -3 & 6 & -3 \end{pmatrix}$$

The adjoint is equal to the determinant times the inverse of the given matrix:

```
Inverse[A] == (1/Det[A]) adjA
```

```
True
```

- A general cofactor matrix

```
Clear[a, A]; A = Array[a## &, {3, 3}]; A // MatrixForm
```

$$\begin{pmatrix} a_{1,1} & a_{1,2} & a_{1,3} \\ a_{2,1} & a_{2,2} & a_{2,3} \\ a_{3,1} & a_{3,2} & a_{3,3} \end{pmatrix}$$

```
CofactorMatrix[A]
```

$\{\{-a_{2,3}\,a_{3,2} + a_{2,2}\,a_{3,3},\ a_{2,3}\,a_{3,1} - a_{2,1}\,a_{3,3},\ -a_{2,2}\,a_{3,1} + a_{2,1}\,a_{3,2}\},$
$\{a_{1,3}\,a_{3,2} - a_{1,2}\,a_{3,3},\ -a_{1,3}\,a_{3,1} + a_{1,1}\,a_{3,3},\ a_{1,2}\,a_{3,1} - a_{1,1}\,a_{3,2}\},$
$\{-a_{1,3}\,a_{2,2} + a_{1,2}\,a_{2,3},\ a_{1,3}\,a_{2,1} - a_{1,1}\,a_{2,3},\ -a_{1,2}\,a_{2,1} + a_{1,1}\,a_{2,2}\}\}$

Manipulation

- Exploring cofactor matrices

```
MinorMatrix[m_List?MatrixQ] := Map[Reverse, Minors[m], {0, 1}]
```

```
CofactorMatrix[m_List?MatrixQ] := MapIndexed[#1 (-1)^(Plus @@ #2) &, MinorMatrix[m], {2}]
```

```
Manipulate[MatrixForm[CofactorMatrix[{{a, 2, 3}, {4, b, 6}, {7, c, 10}}]],
 {a, -3, 3, 1}, {b, -3, 3, 1}, {c, -3, 3, 1}]
```

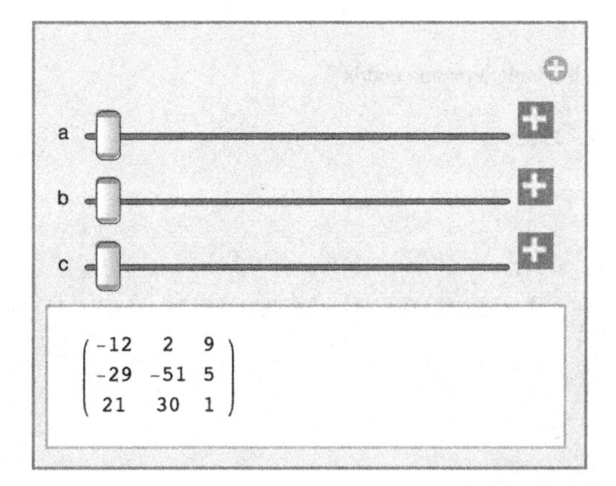

We combine **Manipulate**, **MatrixForm**, and **CofactorMatrix** to explore cofactor matrices. If we let $a = b = c = -3$, for example, the manipulation produces the cofactor matrix of the matrix {{-3, 2, 3}, {4, -3, 6}, {7, -3, 10}}.

Adjoint transformation

If a matrix A represents a linear transformation $T : V \longrightarrow V$ on an inner product space V in an orthonormal basis B, the transpose of A represents a linear transformation $T^* : V \longrightarrow V$ called the *adjoint* of T. The transformations T and T^* are linked by the equation $< T[\mathbf{u}], \mathbf{v} > = < \mathbf{u}, T^*[\mathbf{v}] >$ for all vectors **u** and **v** in V.

Illustration

```
Clear[x, y, z, u, v, w]
```

```
MatrixForm[A = {{2, 1, 7}, {3, 5, 8}, {6, 4, 2}}]
```

$$\begin{pmatrix} 2 & 1 & 7 \\ 3 & 5 & 8 \\ 6 & 4 & 2 \end{pmatrix}$$

```
MatrixForm[B = Transpose[A]]
```

$$\begin{pmatrix} 2 & 3 & 6 \\ 1 & 5 & 4 \\ 7 & 8 & 2 \end{pmatrix}$$

With respect to the Euclidean inner product (the dot product), the transformation T and its adjoint T^* are related by the equation

```
Expand[Dot[A.{x, y, z}, {u, v, w}] == Dot[{x, y, z}, B.{u, v, w}]]
```

```
True
```

Manipulation

- Exploring a transformation and its adjoint by varying the transformation matrix

```
MatrixForm[A = {{a, 1, 7}, {3, 5, b}, {6, c, 2}}]
```

$$\begin{pmatrix} a & 1 & 7 \\ 3 & 5 & b \\ 6 & c & 2 \end{pmatrix}$$

```
Manipulate[Evaluate[Dot[A.{1, 2, 3}, {4, 5, 6}] == Dot[{1, 2, 3}, Transpose[A].{4, 5, 6}]],
  {a, -2, 2, 1}, {b, -5, 5, 1}, {c, -8, 8, 1}]
```

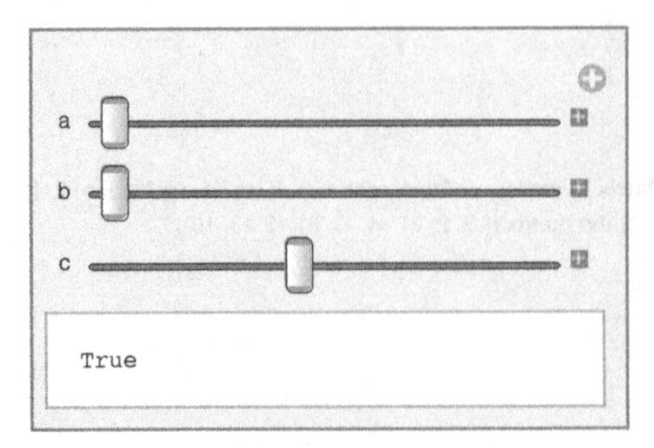

We combine **Manipulate**, **Evaluate**, **Dot,** and **Transpose**, to explore adjoint transformations. The manipulation shows, for example, that the transformation and its transpose determined by $a = -2$, $b = -5$, and $c = -1$ are adjoint.

Adjugate of a matrix

The *adjugate* of a matrix is the transpose of the cofactor matrix of the matrix. On the website http://mathematica.stackexchange.com, a *Mathematica* function for calculating adjugates is defined:

```
adj[m_] :=
  Map[Reverse, Minors[Transpose[m], Length[m] - 1], {0, 1}] *
   Table[(-1)^(i + j), {i, Length[m]}, {j, Length[m]}];
```
(1)

This function is not included in the *Mathematica* repertoire and must to be activated by pressing *Shift* + *Enter*.

Illustration

- Adjugate of a 2-by-2 matrix

```
adj[m_] := Map[Reverse, Minors[Transpose[m], Length[m] - 1], {0, 1}] *
  Table[(-1)^(i + j), {i, Length[m]}, {j, Length[m]}]

A = {{a, b}, {c, d}};

adj[A]

{{d, -b}, {-c, a}}
```

- Adjugate of a numerical 3-by-3 matrix

```
MatrixForm[A = {{6, 1, 9}, {3, 2, 0}, {5, 6, 8}}]
```

$$\begin{pmatrix} 6 & 1 & 9 \\ 3 & 2 & 0 \\ 5 & 6 & 8 \end{pmatrix}$$

```
MatrixForm[adj[A]]
```

$$\begin{pmatrix} 16 & 46 & -18 \\ -24 & 3 & 27 \\ 8 & -31 & 9 \end{pmatrix}$$

```
Inverse[A] == (1/Det[A]) adj[A]

True
```

Affine transformation

An *affine transformation* is an invertible matrix transformation from $\mathbb{R}^2$ to $\mathbb{R}^2$, followed by a translation.

Illustration

If the translation vector is zero, then the affine transformation is simply an invertible matrix transformation.

- An affine transformation defined by an invertible matrix without a translation

```
Clear[A, u]

MatrixForm[A = {{3, 2}, {9, 1}}]
```

$$\begin{pmatrix} 3 & 2 \\ 9 & 1 \end{pmatrix}$$

```
Det[A]

-15

affineA[u_] := A.u

u = {1, 2};

affineA[u]

{7, 11}
```

- An affine transformation defined by an invertible matrix together with a translation

```
Clear[u, v, A]

A = RandomInteger[{0, 9}, {2, 2}];
```

$A = \begin{pmatrix} 3 & 2 \\ 9 & 1 \end{pmatrix};$

```
affine[u_, v_] := A.u + v

u = {1, 2}; v = {3, 4};

affine[u, v]

{10, 15}
```

Mathematica has a built-in function for affine transformations:

- An affine transformation defined with the **AffineTransform** command

```
A = RandomInteger[{0, 9}, {2, 2}];
```

$A = \begin{pmatrix} 3 & 2 \\ 9 & 1 \end{pmatrix}; b = \begin{pmatrix} 3 \\ 4 \end{pmatrix};$

```
tA = AffineTransform[{A, {3, 4}}]
```

$$\text{TransformationFunction}\left[\left(\begin{array}{cc|c} 3 & 2 & 3 \\ 9 & 1 & 4 \\ \hline 0 & 0 & 1 \end{array}\right)\right]$$

```
tA[{1, 2}]
```

{10, 15}

$$B = \begin{pmatrix} 3 & 2 \\ 9 & 1 \end{pmatrix};\ b = \begin{pmatrix} 0 \\ 0 \end{pmatrix};$$

```
tB = AffineTransform[{B, {0, 0}}]
```

$$\text{TransformationFunction}\left[\left(\begin{array}{cc|c} 3 & 2 & 0 \\ 9 & 1 & 0 \\ \hline 0 & 0 & 1 \end{array}\right)\right]$$

```
tB[{1, 2}]
```

{7, 11}

- An affine transformation from $\mathbb{R}^2$ to $\mathbb{R}^2$ defined in two-dimensional matrix notation

$$f[\{x\_, y\_\}] := \begin{pmatrix} 1 & 2 \\ 3 & 4 \end{pmatrix} . \begin{pmatrix} x \\ y \end{pmatrix} + \begin{pmatrix} 5 \\ 6 \end{pmatrix};$$

```
MatrixForm[f[{0, 0}]]
```

$$\begin{pmatrix} 5 \\ 6 \end{pmatrix}$$

```
MatrixForm[f[{3, 4}]]
```

$$\begin{pmatrix} 16 \\ 31 \end{pmatrix}$$

$$\text{MatrixForm}\left[\begin{pmatrix} 5 \\ 6 \end{pmatrix} + \begin{pmatrix} 16 \\ 31 \end{pmatrix} == f[\{0, 0\}] + f[\{3, 4\}]\right]$$

True

- An affine transformation defined with the **AffineTransform** function

$$t = \text{AffineTransform}[\{\{\{a_{1,1}, a_{1,2}\}, \{a_{2,1}, a_{2,2}\}\}, \{b_1, b_2\}\}];$$

```
t[{x, y}]
```

$$\{\{\{0\}, \{0\}\}_1 + x\, a_{1,1} + y\, a_{1,2}, \{\{0\}, \{0\}\}_2 + x\, a_{2,1} + y\, a_{2,2}\}$$

```
A = {{1, 2}, {3, 4}}; b = {5, 6};
```

```
t = AffineTransform[{A, b}];
```

```
t[{7, 8}]
```

```
{28, 59}
```

With the help of homogeneous coordinates, a matrix multiplication and the addition of a translation vector can be combined into a single operation. Here is how. Embed $\mathbb{R}^2$ into $\mathbb{R}^3$ and redefine matrix multiplication.

- Using homogeneous coordinates to combine a matrix multiplication and the addition of a translation vector

```
Clear[A, b, x, y]
```

```
A = {{1, 2}, {3, 4}}; b = {5, 6};
```

```
A.{x, y} + b
```

```
{5 + x + 2 y, 6 + 3 x + 4 y}
```

- Converting the 2-by-2 matrix *A* and the vector **b** to a 3-by-3 matrix hA and a 3-by-3 matrix *hb*

```
MatrixForm[hA = {{1, 2, 0}, {3, 4, 0}, {0, 0, 1}}]
```

$$\begin{pmatrix} 1 & 2 & 0 \\ 3 & 4 & 0 \\ 0 & 0 & 1 \end{pmatrix}$$

```
MatrixForm[hb = {{1, 0, 5}, {0, 1, 6}, {0, 0, 1}}]
```

$$\begin{pmatrix} 1 & 0 & 5 \\ 0 & 1 & 6 \\ 0 & 0 & 1 \end{pmatrix}$$

- Combining the matrix multiplication and translation into two matrix multiplications

```
hb.hA.{x, y, 1}
```

```
{5 + x + 2 y, 6 + 3 x + 4 y, 1}
```

- Projecting the resulting vector from $\mathbb{R}^3$ to $\mathbb{R}^2$

```
p[{u_, v_, w_}] := {u, v}
```

```
p[{5 + x + 2 y, 6 + 3 x + 4 y, 1}]
```

```
{5 + x + 2 y, 6 + 3 x + 4 y}
```

As we can see, the two results are identical:

```
p[{5 + x + 2 y, 6 + 3 x + 4 y, 1}] == A.{x, y} + b
```

```
True
```

- A general affine transformation form defined with the affine transform

```
T = AffineTransform[{{{a1,1, a1,2}, {a2,1, a2,2}}, {b1, b2}}]
```

$$\text{TransformationFunction}\left[\left(\begin{array}{cc|c} a_{1,1} & a_{1,2} & \{5, 6\}_1 \\ a_{2,1} & a_{2,2} & \{5, 6\}_2 \\ \hline 0 & 0 & 1 \end{array}\right)\right]$$

```
T[{x, y}]
```

$\{\{5, 6\}_1 + x\, a_{1,1} + y\, a_{1,2},\ \{5, 6\}_2 + x\, a_{2,1} + y\, a_{2,2}\}$

```
T = AffineTransform[{{{1, 2}, {3, 4}}, {5, 6}}]
```

$$\text{TransformationFunction}\left[\left(\begin{array}{cc|c} 1 & 2 & 5 \\ 3 & 4 & 6 \\ \hline 0 & 0 & 1 \end{array}\right)\right]$$

```
T[{3, 2}]
```

{12, 23}

```
{{1, 2}, {3, 4}}.{3, 2} + {5, 6} == T[{3, 2}]
```

True

A clockwise rotation can be represented both ways, as a matrix multiplication and as an affine transformation.

- A clockwise rotation represented by a matrix multiplication

$$\text{cwr} = \begin{pmatrix} \text{Cos[Pi / 3]} & \text{Sin[Pi / 3]} \\ \text{-Sin[Pi / 3]} & \text{Cos[Pi / 3]} \end{pmatrix}$$

$$\left\{\left\{\frac{1}{2}, \frac{\sqrt{3}}{2}\right\}, \left\{-\frac{\sqrt{3}}{2}, \frac{1}{2}\right\}\right\}$$

```
cwr.{1, 0}
```

$$\left\{\frac{1}{2}, -\frac{\sqrt{3}}{2}\right\}$$

- A clockwise rotation represented by an affine transformation

```
cwt = AffineTransform[{{{Cos[Pi / 3], Sin[Pi / 3]}, {-Sin[Pi / 3], Cos[Pi / 3]}}, {0, 0}}]
```

$$\text{TransformationFunction}\left[\left(\begin{array}{cc|c} \frac{1}{2} & \frac{\sqrt{3}}{2} & 0 \\ -\frac{\sqrt{3}}{2} & \frac{1}{2} & 0 \\ \hline 0 & 0 & 1 \end{array}\right)\right]$$

```
cwt[{1, 0}]
```

$\{\frac{1}{2}, -\frac{\sqrt{3}}{2}\}$

- A counterclockwise rotation represented by a matrix multiplication

```
ccwr = (Cos[Pi/3]  -Sin[Pi/3]
        Sin[Pi/3]   Cos[Pi/3])
```

$\{\{\frac{1}{2}, -\frac{\sqrt{3}}{2}\}, \{\frac{\sqrt{3}}{2}, \frac{1}{2}\}\}$

```
ccwr.{1, 0}
```

$\{\frac{1}{2}, \frac{\sqrt{3}}{2}\}$

- A counterclockwise rotation represented by an affine transformation

```
ccwt = AffineTransform[{{{Cos[Pi/3], Sin[Pi/3]}, {-Sin[Pi/3], Cos[Pi/3]}}, {0, 0}}]
```

$$\text{TransformationFunction}\left[\left(\begin{array}{cc|c} \frac{1}{2} & \frac{\sqrt{3}}{2} & 0 \\ -\frac{\sqrt{3}}{2} & \frac{1}{2} & 0 \\ \hline 0 & 0 & 1 \end{array}\right)\right]$$

```
ccwt[{1, 0}]
```

$\{\frac{1}{2}, -\frac{\sqrt{3}}{2}\}$

- A shear along the *x*-axis

A *shear* along the *x*- and *y*-axis can be represented both ways, as a matrix multiplication and as an affine transformation.

```
s = 5; v = {3, -7};

shear1[s_, v_] := {{1, s}, {0, 1}}.v

shear1[s, v]
```

{-32, -7}

```
shear2 = AffineTransform[{{{1, s}, {0, 1}}, {0, 0}}]
```

$$\text{TransformationFunction}\left[\left(\begin{array}{cc|c} 1 & 5 & 0 \\ 0 & 1 & 0 \\ \hline 0 & 0 & 1 \end{array}\right)\right]$$

```
shear2[{3, -7}]
```

```
{-32, -7}
```

- A shear along the *y*-axis

```
s = 5; v = {3, -7};
```

```
shear3[s_, v_] := {{1, 0}, {s, 1}}.v
```

```
shear3[s, v]
```

```
{3, 8}
```

```
shear4 = AffineTransform[{{{1, 0}, {s, 1}}, {0, 0}}]
```

$$\text{TransformationFunction}\left[\left(\begin{array}{cc|c} 1 & 0 & 0 \\ 5 & 1 & 0 \\ \hline 0 & 0 & 1 \end{array}\right)\right]$$

```
shear4[{3, -7}]
```

```
{3, 8}
```

- A reflection about $y = x$ represented by a matrix and an affine transform

A *reflection* about the lines $y = x$ can also be represented both ways, as a matrix multiplication and an affine transformation.

$$\text{MatrixForm}\left[\text{reflyx} = \begin{pmatrix} 0 & 1 \\ 1 & 0 \end{pmatrix}\right];$$

```
point = {Cos[Pi / 3], Sin[Pi / 3]}
```

$$\left\{\frac{1}{2}, \frac{\sqrt{3}}{2}\right\}$$

```
reflyx.point
```

$$\left\{\frac{\sqrt{3}}{2}, \frac{1}{2}\right\}$$

```
t = AffineTransform[{reflyx, {0, 0}}]
```

$$\text{TransformationFunction}\left[\left(\begin{array}{cc|c} 0 & 1 & 0 \\ 1 & 0 & 0 \\ \hline 0 & 0 & 1 \end{array}\right)\right]$$

```
t[point]
```

$$\left\{\frac{\sqrt{3}}{2}, \frac{1}{2}\right\}$$

Algebraic multiplicity of an eigenvalue

The number of repetitions of a linear factor $(t - \lambda)$ in the factorization of the characteristic polynomial p[A, t] of a matrix A into linear factors is called the *algebraic multiplicity* of the eigenvalue λ.

Illustration

```
MatrixForm[A = UpperTriangularize[RandomInteger[{0, 3}, {5, 5}]]];
```

$$A = \begin{pmatrix} 1 & 3 & 1 & 0 & 0 \\ 0 & 3 & 2 & 0 & 1 \\ 0 & 0 & 0 & 1 & 0 \\ 0 & 0 & 0 & 3 & 1 \\ 0 & 0 & 0 & 0 & 3 \end{pmatrix};$$

```
Clear[t]
```

```
p[A, t] = CharacteristicPolynomial[A, t]
```

$-27\,t + 54\,t^2 - 36\,t^3 + 10\,t^4 - t^5$

```
Factor[p[A, t]]
```

$-(-3 + t)^3\,(-1 + t)\,t$

```
Eigenvalues[A]
```

{3, 3, 3, 1, 0}

The algebraic multiplicity of eigenvalue 1 is 1, and that of the eigenvalues 0 and 3 is 2.

- Algebraic multiplicities of eigenvalues

$$A = \begin{pmatrix} 1 & 2 & 0 & 3 & 1 \\ 0 & 0 & 2 & 0 & 0 \\ 0 & 0 & 0 & 2 & 1 \\ 0 & 0 & 0 & a & a \\ 0 & 0 & 0 & 0 & 3 \end{pmatrix};$$

```
p[A, t] = CharacteristicPolynomial[A, t]
```

$(3 - t)\,\left(a\,t^2 - t^3 - a\,t^3 + t^4\right)$

```
Factor[p[A, t]]
```

```
-(-3 + t) (-1 + t) t^2 (-a + t)
```

```
Eigenvalues[A]
```

```
{3, 1, 0, 0, a}
```

Manipulation

- Algebraic multiplicities of eigenvalues

```
A = {{1, 2, 0, 3}, {0, 0, 2, 0}, {0, 0, a, 2}, {0, 0, a, a}};
```

```
Manipulate[Evaluate[Factor[CharacteristicPolynomial[A, t]]], {a, -2, 2, 1}]
```

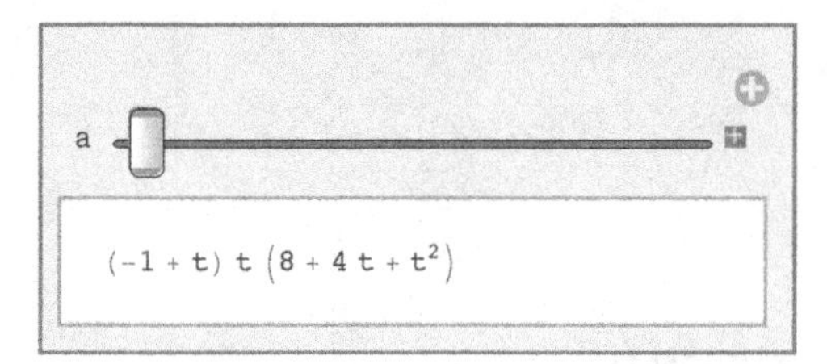

If we combine **Manipulate**, **Evaluate**, **Factor**, and **CharacteristicPolynomial** and let a = - 2, the manipulation displays the factored characteristic polynomial of the generated matrix and implies that the multiplicities of the eigenvalues of the matrix are 1, 1, and 2.

Angle

An *angle* is a measure of revolution, expressed in either degrees or radians. An angle θ between two vectors **u** and **v,** expressed in radians, is the value of the function **ArcCos**[θ] where **Cos**[θ] is the cosine determined by **u** and **v**.

1 revolution = 360 degrees = 2 π radians

```
f[degrees_] := (2 π/360) degrees; g[radians_] := (360/(2 π)) radians;
```

```
{f[360], g[2 π]}
```

```
{2 π, 360}
```

Illustration

- Angle between two vectors with respect to the Euclidean norm

```
cos[u_, v_] := Dot[u, v] / (Sqrt[Dot[u, u]] Sqrt[Dot[v, v]])
```

```
u = {3, 4}; v = {1, 5};
```

```
cos[u, v]
```

$$\frac{23}{5\sqrt{26}}$$

```
radians = N[ArcCos[cos[u, v]]]
```

0.446106

```
degrees = g[radians]
```

25.56

- Angle between two vectors with respect to a non-Euclidean norm

```
A = {{5, 1}, {2, 7}};
```

```
MatrixForm[AAt = A.Transpose[A]]
```

$$\begin{pmatrix} 26 & 17 \\ 17 & 53 \end{pmatrix}$$

```
{SymmetricMatrixQ[AAt], PositiveDefiniteMatrixQ[AAt]}
```

{True, True}

The following command defines an inner product between vectors **u** and **v**:

```
⟨u_, v_⟩ := u.AAt.v
```

$$\text{cos[u\_, v\_]} := \frac{\langle u, v\rangle}{\sqrt{\langle u, v\rangle}\ \sqrt{\langle v, v\rangle}}$$

```
u = {3, 2}; v = {1, 9};
```

```
cos[u, v]
```

$$\sqrt{\frac{61}{185}}$$

```
radians = N[ArcCos[cos[u, v]]]
```

0.959144

```
degrees = g[radians]
```

54.9549

- The angle between two vectors in $\mathbb{R}^3$

```
u = {1, 2, 3}; v = {4, 5, 6};
```

```
cosine = (Dot[u, v]) / (Norm[u] Norm[v]);
```

```
angle = N[ArcCos[cosine]]
```

```
0.225726
```

This tells us that the angle between the vectors **u** and **v** is 0.225726 radians. In degrees, we get

```
angledeg = g[angle]
```

```
12.9332
```

Area of a parallelogram

If **u** and **v** are two vectors in $\mathbb{R}^3$, then the Euclidean norm **Norm[Cross[**u, v**]]** of the cross product of **u** and **v** is the *area of the parallelogram* determined by **u** and **v**.

Illustration

- Area of a parallelogram

```
u = {1, 2, 3}; v = {4, 5, 6};
```

```
area = Norm[Cross[u, v]]
```

$3\sqrt{6}$

Manipulation

- Exploring the areas of parallelograms

```
Manipulate[Norm[Cross[{1, a, 3}, {b, 5, 6}]], {a, -3, 3, 1}, {b, -2, 2, 1}]
```

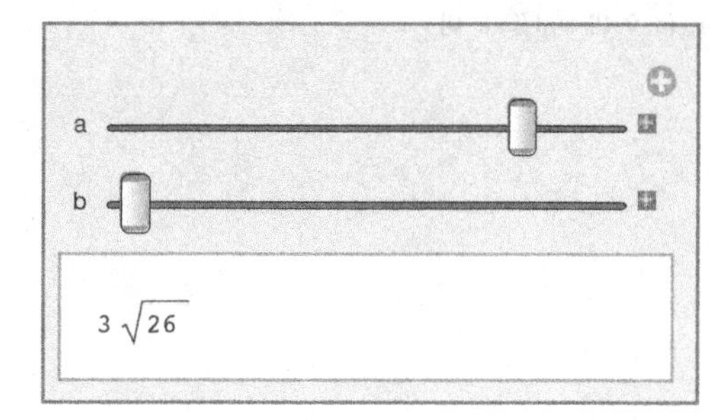

We combine **Manipulate**, **Norm**, and **Cross** to explore the area of parallelograms. If we let $a = 2$ and $b = -2$, the manipulation produces the area of the parallelogram determined by the vectors {1, 2, 3} and {-2, 5, 6}.

Area of a triangle

The *area of the triangle* described by three points $P(p_1, p_2, p_3)$, $Q(q_1, q_2, q_3)$, and $R(r_1, r_2, r_3)$ in $\mathbb{R}^3$ is half the area of the parallelogram determined by vectors

$u_{PQ} = \{p_1 - q_1, p_2 - q_2, q_3 - q_3\}$ and $v_{PR} = \{p_1 - r_1, p_2 - r_2, p_3 - r_3\}$

Illustration

- Area of a triangle described by the points $P_1 = \{1, 2, 3\}$, $P_2 = \{-2, 1, 4\}$, and $P_3 = \{5, 4, 2\}$

```
P1 = {1, 2, 3}; P2 = {-2, 1, 4}; P3 = {5, 4, 2};
```

```
u = P1 - P2;  v = P1 - P3;
```

$$\text{area} = \frac{1}{2}\,\text{Norm[Cross[u, v]]}$$

$$\sqrt{\frac{3}{2}}$$

In the plane, the formula for the area of a triangle is half the product of the base and height of the triangle.

- Area = $\frac{1}{2}$(Base × Height)

```
v1 = {0, 0}; v2 = {6, 0}; v3 = {3, 2};
```

$$\text{area} = \frac{1}{2}\,(6 \times 2)$$

6

By putting the z-coordinate equal to zero, we can use the norm and cross product functions to find the areas of triangles in the plane.

- Area = half the norm of the cross product of the vectors {6, 0, 0} and {3, 2, 0}

$$\text{area} = \frac{1}{2}\,\text{Norm[Cross[\{6, 0, 0\}, \{3, 2, 0\}]]}$$

6

Manipulation

- Exploring the area of triangles

```
Manipulate[1/2 Norm[Cross[{6, a, 0}, {3, 2 b, 0}]], {a, -2, 2, 1}, {b, -2, 2, 1}]
```

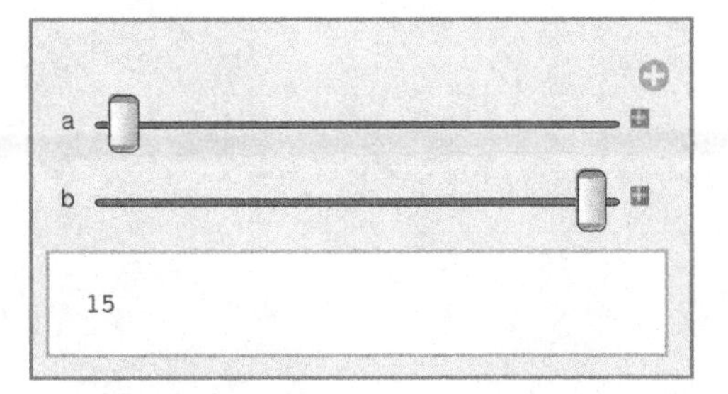

We combine **Manipulate**, **Norm**, and **Cross** to explore the areas of triangles. If we let a = - 2 and b = 2, for example, the manipulation shows that the area of the triangle determined by the vectors {6, -2, 0} and {3, 4, 0} is 15.

Array

Mathematica handles huge arrays of numeric, symbolic, textual, or any other data, with any dimension or structure. Vectors (lists of data) and matrices (two-dimensional lists of data) are special cases of *arrays*.

Illustration

- A one-dimensional array (vector in $\mathbb{R}^5$) generated by the squaring function

```
v = Array[#^2 &, 5]
```

```
{1, 4, 9, 16, 25}
```

- A two-dimensional array (a 3-by-4 matrix) generated by the square-root function

```
MatrixForm[A = Array[Sqrt[#] &, {3, 4}]]
```

$$\begin{pmatrix} 1 & 1 & 1 & 1 \\ \sqrt{2} & \sqrt{2} & \sqrt{2} & \sqrt{2} \\ \sqrt{3} & \sqrt{3} & \sqrt{3} & \sqrt{3} \end{pmatrix}$$

- A 2-by-2 array with polynomial elements

```
Clear[t]
```

```
A11 = RandomInteger[{0, 9}, {2, 2}]
```

```
A11 = {{2, 2}, {6, 5}};
```

```
cp11 = CharacteristicPolynomial[A11, t]
```

$-2 - 7\,t + t^2$

```
A12 = RandomInteger[{0, 9}, {2, 2}]

A12 = {{1, 5}, {9, 5}};

cp12 = CharacteristicPolynomial[A12, t]

-40 - 6 t + t^2

A21 = RandomInteger[{0, 9}, {2, 2}]

A21 = {{2, 6}, {0, 4}};

cp21 = CharacteristicPolynomial[A21, t]

8 - 6 t + t^2

A22 = RandomInteger[{0, 9}, {2, 2}]

A22 = {{2, 9}, {0, 0}};

cp22 = CharacteristicPolynomial[A22, t];

MatrixForm[A = {{cp11, cp12}, {cp21, cp22}}]
```

$$\begin{pmatrix} -2 - 7t + t^2 & -40 - 6t + t^2 \\ 8 - 6t + t^2 & -2t + t^2 \end{pmatrix}$$

Arrow

Vectors are frequently represented graphically by arrows in a Cartesian coordinate system to illustrate their length and direction. The *Mathematica* **Arrow** function can be used to represent *arrows* in two- and three-dimensional coordinate systems.

Illustration

- An arrow in $\mathbb{R}^2$

```
Graphics[Arrow[{{0, 0}, {1, .2}}], Axes → True]
```

0.20
0.15
0.10
0.05
0.2 0.4 0.6 0.8 1.0

- An arrow in $\mathbb{R}^3$

```
Graphics3D[Arrow[{{.1, .2, .25}, {.2, .75, 0}}], Axes → True]
```

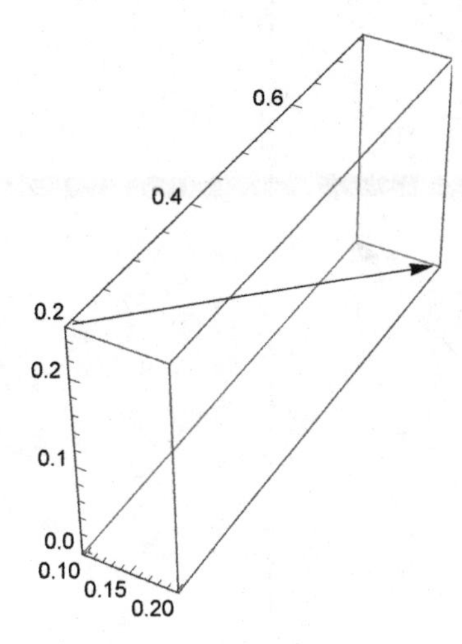

Manipulation

- Arrows in $\mathbb{R}^2$

```
Manipulate[Graphics[Arrow[{{a, b}, {1, .2}}], Axes → True], {a, -2, 2}, {b, -3, 3}]
```

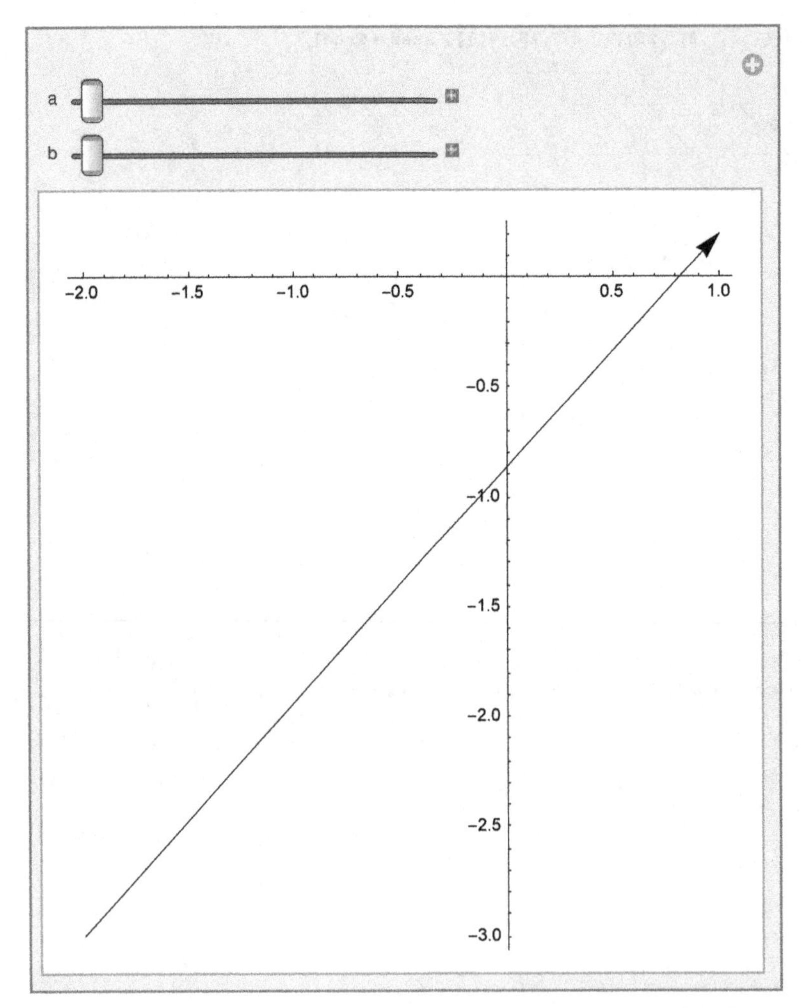

We combine **Manipulate**, **Graphics**, and **Arrow** to explore the representation of vectors as arrows. If we let $a = -2$ and $b = -3$, for example, the manipulation displays an arrow in $\mathbb{R}^2$ starting at {-2, -3} and ending at {1, .2}.

Augmented matrix

The *augmented matrix* of a linear system is the matrix of the coefficients of the variables of the system and the vector of constants of the system.

Illustration

- The augmented matrix of a linear system in three variables and two equations

```
system = {3 x + 5 y - z == 1, x - 2 y + 4 z == 5};
```

```
MatrixForm[augA = {{3, 5, -1, 1}, {1, -2, 4, 5}}]
```

$$\begin{pmatrix} 3 & 5 & -1 & 1 \\ 1 & -2 & 4 & 5 \end{pmatrix}$$

- Combining the coefficient matrix and a constant vector using **ArrayFlatten**

```
MatrixForm[A = {{3, 5, -1}, {1, -2, 4}}]
```

$$\begin{pmatrix} 3 & 5 & -1 \\ 1 & -2 & 4 \end{pmatrix}$$

```
MatrixForm[v = {{1}, {5}}]
```

$$\begin{pmatrix} 1 \\ 5 \end{pmatrix}$$

```
MatrixForm[augA = ArrayFlatten[{{A, v}}]]
```

$$\begin{pmatrix} 3 & 5 & -1 & 1 \\ 1 & -2 & 4 & 5 \end{pmatrix}$$

- Combining the coefficient matrix and a constant vector using **CoefficientArrays** and **ArrayFlatten**

```
system = {3 x + 5 y - z == 1, x - 2 y + 4 z == 5};
```

```
A = Normal[CoefficientArrays[system, {x, y, z}]][[2]];
```

```
v = {{1}, {5}};
```

```
MatrixForm[augA = ArrayFlatten[{{A, v}}]]
```

$$\begin{pmatrix} 3 & 5 & -1 & 1 \\ 1 & -2 & 4 & 5 \end{pmatrix}$$

- Combining the coefficient matrix and a constant vector using **Join**

```
system = {3 x + 5 y - z == 1, x - 2 y + 4 z == 5};
```

```
A = {{3, 5, -1}, {1, -2, 4}};
```

```
v = {{1}, {5}};
```

```
MatrixForm[Join[A, v, 2]]
```

$$\begin{pmatrix} 3 & 5 & -1 & 1 \\ 1 & -2 & 4 & 5 \end{pmatrix}$$

The **ClassroomUtilities** add-on package for *Mathematica* contains two elegant converters for alternating between linear systems and augmented matrices.

- Using the **ClassroomUtilities**

```
Needs["ClassroomUtilities`"]
```

```
eqns = {3 x + y == 4, x - y == 1}; vars = {x, y};
```

```
MatrixForm[CreateAugmentedMatrix[eqns, vars]]
```

$$\begin{pmatrix} 3 & 1 & -4 \\ 1 & -1 & -1 \end{pmatrix}$$

```
Clear[x, y]
```

```
A = {{3, 1, -4}, {1, -1, -1}}; vars = {x, y};
```

```
CreateEquations[A, vars]
```

```
{3 x + y == -4, x - y == -1}
```

Manipulation

- Linear systems and their augmented matrices

```
Manipulate[{{3 x + 5 y - z c == a, x - 2 y + 4 z == b},
  MatrixForm[{{3, 5, -c, a}, {1, -2, 4, b}}]}, {a, 0, 5, 1}, {b, 0, 5, 1}, {c, -5, 5, 1}]
```

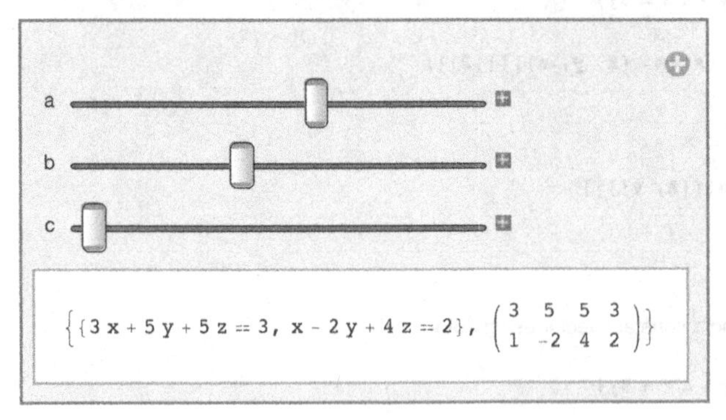

We use **Manipulate** and **MatrixForm** to explore the connection between linear systems and their augmented matrices. If we let $a = 3$, $b = 2$, and $c = - 5$, for example, the manipulation displays the resulting linear system and its augmented matrix.

B

Back substitution

A linear system $A\mathbf{v} = \mathbf{b}$ can sometimes be solved by decomposing the coefficient matrix A into a product LU, where L is a lower-triangular and U is an upper-triangular matrix. The system can then be solved by solving the systems $L\mathbf{w} = \mathbf{b}$ and $U\mathbf{v} = \mathbf{w}$. Since U is upper-triangular, the system $U\mathbf{v} = \mathbf{w}$ can be solved by *back substitution*. (The associated system $L\mathbf{w} = \mathbf{b}$ is solved by *forward substitution*.)

Illustration

- Solving a linear system in three equations and three variables

```
system = {eq1, eq2, eq3} = {x - 2 y + 5 z == 12, 4 y + 3 z == 5, 2 z == 4};
```

```
solutionz = Reduce[eq3]
```

```
z == 2
```

```
solutiony = Reduce[eq2 /. {z → 2}]
```

$$y == -\frac{1}{4}$$

```
solutionx = Reduce[eq1 /. {z → 2, y → -1/4}]
```

$$x == \frac{3}{2}$$

```
Flatten[Reduce[system, {x, y, z}]] == (solutionx && solutiony && solutionz)
```

```
True
```

Band matrix

A *band matrix* is a sparse matrix whose nonzero elements occur only on the main diagonal and on zero or more diagonals on either side of the main diagonal.

Illustration

- A diagonal matrix is a band matrix

```
MatrixForm[A = DiagonalMatrix[{1, 2, 3, 0, 5, 5}]];
```

$$A = \begin{pmatrix} 1 & 0 & 0 & 0 & 0 & 0 \\ 0 & 2 & 0 & 0 & 0 & 0 \\ 0 & 0 & 3 & 0 & 0 & 0 \\ 0 & 0 & 0 & 0 & 0 & 0 \\ 0 & 0 & 0 & 0 & 5 & 0 \\ 0 & 0 & 0 & 0 & 0 & 5 \end{pmatrix};$$

- A Hessenberg matrix is a band matrix

```
A = {{1, 2, 3, 4}, {5, 6, 7, 8}, {9, 10, 11, 12}, {13., 14, 15, 16}};

{p, h} = HessenbergDecomposition[A];

MatrixForm[Chop[h]]
```

$$\begin{pmatrix} 1. & -5.3669 & 0.443129 & 0 \\ -16.5831 & 33.0873 & -9.55746 & 0 \\ 0 & -2.20899 & -0.0872727 & 0 \\ 0 & 0 & 0 & 0 \end{pmatrix}$$

- A shear matrix is a band matrix

```
A = {{1, a}, {b, 1}};

A.{x, y}

{x + a y, b x + y}
```

- Two-dimensional display of a band matrix

```
diagonal = {0, 2, 1, 5, 8, 0, 8, 9};

A = SparseArray[DiagonalMatrix[diagonal]]
```

```
A // MatrixForm
```

$$\begin{pmatrix} 0 & 0 & 0 & 0 & 0 & 0 & 0 & 0 \\ 0 & 2 & 0 & 0 & 0 & 0 & 0 & 0 \\ 0 & 0 & 1 & 0 & 0 & 0 & 0 & 0 \\ 0 & 0 & 0 & 5 & 0 & 0 & 0 & 0 \\ 0 & 0 & 0 & 0 & 8 & 0 & 0 & 0 \\ 0 & 0 & 0 & 0 & 0 & 0 & 0 & 0 \\ 0 & 0 & 0 & 0 & 0 & 0 & 8 & 0 \\ 0 & 0 & 0 & 0 & 0 & 0 & 0 & 9 \end{pmatrix}$$

- A Jordan block is a band matrix

```
A = {{27, 48, 81}, {-6, 0, 0}, {1, 0, 3}};

{s, j} = JordanDecomposition[A];

MatrixForm[j]
```

$$\begin{pmatrix} 6 & 0 & 0 \\ 0 & 12 & 1 \\ 0 & 0 & 12 \end{pmatrix}$$

Basic variable of a linear system

The variables of a linear system $A\mathbf{v} = \mathbf{b}$ corresponding to the pivot columns of the coefficient matrix A are the *basic variables* of the system.

Illustration

- Basic variables of a linear system

```
system = {3 x + 4 y - z == 9, x + y + 4 z == 1};

A = {{3, 4, -1}, {1, 1, 4}};

MatrixForm[RowReduce[A]]
```

$$\begin{pmatrix} 1 & 0 & 17 \\ 0 & 1 & -13 \end{pmatrix}$$

Since the first and second columns of A are the pivot columns of the given linear system, the variables x and y are basic variables of the system.

Basis of a vector space

A *basis* of a vector space is a linearly independent subset of vectors that spans the space.

Illustration

- A basis for $\mathbb{R}^3$

```
basis = {{0, 3, 2}, {5, 8, 1}, {5, 0, 6}}
```

  - Linear independence

```
Solve[a {0, 3, 2} + b {5, 8, 1} + c {5, 0, 6} == {0, 0, 0}]

{{a → 0, b → 0, c → 0}}
```

  - Span

```
span = Solve[a {0, 3, 2} + b {5, 8, 1} + c {5, 0, 6} == {x, y, z}, {a, b, c}]
```

$$\left\{\left\{a \to \frac{1}{155}(-48 x + 25 y + 40 z),\ b \to \frac{1}{155}(18 x + 10 y - 15 z),\ c \to \frac{1}{155}(13 x - 10 y + 15 z)\right\}\right\}$$

```
span /. {x → 1, y → 1, z → 1}
```

$$\left\{\left\{a \to \frac{17}{155},\ b \to \frac{13}{155},\ c \to \frac{18}{155}\right\}\right\}$$

$$\text{coordinates} = \left\{\left\{a \to \frac{17}{155},\ b \to \frac{13}{155},\ c \to \frac{18}{155}\right\}\right\}$$

$$\left\{\left\{a \to \frac{17}{155},\ b \to \frac{13}{155},\ c \to \frac{18}{155}\right\}\right\}$$

```
{1, 1, 1} == a {0, 3, 2} + b {5, 8, 1} + c {5, 0, 6} /. coordinates
```

```
{True}
```

The columns or rows of an invertible matrix are linearly independent and every vector in the space can be written as a unique linear combination of the columns or the rows of the matrix, provided that the order of the columns or rows is respected. Therefore the columns or rows form a basis of the column space or row space of the matrix.

- An invertible matrix considered as a basis for $\mathbb{R}^3$

```
MatrixForm[A = RandomInteger[{0, 9}, {3, 3}]];
```

$$A = \begin{pmatrix} 5 & 8 & 5 \\ 9 & 9 & 0 \\ 8 & 0 & 6 \end{pmatrix};$$

```
Det[A]
```

```
-522
```

```
Solve[{x, y, z} == a A[[1]] + b A[[2]] + c A[[3]], {x, y, z}]
```

```
{{x → 5 a + 9 b + 8 c, y → 8 a + 9 b, z → 5 a + 6 c}}
```

```
{x, y, z} /. {{x → 5 a + 9 b + 8 c, y → 8 a + 9 b, z → 5 a + 6 c}}
```

```
{{5 a + 9 b + 8 c, 8 a + 9 b, 5 a + 6 c}}
```

The result shows that every vector {*x*, *y*, *z*} in $\mathbb{R}^3$ can be written as a linear combination of the rows of the matrix *A*. To show that the rows form a basis, the uniqueness of the linear combination still needs to be verified.

- The rank (number of linearly independent rows or columns) of a matrix

```
A = {{1, 2, 3}, {4, 5, 6}, {7, 8, 10}};
```

```
MatrixRank[A]
```

3

This tells us that the three rows of the matrix m are linearly independent. So are the three columns.

```
lc = {x, y, z} == a A[[1]] + b A[[2]] + c A[[3]]
```

{x, y, z} == {a + 4 b + 7 c, 2 a + 5 b + 8 c, 3 a + 6 b + 10 c}

```
Solve[lc, {x, y, z}]
```

{{x → a + 4 b + 7 c, y → 2 a + 5 b + 8 c, z → 3 a + 6 b + 10 c}}

Since the rows of the matrix *A* are linearly independent, each vector {*x*, *y*, *z*} in the space $\mathbb{R}^3$ can be written as a unique linear combination of the rows of *A*. Randomly generated *n*-by-*n* matrices with real entries are often (usually) invertible. Their rows and columns can therefore be used as bases for $\mathbb{R}^n$.

- A basis for $\mathbb{R}^5$ derived from randomly generated matrices

```
MatrixForm[A = RandomInteger[{0, 9}, {5, 5}]];
```

$$A = \begin{pmatrix} 9 & 7 & 9 & 3 & 5 \\ 3 & 9 & 2 & 5 & 7 \\ 4 & 7 & 5 & 6 & 1 \\ 7 & 3 & 4 & 4 & 1 \\ 1 & 2 & 5 & 3 & 1 \end{pmatrix};$$

```
MatrixRank[A]
```

5

```
Reduce[a A[[1]] + b A[[2]] + c A[[3]] + d A[[4]] + e A[[5]] == 0, {a, b, c, d, e}]
```

a == 0 && b == 0 && c == 0 && d == 0 && e == 0

This shows that the rows of the matrix *A* are linearly independent and form a basis for $\mathbb{R}^5$. Similarly, the calculation

```
Reduce[a A[[All,1]] + b A[[All,2]] + c A[[All,3]] + d A[[All,4]] + e A[[All,5]] == 0, {a, b, c, d, e}]
```

a == 0 && b == 0 && c == 0 && d == 0 && e == 0

shows that the columns of *A* are linearly independent and also form a basis for $\mathbb{R}^5$.

Manipulation

- Linearly dependent and independent vectors

```
Manipulate[Evaluate[Reduce[a {1, 2, 3} + b {4, 5, 6 } + c {7, 8, 9 d} == {0, 0, 0}, {a, b, c}]],
 {d, -2, 2, 1}]
```

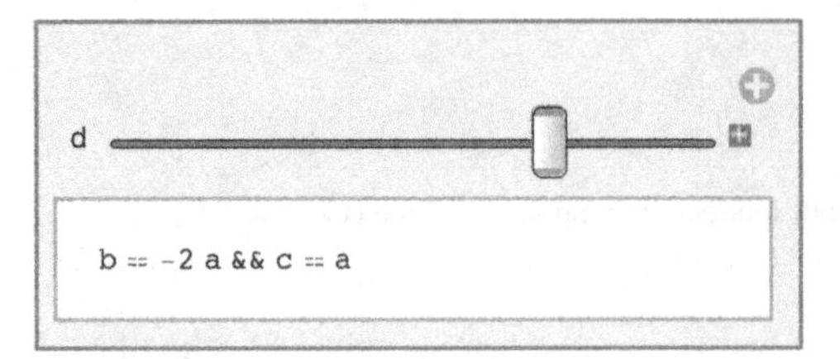

We combine **Manipulate** and **Reduce** to explore the linear dependence and independence of vectors in $\mathbb{R}^3$. The displayed window shows, for example, that if we let $d = 1$, the generated vectors are linearly dependent. On the other hand, if we let $d = 2$, the generated vectors are linearly independent.

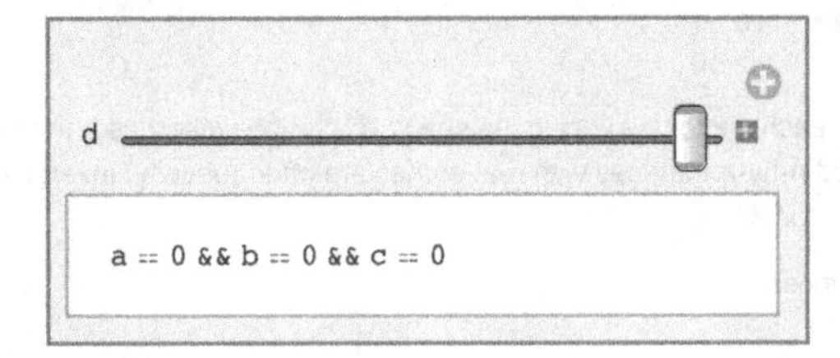

- Exploring the invertibility of a matrix

```
Manipulate[Inverse[{{a, b}, {2, 3}}], {a, -2, 2, 1}, {b, -2, 2, 1}]
```

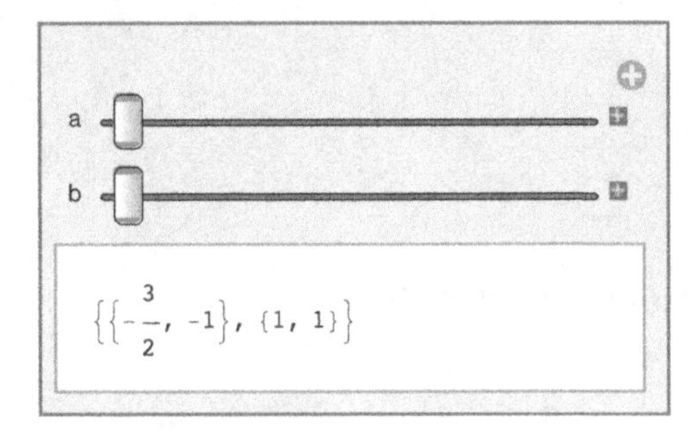

We combine **Manipulate** and **Inverse** to explore the invertibility of matrices. If we let $a = b = -2$, for example, the window displays the inverse of the generated matrix.

Bijective linear transformation

A linear transformation $T : V \longrightarrow W$ from a vector space V to a vector space W is bijective if it is both injective (one-to-one) and surjective (onto). Bijective transformations are invertible in the sense that there exists a linear transformation $S : W \longrightarrow V$ for which $S[T[\mathbf{v}]] = \mathbf{v}$ and $T[S[\mathbf{w}]] = \mathbf{w}$ for all $\mathbf{v}$ in V and $\mathbf{w}$ in W.

Illustration

- A bijective linear transformation from $\mathbb{R}^2$ to $\mathbb{R}^2$

```
T[{x_, y_}] := {{Cos[θ], Sin[θ]}, {-Sin[θ], Cos[θ]}}.{x, y}
```

```
S[{x_, y_}] := {{Cos[θ], -Sin[θ]}, {Sin[θ], Cos[θ]}}.{x, y}

θ = π / 3; x = 1; y = 1;

{T[{x, y}], S[{x, y}]}
```

$$\left\{\left\{\frac{1}{2}+\frac{\sqrt{3}}{2}, \frac{1}{2}-\frac{\sqrt{3}}{2}\right\}, \left\{\frac{1}{2}-\frac{\sqrt{3}}{2}, \frac{1}{2}+\frac{\sqrt{3}}{2}\right\}\right\}$$

```
Simplify[Composition[S, T][{x, y}] == {x, y}]

True

Simplify[Composition[T, S][{x, y}] == {x, y}]

True
```

Bilinear functional

If V is a real vector space, then a function T[**u**, **v**] from $V \times V$ with values in the real vector space ℝ is a (real) bilinear functional if it is linear both in **u** and **v**. A similar definition leads to the idea of a complex bilinear functional. In the context of linear functionals, the values T[*u*, **v**] are usually written as ⟨**u**, **v**⟩. Bilinearity is defined by the following properties:

Properties of bilinear functionals

$$\langle a\,\mathbf{u} + b\,\mathbf{v}, \mathbf{w}\rangle = a\,\langle \mathbf{u}, \mathbf{w}\rangle + b\,\langle \mathbf{u}, \mathbf{w}\rangle \tag{1}$$

$$\langle \mathbf{u}, c\,\mathbf{v} + d\,\mathbf{w}\rangle = c\,\langle \mathbf{u}, \mathbf{v}\rangle + d\,\langle \mathbf{u}, \mathbf{w}\rangle \tag{2}$$

Illustration

- The trace as a bilinear functional on 2-by-2 real matrices

```
Clear[a, b, c, d, e, f, g, h]

A = {{a, b}, {c, d}}; B = {{e, f}, {g, h}};

{Tr[A], Tr[B]}

{a + d, e + h}

Tr[3 A + 5 B] == Expand[3 Tr[A] + 5 Tr[B]]

True
```

- Multiplication as a bilinear functional on $\mathbb{R}^2$

```
Clear[x, y, z, a, b, c, d]
```

```
T[x_, y_] := x y

left = Expand[T[a x + b y, z]] == a T[x, z] + b T[y, z];

right = Expand[T[x, c y + d z]] == c T[x, y] + d T[x, z];

a = b = 1; x = 5; y = -6; z = 3;

{T[a x + b y, z], T[x, z] + b T[y, z]}

{-3, -3}

c = d = 1; x = 2; y = 7; z = -9;

{T[x, c y + d z], c T[x, y] + d T[x, z]}

{-4, -4}
```

- The dot product as a bilinear functional on $\mathbb{R}^4$

```
Clear[x, y, z, a, b, c, d, e, f, g, h, i, j, k, l]

cr = CharacterRange["a", "l"]

{a, b, c, d, e, f, g, h, i, j, k, l}

x = {a, b, c, d}; y = {e, f, g, h}; z = {i, j, k, l};

Expand[Dot[5 x + 7 y, z] == 5 Dot[x, z] + 7 Dot[y, z]]

True
```

C

Cartesian product of vector spaces

The *Cartesian product* $U \times V$ of two finite-dimensional vector spaces *U* and *V* consists of all pairs {**u**, **v**} of vectors **u** in U and **v** in *V*. The vector space operations are defined componentwise:

$$\{u_1, v_1\} + \{u_2, v_2\} = \{u_1 + u_2, v_1 + v_2\} \quad (1)$$

$$a\{u, v\} = \{a\,u, a\,v\} \quad (2)$$

Illustration

- The Cartesian product of two vector spaces

```
basisU = {{1, 0}, {0, 1}}

{{1, 0}, {0, 1}}

basisV = {{1, 2, 3}, {0, 4, 1}}

{{1, 2, 3}, {0, 4, 1}}

basisUV = Tuples[{basisU, basisV}]

{{{1, 0}, {1, 2, 3}}, {{1, 0}, {0, 4, 1}}, {{0, 1}, {1, 2, 3}}, {{0, 1}, {0, 4, 1}}}
```

Cauchy–Schwarz inequality

If **u** and **v** are two vectors in an inner product space *V*, then the Cauchy–Schwarz *inequality* states that for all vectors **u** and **v** in *V*,

$$|\langle u, v\rangle|^2 \leq \mathrm{Dot}[\langle u, u\rangle, \langle v, v\rangle] \quad (1)$$

The bilinear functional <**u**, **v**> is the inner product of the space *V*. The inequality becomes an equality if and only if **u** and **v** are linearly dependent.

Illustration

- The Cauchy–Schwarz inequality of two vectors in $\mathbb{R}^3$

```
u = {1, 2, 3}; v = {-4, 5, 6};

Abs[Dot[u, v]]^2 ≤ Dot[u, u] Dot[v, v]

True
```

```
{Abs[Dot[u, v]]^2, Dot[u, u] Dot[v, v]}
```

```
{576, 1078}
```

If **w** is a multiple of **u**, for example, then the inequality becomes an equality:

```
w = 3 u;
```

```
Abs[Dot[u, w]]^2 == Dot[u, u] Dot[w, w]
```

```
True
```

- The Cauchy–Schwarz inequality of two vectors in $\mathbb{C}^2$

```
u = {3 + I, -5}; v = {5 - 2 I, 6 + I};
```

```
Abs[Dot[u, Conjugate[v]]]^2 ≤ Dot[u, Conjugate[u]] Dot[v, Conjugate[v]]
```

```
True
```

Manipulation

- Cauchy–Schwarz inequalities

```
u = {b, 2, 3}; v = {a, 5, 6};
```

```
Manipulate[Evaluate[Abs[Dot[u, v]]^2 ≤ Dot[u, u] Dot[v, v]], {a, -2, 2}, {b, -5, 5}]
```

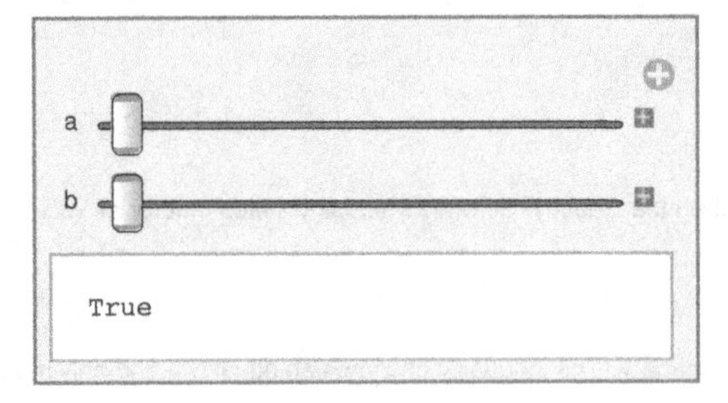

We can combine **Manipulate**, **Abs**, and **Dot** to explore the Cauchy–Schwarz inequality. If we assign the values a = - 2 and b = - 5, for example, the manipulation shows that the Cauchy–Schwarz inequality holds for the vectors {- 5, 2, 3} and {- 2, 5, 6}.

Cayley–Hamilton theorem

One of the best-known properties of characteristic polynomials is that all square real or complex matrices satisfy their characteristic polynomials. This result is known as the *Cayley–Hamilton theorem*.

Illustration

- The Cayley–Hamilton theorem illustrated with a 3-by-3 matrix

```
A = {{2, 1, 1}, {4, 1, 7}, {5, 3, 0}};
```

$$A = \begin{pmatrix} 2 & 1 & 1 \\ 4 & 1 & 7 \\ 5 & 3 & 0 \end{pmatrix};$$

```
characteristicPolynomialA = Det[A - t IdentityMatrix[3]]
```

$28\,t + 3\,t^2 - t^3$

```
charPolyA = -MatrixPower[A, 3] + 3 MatrixPower[A, 2] + 28 A
```

```
{{0, 0, 0}, {0, 0, 0}, {0, 0, 0}}
```

This shows that the matrix *A* satisfies of its own characteristic polynomial.

Manipulation

- The Cayley–Hamilton illustrated with **Manipulate**

$$A = \begin{pmatrix} 2 & 1 & 1 \\ a & 1 & 7 \\ 5 & 3 & 0 \end{pmatrix};$$

```
cpA = CharacteristicPolynomial[A, t]
```

$-12 + 3\,a + 24\,t + a\,t + 3\,t^2 - t^3$

```
Manipulate[
 Evaluate[-MatrixPower[A, 3] + 3 MatrixPower[A, 2] + (24 + a) A - (12 - 3 a) IdentityMatrix[3]],
 {a, -5, 5, 1}]
```

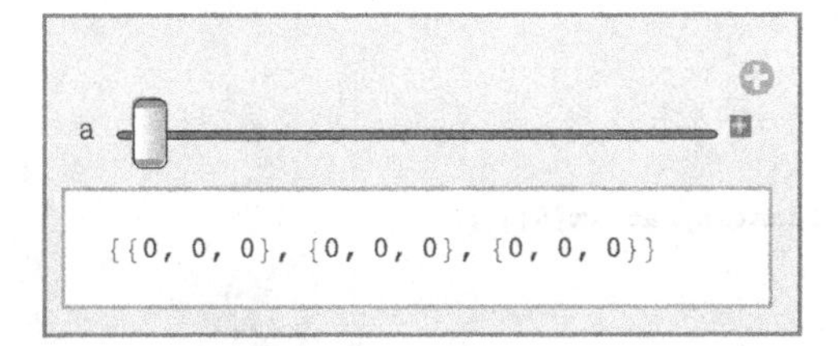

We can use **Manipulate**, **MatrixPower**, and **IdentityMatrix** to show that the Cayley–Hamilton theorem holds for the generated matrices. If we let a = - 5, for example, the manipulation shows that the Cayley–Hamilton theorem holds for the generated matrix.

Change-of-basis matrix

See Coordinate conversion matrix

Characteristic polynomial

The *characteristic polynomial* p[t] of an *n*-by-*n* real or complex matrix *A* in the variable *t* is the polynomial

```
p[t] = Det[A - t IdentityMatrix[n]]
```
(1)

The roots of p[t] are the eigenvalues of the matrix *A*.

Illustration

- The characteristic polynomial of a 5-by-5 matrix

```
Clear[A, t]

A = RandomInteger[{0, 9}, {5, 5}];
```

$$A = \begin{pmatrix} 4 & 3 & 8 & 1 & 9 \\ 1 & 7 & 4 & 9 & 8 \\ 5 & 3 & 8 & 5 & 1 \\ 2 & 3 & 6 & 0 & 5 \\ 4 & 9 & 8 & 6 & 8 \end{pmatrix};$$

```
CharacteristicPolynomial[A, t]
```

$3206 - 851\,t - 675\,t^2 - 8\,t^3 + 27\,t^4 - t^5$

- Using determinants to find the characteristic polynomial of a 5-by-5 matrix

```
Det[A - t IdentityMatrix[5]]
```

$3206 - 851\,t - 675\,t^2 - 8\,t^3 + 27\,t^4 - t^5$

The two calculations show that

```
CharacteristicPolynomial[A, t] == Det[A - t IdentityMatrix[5]]

True
```

Cholesky decomposition

The *Cholesky decomposition* of a square matrix is a decomposition of a Hermitian, positive-definite matrix into a product of a lower-triangular matrix and its conjugate transpose.

Illustration

- Cholesky decomposition of a 2-by-2 symmetric positive-definite matrix

```
MatrixForm[A = {{2, 1}, {1, 2}}]
```

$$\begin{pmatrix} 2 & 1 \\ 1 & 2 \end{pmatrix}$$

```
{HermitianMatrixQ[A], PositiveDefiniteMatrixQ[A]}
```

```
{True, True}
```

```
cdA = CholeskyDecomposition[A]
```

$$\left\{\left\{\sqrt{2}, \frac{1}{\sqrt{2}}\right\}, \left\{0, \sqrt{\frac{3}{2}}\right\}\right\}$$

```
ConjugateTranspose[cdA].cdA
```

```
{{2, 1}, {1, 2}}
```

- Cholesky decomposition of a 3-by-3 symmetric positive-definite matrix

$$A = \begin{pmatrix} 2 & -1 & 0 \\ -1 & 2 & 1 \\ 0 & 1 & 2 \end{pmatrix};$$

```
{PositiveDefiniteMatrixQ[A], HermitianMatrixQ[A]}
```

```
{True, True}
```

```
MatrixForm[cdA = CholeskyDecomposition[A]]
```

$$\begin{pmatrix} \sqrt{2} & -\frac{1}{\sqrt{2}} & 0 \\ 0 & \sqrt{\frac{3}{2}} & \sqrt{\frac{2}{3}} \\ 0 & 0 & \frac{2}{\sqrt{3}} \end{pmatrix}$$

```
MatrixForm[ConjugateTranspose[cdA].cdA]
```

$$\begin{pmatrix} 2 & -1 & 0 \\ -1 & 2 & 1 \\ 0 & 1 & 2 \end{pmatrix}$$

Codimension of a vector subspace

If *W* is a subspace of a finite-dimensional vector space *V*, then the *codimension* of the subspace *W* in the space *V* is the difference *dim*[*V*] - *dim*[*W*] between the dimension of the space *V* and the dimension of the subspace *W*.

Illustration

- A subspace of $\mathbb{R}^3$ with codimension 2 in $\mathbb{R}^3$

```
V = ℝ^3; W = {{a, 0, b} : a, b ∈ Reals};
```

The dimension of *V* is 3 and the dimension of the subspace *W* is 2. Hence the codimension of *W* is 1.

- A subspace of $\mathbb{R}^5$ with codimension 3 in $\mathbb{R}^5$

```
V = ℝ^5; W = span[{{1, 0, 1, 0, 1}, {0, 1, 1, 0, 0}}]

W = {a {1, 0, 1, 0, 1} + b {0, 1, 1, 0, 0} : a, b ∈ ℝ} = {a, b, b, 0, a} : a, b ∈ ℝ}
```

Since *W* is a subspace of $\mathbb{R}^5$ of dimension 2, its codimension in $\mathbb{R}^5$ is 3.

Codomain of a linear transformation

The *codomain* of a linear transformation *T* is the vector space in which the transformation takes its values.

The notation $T : A \longrightarrow B$ identifies the vector space *A* as the domain of *T* and the vector space *B* as its codomain. The *range* of *T* is the subspace of the codomain of *T* consisting of all values of *T*. If the range of *T* coincides with the codomain of *T*, the transformation *T* is said to be *onto* or *surjective*. If there exists a transformation $S : B \longrightarrow A$ for which the composition $(S \cdot T)$ of *S* and *T* is the identity transformation $idA : A \longrightarrow A$ on *A* and the composition $(T \cdot S)$ is the identity transformation $idB : B \longrightarrow B$, then the transformation *T* is said to be *one-to-one* or *injective*. If *T* is both onto and one-to-one, it is said to be *bijective*.

Illustration

- An injective linear transformation $T: \mathbb{R}^2 \longrightarrow \mathbb{R}^3$

```
T[{x_, y_}] := {x + y, 2 x - y, y};

T[{1, 2}]

{3, 0, 2}
```

The codomain of the transformation *T* is $\mathbb{R}^3$. However, not every vector in $\mathbb{R}^3$ is a value of *T*. For example, the vector {*5, 5, 5*} in $\mathbb{R}^3$ is not a value of *T*.

```
Clear[x, y, a, b, c, d]

Solve[{x + y, 2 x - y, y} == {5, 5, 5}, {x, y}]

{}
```

- A surjective linear transformation $T: \mathbb{R}^3 \longrightarrow \mathbb{R}^2$

```
T[{x_, y_, z_}] := {y, x};
```

```
T[{-2, 3, 5}]
```

```
{3, -2}
```

The codomain of the transformation T is $\mathbb{R}^2$. And, any vector $\{x, y\}$ in $\mathbb{R}^2$ is a value of T.

```
Solve[{x, y} == T[{a, b, c}], {a, b, c}]
```

Solve::svars : Equations may not give solutions for all "solve" variables. ≫

```
{{a → y, b → x}}
```

```
T[{a, b, c}] /. {a → y, b → x}
```

```
{x, y}
```

- A bijective linear transformation $T : \mathbb{R}[t, 3] \longrightarrow \mathbb{R}^4$

```
T[a_ + b_ t_ + c_ t_^2 + d_ t_^3] := {a, b, c, d}
```

```
S[{a_, b_, c_, d_}] := a + b t + c t^2 + d t^3
```

```
Composition[S, T][a + b t + c t^2 + d t^3]
```

```
a + b t + c t^2 + d t^3
```

```
Composition[T, S][{a, b, c, d}]
```

```
{a, b, c, d}
```

Cofactor matrix

The *cofactor matrix* of a square matrix A is the matrix of cofactors of A. The cofactors *cfAij* are $(-1)^{i+j}$ times the determinants of the submatrices *Aij* obtained from A by deleting the ith rows and j^{th} columns of A. The cofactor matrix is also referred to as the *minor matrix*. It can be used to find the inverse of A.

Illustration

- Cofactor matrix of a 3-by-3 matrix

$$A = \begin{pmatrix} 4 & 5 & 2 \\ 1 & 4 & 6 \\ 7 & 0 & 6 \end{pmatrix};$$

$$\text{cfA11} = (-1)^{1+1}\,\text{Det}\left[\begin{pmatrix} 4 & 6 \\ 0 & 6 \end{pmatrix}\right];\ \text{cfA12} = (-1)^{1+2}\,\text{Det}\left[\begin{pmatrix} 1 & 6 \\ 7 & 6 \end{pmatrix}\right];\ \text{cfA13} = (-1)^{1+3}\,\text{Det}\left[\begin{pmatrix} 1 & 4 \\ 7 & 0 \end{pmatrix}\right];$$

$$\text{cfA21} = (-1)^{2+1}\,\text{Det}\left[\begin{pmatrix} 5 & 2 \\ 0 & 6 \end{pmatrix}\right];\ \text{cfA22} = (-1)^{2+2}\,\text{Det}\left[\begin{pmatrix} 4 & 2 \\ 7 & 6 \end{pmatrix}\right];\ \text{cfA23} = (-1)^{2+3}\,\text{Det}\left[\begin{pmatrix} 4 & 5 \\ 7 & 0 \end{pmatrix}\right];$$

```
cfA31 = (-1)^(3+1) Det[(5 2; 4 6)]; cfA32 = (-1)^(3+2) Det[(4 2; 1 6)]; cfA33 = (-1)^(3+3) Det[(4 5; 1 4)];
```

```
MatrixForm[cfA = {{cfA11, cfA12, cfA13}, {cfA21, cfA22, cfA23}, {cfA31, cfA32, cfA33}}]
```

$$\begin{pmatrix} 24 & 36 & -28 \\ -30 & 10 & 35 \\ 22 & -22 & 11 \end{pmatrix}$$

```
(1/Det[A]) Transpose[cfA] == Inverse[A]
```

```
True
```

The following *Mathematica* definition can be used to calculate the cofactors of a given matrix:

```
Clear[A, i, j]
```

$$A = \begin{pmatrix} 4 & 5 & 2 \\ 1 & 4 & 6 \\ 7 & 0 & 6 \end{pmatrix};$$

```
Cofactor[m_List?MatrixQ, {i_Integer, j_Integer}] :=
  (-1)^(i + j) Det[Drop[Transpose[Drop[Transpose[m], {j}]], {i}]]
```

The Cofactor command is a defined command and needs to be activated before it can be used by typing *Shift + Enter*.

```
CfA = MatrixForm[Table[Cofactor[A, {i, j}], {i, 1, 3}, {j, 1, 3}]];
```

```
MatrixForm[cfA] == CfA
```

```
True
```

The next *Mathematica* definitions can be used to calculate the cofactor matrix in one step:

```
MinorMatrix[m_List?MatrixQ] := Map[Reverse, Minors[m], {0, 1}]
```

```
CofactorMatrix[m_List?MatrixQ] := MapIndexed[#1 (-1)^(Plus @@ #2) &, MinorMatrix[m], {2}]
```

Both commands are defined commands and must be activated by typing *Shift + Enter* before they can be used. The commands are defined in *MathWorld* at http://mathworld.wolfram.com/Cofactor.html.

```
MatrixForm[MinorMatrix[A]]
```

$$\begin{pmatrix} 24 & -36 & -28 \\ 30 & 10 & -35 \\ 22 & 22 & 11 \end{pmatrix}$$

Column space

The *column space* of a matrix A is the set of all linear combinations of the columns of A. The column space of the matrix tells us if and when a linear system $A\mathbf{v} = \mathbf{b}$ has a solution $\mathbf{v}$.

Illustration

- A basis for the column space of a 3-by-4 matrix

```
MatrixForm[A = {{1, 2, 3, 4}, {5, 6, 7, 8}, {0, 1, 3, 0}}]
```

$$\begin{pmatrix} 1 & 2 & 3 & 4 \\ 5 & 6 & 7 & 8 \\ 0 & 1 & 3 & 0 \end{pmatrix}$$

```
MatrixForm[B = RowReduce[A]]
```

$$\begin{pmatrix} 1 & 0 & 0 & -5 \\ 0 & 1 & 0 & 9 \\ 0 & 0 & 1 & -3 \end{pmatrix}$$

The reduction shows that the first three columns of the matrix B are pivot columns. They are linearly independent and span the column space of *A*.

- Linear independence

```
Clear[a, b, c]
```

$$\text{Solve}\left[a \begin{pmatrix} 1 \\ 5 \\ 0 \end{pmatrix} + b \begin{pmatrix} 2 \\ 6 \\ 1 \end{pmatrix} + c \begin{pmatrix} 3 \\ 7 \\ 3 \end{pmatrix} == \begin{pmatrix} 0 \\ 0 \\ 0 \end{pmatrix}, \{a, b, c\}\right]$$

```
{{a → 0, b → 0, c → 0}}
```

Thus the zero vector is a linear combination of the first three columns if and only if the coefficients *a*, *b*, and *c* are zero.

The fourth column of *A* is a linear combination of the first three:

$$\text{Solve}\left[a \begin{pmatrix} 1 \\ 5 \\ 0 \end{pmatrix} + b \begin{pmatrix} 2 \\ 6 \\ 1 \end{pmatrix} + c \begin{pmatrix} 3 \\ 7 \\ 3 \end{pmatrix} == \begin{pmatrix} 4 \\ 8 \\ 0 \end{pmatrix}, \{a, b, c\}\right]$$

```
{{a → -5, b → 9, c → -3}}
```

$$-5 \begin{pmatrix} 1 \\ 5 \\ 0 \end{pmatrix} + 9 \begin{pmatrix} 2 \\ 6 \\ 1 \end{pmatrix} - 3 \begin{pmatrix} 3 \\ 7 \\ 3 \end{pmatrix} == \begin{pmatrix} 4 \\ 8 \\ 0 \end{pmatrix}$$

```
True
```

- Spanning

```
Clear[a, b, c, x, y, z]
```

$$\text{Solve}\left[a \begin{pmatrix} 1 \\ 5 \\ 0 \end{pmatrix} + b \begin{pmatrix} 2 \\ 6 \\ 1 \end{pmatrix} + c \begin{pmatrix} 3 \\ 7 \\ 3 \end{pmatrix} == \begin{pmatrix} x \\ y \\ z \end{pmatrix}, \{a, b, c\}\right]$$

$$\left\{\left\{a \to \frac{1}{4}(-11\,x + 3\,y + 4\,z),\ b \to \frac{1}{4}(15\,x - 3\,y - 8\,z),\ c \to \frac{1}{4}(-5\,x + y + 4\,z)\right\}\right\}$$

This shows that every vector {x. y, z} in the column space of the matrix *A* can be constructed as a linear combination of the first three columns of the matrix *A*.

- The column space of a 3-by-5 matrix determined by the pivot columns of the matrix

```
MatrixForm[Transpose[A = RandomInteger[{0, 9}, {5, 3}]]]
```

$$A = \begin{pmatrix} 0 & 4 & 8 & 3 & 4 \\ 8 & 5 & 0 & 6 & 9 \\ 2 & 7 & 3 & 8 & 1 \end{pmatrix};$$

```
MatrixForm[Transpose[A]]
```

$$\begin{pmatrix} 0 & 4 & 8 & 3 & 4 \\ 8 & 5 & 0 & 6 & 9 \\ 2 & 7 & 3 & 8 & 1 \end{pmatrix}$$

```
MatrixForm[B = RowReduce[Transpose[A]]]
```

$$\begin{pmatrix} 1 & 0 & 0 & -\frac{11}{272} & \frac{26}{17} \\ 0 & 1 & 0 & \frac{43}{34} & -\frac{11}{17} \\ 0 & 0 & 1 & -\frac{35}{136} & \frac{14}{17} \end{pmatrix}$$

This shows that the first three columns of *A* are the pivot columns of *A*. They are linearly independent and span the column space of *A*.

- Linear independence

$$\text{Solve}[a\,B_{[[All,1]]} + b\,B_{[[All,2]]} + c\,B_{[[All,3]]} == \{0, 0, 0\}, \{a, b, c\}]$$

```
{{a → 0, b → 0, c → 0}}
```

- The third column as a linear combination of the first three columns

```
Clear[a, b, c]
```

$$\text{Solve}[a\,B_{[[All,1]]} + b\,B_{[[All,2]]} + c\,B_{[[All,3]]} == \{-11/272, 43/34, -35/136\}, \{a, b, c\}]$$

$$\left\{\left\{a \to -\frac{11}{272},\ b \to \frac{43}{34},\ c \to -\frac{35}{136}\right\}\right\}$$

- The fourth column as a linear combination of the first three columns

```
Clear[a, b, c]
```

```
Solve[a B[[All,1]] + b B[[All,2]] + c B[[All,3]] == {26/17, -11/17, 14/17}, {a, b, c}]
```

$\{\{a \to \frac{26}{17}, b \to -\frac{11}{17}, c \to \frac{14}{17}\}\}$

Column vector

A *column vector* is a vertical list of scalars. In *Mathematica*, column vectors are represented as lists of singleton lists.

Illustration

- A column vector of height 3 in standard form

Unless required to do otherwise, *Mathematica* outputs column vectors in standard form.

```
columnvector = {{1}, {2}, {3}}
```

```
{{1}, {2}, {3}}
```

- A column vector of height 3 in traditional form.

```
MatrixForm[columnvector]
```

$$\begin{pmatrix} 1 \\ 2 \\ 3 \end{pmatrix}$$

Companion matrix

The (Frobenius) *companion matrix* of the monic polynomial

$$p[t] = a_0 + a_1 t + \cdots + a_{n-1} t^{n-1} + t^n \quad (1)$$

is the square matrix

$$C[p] = \begin{pmatrix} 0 & 0 & \ldots & 0 & -a_0 \\ 1 & 0 & \ldots & 0 & -a_1 \\ 0 & 1 & \ldots & 0 & -a_2 \\ \vdots & \vdots & \ddots & \vdots & \vdots \\ 0 & 0 & \square & 1 & -a_{n-1} \end{pmatrix} \quad (2)$$

The characteristic polynomial and the minimal polynomial of C[p] are both equal to p[t]. Sometimes the companion matrix is defined as the transpose of C[p].

Illustration

- Companion matrix of a monic polynomial of degree 3

```
p = t^3 - 4 t^2 + 5;
```

The following defined *Mathematica* function can be used to calculate the companion matrix of a polynomial:

```
CompanionMatrix[p_, x_] := Module[{n, w = CoefficientList[p, x]}, w = -w / Last[w];
  n = Length[w] - 1;
  SparseArray[{{i_, n} :> w[[i]], {i_, j_} /; i == j + 1 -> 1}, {n, n}]]
```

The command is defined in *MathWorld* and can be found at http://mathworld.wolfram.com/CompanionMatrix.html.

```
MatrixForm[Normal[CompanionMatrix[p, t]]]
```

$$\begin{pmatrix} 0 & 0 & -5 \\ 1 & 0 & 0 \\ 0 & 1 & 4 \end{pmatrix}$$

- Companion matrix of a monic polynomial of degree 4

```
p = 3 + 5 t - 7 t^3 + t^4;
```

```
MatrixForm[A = Normal[CompanionMatrix[p, t]]]
```

$$\begin{pmatrix} 0 & 0 & 0 & -3 \\ 1 & 0 & 0 & -5 \\ 0 & 1 & 0 & 0 \\ 0 & 0 & 1 & 7 \end{pmatrix}$$

Complex conjugate

The *complex conjugate* of the complex number ($a + b\,i$) is the number ($a - b\,i$).

Illustration

- The complex conjugates of five complex numbers

```
numbers = {{3 - 7 I}, {5 I}, {9}, {-2 + 4 I}, {8 + I}}
```

```
{{3 - 7 i}, {5 i}, {9}, {-2 + 4 i}, {8 + i}}
```

```
conjugates = Conjugate[numbers]
```

```
{{3 + 7 i}, {-5 i}, {9}, {-2 - 4 i}, {8 - i}}
```

- The conjugates of a product and a quotient of complex numbers

```
numbers = {{3 - 7 I} {5 I}, {9} / {-2 + 4 I}}
```

$$\left\{\{35 + 15\,i\},\ \left\{-\frac{9}{10} - \frac{9\,i}{5}\right\}\right\}$$

```
conjugates = Conjugate[numbers]
```

$\left\{\{35 - 15\,i\}, \left\{-\frac{9}{10} + \frac{9\,i}{5}\right\}\right\}$

Composition of linear transformations

If T is a linear transformation from $\mathbb{R}^n$ to $\mathbb{R}^m$ and S is a linear transformation from $\mathbb{R}^m$ to $\mathbb{R}^p$, then the *composition* of T and S, denoted by $(S \circ T)$, is defined by $S(T(\mathbf{x}))$ for all $\boldsymbol{x}$ in $\mathbb{R}^n$. The composite transformation $(S \circ T)$ is represented by the matrix product ST of the matrices S and T provided that the domain basis of S is the same as the codomain basis of T.

Illustration

- The composition function

```
Composition[f, g][x, y] == f[g[x, y]]
```

```
True
```

- Composition of a linear transformation from $\mathbb{R}^3$ to $\mathbb{R}^2$ with a linear transformation from $\mathbb{R}^2$ to $\mathbb{R}^2$

```
T[{x_, y_, z_}] := {3 x + 2 y - 4 z, x - 5 y + 3 x}
```

```
S[{u_, v_}] := {4 u - 2 v, 2 u + v}
```

```
T[{1, 2, 3}]
```

```
{-5, -6}
```

```
S[{-5, -6}]
```

```
{-8, -16}
```

The matrices representing the transformations T, S, and the composition of T and S are as follows:

```
mT= Transpose[{T[{1, 0, 0}], T[{0, 1, 0}], T[{0, 0, 1}]}]
```

```
{{3, 2, -4}, {4, -5, 0}}
```

```
mS = Transpose[{S[{1, 0}], S[{0, 1}]}]
```

```
{{4, -2}, {2, 1}}
```

```
mP = mS.mT
```

```
{{4, 18, -16}, {10, -1, -8}}
```

```
mP.{1, 2, 3}
```

```
{ 8, - 16}
```

This last operation shows that the matrix *mP* produces the same result as the composition of the functions *T* and *S*.

Condition number of a matrix

Various definitions of *condition numbers* of matrices are used to measure the impact of a relatively small change in inputs on the outputs. For example, the condition numbers associated with a linear equation $A\mathbf{v} = \mathbf{b}$ measure how inaccurate the solution **v** will be after a small change in the coefficient matrix *A*. Thus the condition number of a matrix *A* relative to a matrix norm is defined as the norm of *A* divided by the norm of the inverse of *A*.

Illustration

- Condition numbers of Hilbert matrices

```
hilbert = Table[HilbertMatrix[n], {n, 1, 5}];
```

```
conditionnumbers = Table[LinearAlgebra`MatrixConditionNumber[hilbert[[n]]], {n, 1, 5}]
```

```
{1, 27, 748, 28 375, 943 656}
```

This example shows the rapid growth of the condition numbers of even small Hilbert matrices.

Another measure used to calculate the condition number of a matrix is the two-norm consisting of the ratio of the largest to the smallest singular value of the matrix.

```
{u, w, v} = SingularValueDecomposition[hilbert[[5]]];
```

```
MatrixForm[N[w]]
```

$$\begin{pmatrix} 1.56705 & 0. & 0. & 0. & 0. \\ 0. & 0.208534 & 0. & 0. & 0. \\ 0. & 0. & 0.0114075 & 0. & 0. \\ 0. & 0. & 0. & 0.000305898 & 0. \\ 0. & 0. & 0. & 0. & 3.28793 \times 10^{-6} \end{pmatrix}$$

In the case of the 5-by-5 Hilbert matrix, the condition number based on the two-norm is therefore

```
N[w][[1]][[1]]/N[w][[5]][[5]]
```

```
476 607.
```

Congruence transformation

For any *n*-by-*n* symmetric matrix *A* and any nonsingular *n*-by-*n* matrix *B*, the transformation $A \longrightarrow B^T AB$ is a *congruence transformation*. Congruence transformations preserve the number of positive, negative, and zero eigenvalues.

Illustration

- Congruence transformation of a 5-by-5 symmetric matrix

$$A = \begin{pmatrix} 0 & 2 & 0 & 3 & 0 \\ 2 & 2 & 0 & 1 & 0 \\ 0 & 0 & 0 & 2 & 0 \\ 3 & 1 & 2 & 0 & 0 \\ 0 & 0 & 0 & 0 & 0 \end{pmatrix};$$

```
SymmetricMatrixQ[A]
```

```
True
```

```
N[Eigenvalues[A]]
```

```
{4.93812, -3.65125, 1.36381, -0.650674, 0.}
```

```
B = {{5, 3, 3, 9, 5}, {5, 6, 4, 4, 9}, {2, 0, 3, 5, 6}, {5, 7, 0, 1, 9}, {8, 5, 6, 5, 0}};
```

```
Det[B]
```

```
3728
```

```
N[Eigenvalues[Transpose[B].A.B]]
```

```
{2299.83, - 226.481, - 6.48762, 5.13604, 0.}
```

Both *A* and $B^T AB$ have two positive eigenvalues, two negative eigenvalues, and one zero eigenvalue.

Congruent symmetric matrices

Two symmetric matrices *A* and *B* are said to be *congruent* if there exists an orthogonal matrix *Q* for which $A = Q^T BQ$.

Illustration

- Two congruent symmetric matrices

```
A = {{3, -1}, {-1, 3}}; B = {{2, 0}, {0, 4}};
```

```
Solve[{{a, b}, {b, c}}.B.{{a, b}, {b, c}} == {{3, -1}, {-1, 3}}];
```

$$Q = \{\{a, b\}, \{b, c\}\} \;/.\; \left\{a \to -\frac{1}{\sqrt{2}},\ c \to \frac{1}{\sqrt{2}},\ b \to -\frac{1}{\sqrt{2}}\right\}$$

$$\left\{\left\{-\frac{1}{\sqrt{2}}, -\frac{1}{\sqrt{2}}\right\}, \left\{-\frac{1}{\sqrt{2}}, \frac{1}{\sqrt{2}}\right\}\right\}$$

```
Q.Transpose[Q] == IdentityMatrix[2]
```

True

```
A == Transpose[Q].B.Q
```

True

- Congruence and quadratic forms

The previous calculations show that following two matrices are congruent:

$A = \begin{pmatrix} 3 & -1 \\ -1 & 3 \end{pmatrix}$ and $B = \begin{pmatrix} 2 & 0 \\ 0 & 4 \end{pmatrix}$

- Plotting q[x, y] and q[u, v]

```
q[x_, y_] := {x, y}.A.{x, y}
```

```
plot1 = Plot3D[q[x, y], {x, -5, 5}, {y, -5, 5}]
```

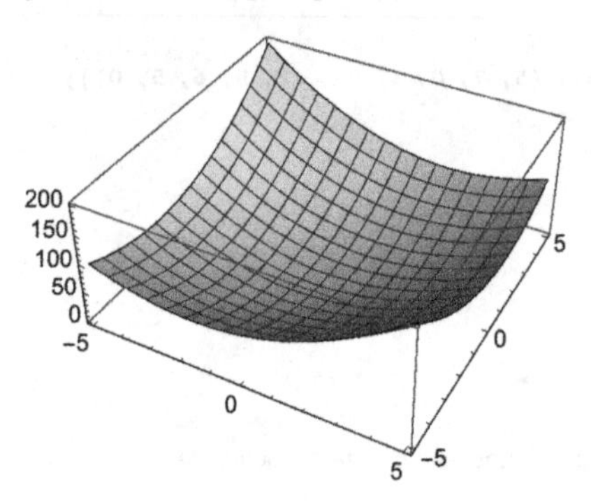


```
q[u_, v_] := {u, v}.B.{u, v}
```

```
plot2 = Plot3D[q[u, v], {u, -5, 5}, {v, -5, 5}]
```

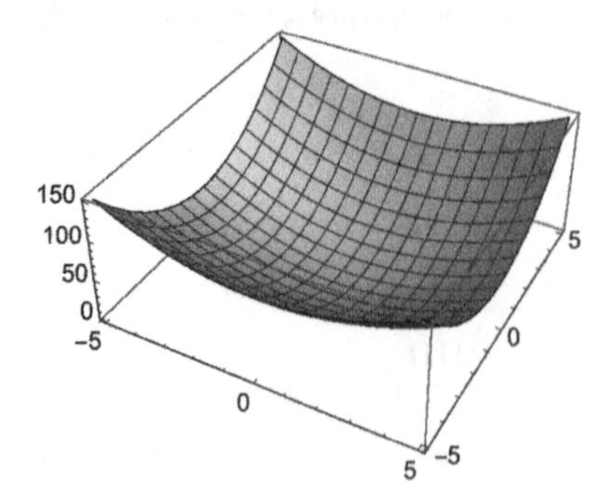

Conjugate transpose

The *conjugate transpose* of a complex matrix is the result of transposing the matrix and replacing its elements by their conjugates.

Illustration

- The conjugate transpose of a 2-by-2 complex matrix

```
MatrixForm[A = {{3 + I, -5}, {-I, 4 - I}}]
```

$$\begin{pmatrix} 3+i & -5 \\ -i & 4-i \end{pmatrix}$$

```
MatrixForm[ConjugateTranspose[A]]
```

$$\begin{pmatrix} 3-i & i \\ -5 & 4+i \end{pmatrix}$$

- The conjugate transpose of a 3-by-3 real matrix

```
MatrixForm[A = RandomReal[{0, 9}, {3, 3}]];
```

$$A = \begin{pmatrix} 5.77847 & 3.40248 & 8.44687 \\ 5.34368 & 6.38133 & 1.38616 \\ 0.838548 & 0.3028 & 8.92482 \end{pmatrix}$$

```
MatrixForm[ConjugateTranspose[A] == Transpose[A]]
```

```
True
```

Consistent linear system

A linear system is *consistent* if it has a solution. A consistent system can have either one solution or infinitely many solutions. In the latter case, the system is said to be underdetermined.

Illustration

- A consistent linear system in two equations and two variables

```
system = {3 x - 5 y == 1, x + y == 10};
```

```
solution = Solve[system]
```

$$\left\{\left\{x \to \frac{51}{8},\ y \to \frac{29}{8}\right\}\right\}$$

```
system /. solution
```

```
{{True, True}}
```

- A consistent linear system in two equations and three variables

```
system = {3 x - 5 y + z == 1, x + y == 10};

solution = Solve[system]

{{y → 10 - x, z → 51 - 8 x}}

Simplify[system /. solution]

{{True, True}}
```

- A consistent linear system in three equations and two variables

```
system = {3 x - 5 y == 1, x + y == 10, 4 x - 4 y == 11};

solution = Solve[system]
```

$$\left\{\left\{x \to \frac{51}{8},\ y \to \frac{29}{8}\right\}\right\}$$

```
system /. solution

{{True, True, True}}
```

- A consistent underdetermined linear system

```
Clear[x, y, z, b]

A = {{1, 1, 1}, {1, 1, 3}}; b = {1, 4};

LinearSolve[A, b]
```

$$\left\{-\frac{1}{2},\ 0,\ \frac{3}{2}\right\}$$

```
Reduce[{x + y + z == 1, x + y + 3 z == 4}, {x, y}]
```

$$z == \frac{3}{2}\ \&\&\ y == -\frac{1}{2} - x$$

As we can see, the **LinearSolve** command produces one solution, whereas the **Reduce** command produces all solutions.

Manipulation

- Consistent and non-consistent linear systems

```
Clear[x, y, a, b]

system = {x + a y == 1, x - b y == 2};
```

```
Manipulate[Evaluate[Reduce[system, {x, y}]], {a, -5, 5, 1}, {b, 0, 10, 1}]
```

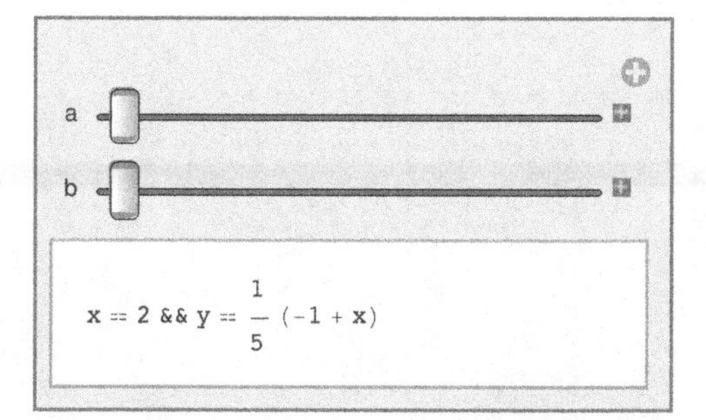

We can combine **Manipulate** and **Reduce** to explore the consistency of linear systems. If we let $a = -5$ and $b = 0$, the manipulation shows that the resulting system is consistent. Other assignments produce inconsistent systems. For example, the assignment $a = b = 0$ produces an inconsistent system.

Contraction along a coordinate axis

For any real number $0 < s < 1$, a left-multiplication of a vector **v** in $\mathbb{R}^2$ by the matrix

```
MatrixForm[A = {{s, 0}, {0, 1}}]
```

$$\begin{pmatrix} \frac{1}{2} & 0 \\ 0 & 1 \end{pmatrix}$$

is a *contraction along the x-axis.* A left-multiplication by the matrix

```
MatrixForm[B = {{1, 0}, {0, s}}]
```

$$\begin{pmatrix} 1 & 0 \\ 0 & \frac{1}{2} \end{pmatrix}$$

is a *contraction along the y-axis.*

Illustration

- Contractions along the coordinate axes

```
Clear[A, B, s]
```

```
MatrixForm[A = {{s, 0}, {0, 1}}]
```

$$\begin{pmatrix} s & 0 \\ 0 & 1 \end{pmatrix}$$

```
MatrixForm[B = {{1, 0}, {0, s}}]
```

$$\begin{pmatrix} 1 & 0 \\ 0 & s \end{pmatrix}$$

```
v = {3, 5}; s = 1/2;
```

```
{A.v, B.v}
```

$$\left\{\left\{\frac{3}{2}, 5\right\}, \left\{3, \frac{5}{2}\right\}\right\}$$

Coordinate conversion matrix

Let **v** be a vector in an n-dimensional real or complex vector space V, let $[\mathbf{v}]_{B_1}$ be the coordinate vector of **v** in a basis B_1, and let $[\mathbf{v}]_{B_1}$ be the coordinate vector of **v** in a basis B_2. Then there exists an invertible matrix P for which $P[\mathbf{v}]_{B_1} = [\mathbf{v}]_{B_2}$ and $P^{-1}[\mathbf{v}]_{B_2} = P[\mathbf{v}]_{B_1}$. The matrix P is called a *coordinate conversion matrix.*

Illustration

- A 2-by-2 coordinate conversion matrix from the basis B_1 to the basis B_2 for $\mathbb{R}^2$

```
Clear[P, v, a, b]
```

```
B1 = {{1, 0}, {0, 1}}; B2 = {{2, 1}, {1, 3}}; v = vB1 = {5, -3};
```

```
Reduce[{5, -3} == a {2, 1} + b {1, 3}, {a, b}]
```

$$a == \frac{18}{5} \&\& b == -\frac{11}{5}$$

```
vB2 = {a, b} /. {a → 18/5, b → -11/5}
```

$$\left\{\frac{18}{5}, -\frac{11}{5}\right\}$$

```
Reduce[{1, 0} == a {2, 1} + b {1, 3}, {a, b}]
```

$$a == \frac{3}{5} \&\& b == -\frac{1}{5}$$

```
Reduce[{0, 1} == a {2, 1} + b {1, 3}, {a, b}]
```

$$a == -\frac{1}{5} \&\& b == \frac{2}{5}$$

$$\mathbf{P = \left\{\left\{\frac{3}{5}, \frac{-1}{5}\right\}, \left\{\frac{-1}{5}, \frac{2}{5}\right\}\right\};}$$

```
P.{5, -3}
```

$$\left\{\frac{18}{5}, -\frac{11}{5}\right\}$$

$$\mathbf{Inverse[P].\left\{\frac{18}{5}, -\frac{11}{5}\right\}}$$

```
{5, -3}
```

- A coordinate conversion matrix for two bases for $\mathbb{R}^3$

```
Clear[a, b, c]
```

$$\mathbf{B_1 = \{\{1, 2, 3\}, \{0, 1, 0\}, \{1, 0, 1\}\}; B_2 = \{\{0, 1, 1\}, \{1, 1, 0\}, \{1, 0, 1\}\};}$$

The columns of the coordinate conversion matrix from B_1 to B_2 are the coordinate vectors of the B_1 basis vectors in the basis B_2.

```
Reduce[{1, 2, 3} == a {0, 1, 1} + b {1, 1, 0} + c {1, 0, 1}, {a, b, c}]
```

```
a == 2 && b == 0 && c == 1
```

```
Reduce[{0, 1, 0} == a {0, 1, 1} + b {1, 1, 0} + c {1, 0, 1}, {a, b, c}]
```

$$a == \frac{1}{2} \;\&\&\; b == \frac{1}{2} \;\&\&\; c == -\frac{1}{2}$$

```
Reduce[{1, 0, 1} == a {0, 1, 1} + b {1, 1, 0} + c {1, 0, 1}, {a, b, c}]
```

```
a == 0 && b == 0 && c == 1
```

$$\mathbf{P = Transpose\left[\left\{\{2, 0, 1\}, \left\{\frac{1}{2}, \frac{1}{2}, \frac{-1}{2}\right\}, \{0, 0, 1\}\right\}\right];}$$

```
v = {5, 4, -2};
```

```
Reduce[v == a {1, 2, 3} + b {0, 1, 0} + c {1, 0, 1}, {a, b, c}]
```

$$a == -\frac{7}{2} \;\&\&\; b == 11 \;\&\&\; c == \frac{17}{2}$$

```
Reduce[v == d {0, 1, 1} + e {1, 1, 0} + f {1, 0, 1}, {d, e, f}]
```

$$d == -\frac{3}{2} \;\&\&\; e == \frac{11}{2} \;\&\&\; f == -\frac{1}{2}$$

```
P.{-7/2, 11, 17/2} == {-3/2, 11/2, -1/2}
```

True

Coordinate system

Vectors in $\mathbb{R}^2$ and $\mathbb{R}^3$ are often represented as arrows in a Cartesian *coordinate system*. The system consists of two perpendicular lines that cross at a point {0, 0} referred to as the *origin* of the system. The points along the horizontal axis (the *x*-axis) are labeled with the real numbers in their natural ordering, with the number 0 at the origin, positive real numbers to the right, and negative real numbers to the left of the origin. The points along the vertical axis (the *y*-axis) are also labeled with the real numbers in their natural ordering, with the number 0 at the origin. Positive real numbers are placed above the *x*-axis and negative ones below.

Illustration

- A Cartesian coordinate system for $\mathbb{R}^2$

```
Plot[{}, {x, -10, 10}, Axes → True]
```

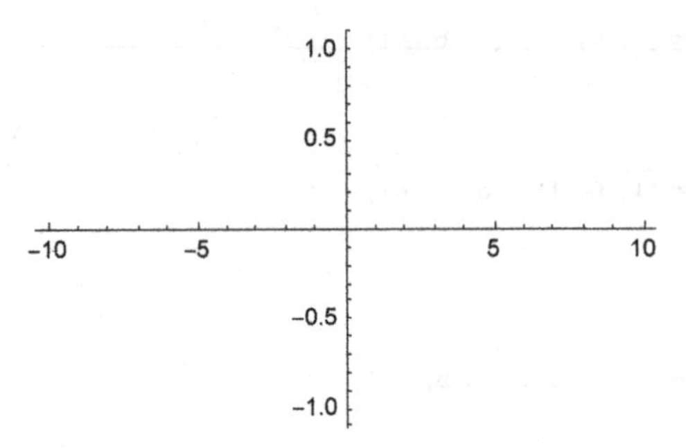

The idea of a graph in a Cartesian coordinate system can be extended to $\mathbb{R}^3$.

- A Cartesian coordinate system for $\mathbb{R}^3$

```
Plot3D[{{}}, {x, -10, 10}, {y, -10, 10}, Axes → True]
```

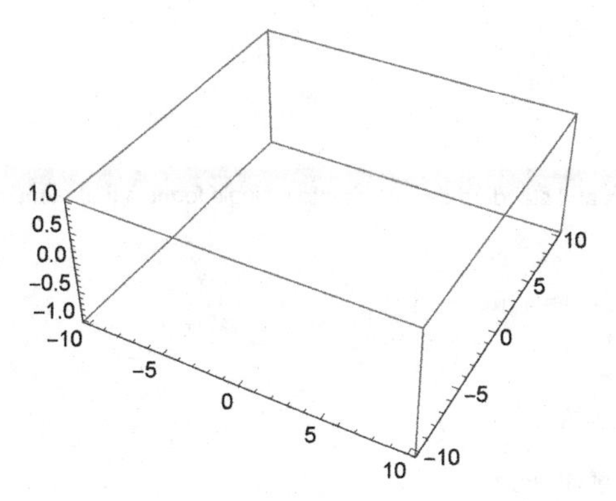

Coordinate vector

The *coordinate vector* v_B in a basis $B = \{\mathbf{b}_1, \ldots, \mathbf{b}_n\}$ of a vector v in the standard basis of an n-dimensional real or complex vector space V is the vector $\{a_1, \ldots, a_n\}$ in $\mathbb{R}^n$ or $\mathbb{C}^n$ with the coordinates $a_1, \ldots, a_n$ for which $v = a_1\,\mathbf{b}_1 + \cdots + a_n\,\mathbf{b}_n$.

Illustration

- A coordinate vector in $\mathbb{R}^2$

```
Quit[]
```

```
v = {3, -5}; B = {{1, 2}, {-3, 4}};
```

```
Reduce[v == a1 B[[1]] + a2 B[[2]], {a1, a2}]
```

$a_1 == -\frac{3}{10}$ && $a_2 == -\frac{11}{10}$

$$\mathbf{v_B} = \left\{\frac{-3}{10}, \frac{-11}{10}\right\};$$

The vector $\mathbf{v}_B$ is the coordinate vector of **v** in the basis B, because the elements of $\mathbf{v}_B$ are the coefficients, a_1 and a_2, needed to express **v** as a linear combination of the basis vectors b_1 and b_2.

- A coordinate vector in $\mathbb{R}^3$

```
v = {3, 4, 5}; B = {{1, 1, 1}, {1, -1, 1}, {0, 0, 1}};
```

```
Reduce[v == a1 B[[1]] + a2 B[[2]] + a3 B[[3]], {a1, a2, a3}]
```

$a_1 == \frac{7}{2}$ && $a_2 == -\frac{1}{2}$ && $a_3 == 2$

```
vB = {7/2, -1/2, 2};
```

Correlation coefficient

The (Pearson) *correlation coefficient* combines covariances and standard deviations into a single formula that measures the degree to which two vectors in $\mathbb{R}^n$ are correlated.

$$\text{Covariance}[x, y] / (\text{StandardDeviation}[x]\ \text{StandardDeviation}[y]) = \frac{\text{Dot}[x, y]}{\text{Norm}[x]\ \text{Norm}[y]} \tag{1}$$

Illustration

- The correlation coefficient of two data sets as a cosine of an angle

```
x = {2, 6, 3, 1, 8, 9}; y = {6, 2, 0, -3, 9, 2}; mx = Mean[x]; my = Mean[y];

pcx = 1/6 Table[(x[[i]] - mx), {i, 1, 6}];

pcy = 1/6 Table[(y[[i]] - my), {i, 1, 6}];

pcv[x, y] = 6 Total[pcx pcy]

lhs = (1/6) (1/5) pcv[x, y] / (StandardDeviation[pcx] StandardDeviation[pcy])

rhs = Dot[pcx, pcy] / (Norm[pcx] Norm[pcy])

lhs == rhs
```

Correlation matrix

A *correlation matrix* is an *m*-by-*m* matrix whose elements are the pairwise correlation coefficients of *m* vectors in $\mathbb{R}^n$.

Illustration

- The correlation matrix of three data sets

```
x = {2, 6, 3, 1, 8}; y = {6, 2, 0, -3, 9}; z = {1, 2, 1, 2, -5};

lhs = MatrixForm[N[Correlation[Transpose[{x, y, z}]]]]

rhs = MatrixForm[N[
   {{Correlation[x, x], Correlation[x, y], Correlation[x, z]},
    {Correlation[y, x], Correlation[y, y], Correlation[y, z]},
    {Correlation[z, x], Correlation[z, y], Correlation[z, z]}}]]
```

```
lhs == rhs
```

Cosine of an angle

The *cosine of an angle* between two vectors **u** and **v** in a vector space V, equipped with an inner product $\langle \mathbf{u}, \mathbf{v} \rangle$, is the scalar defined by the formula

$$\text{cosine}[u,\ v] = \frac{\langle u,\ v\rangle}{\sqrt{\langle u,\ u\rangle}\ \sqrt{\langle v,\ v\rangle}} \tag{1}$$

Illustration

The dot product and Euclidean norm of a vector can be used to find the cosine of the angle between two vectors.

- The cosine of the angle between two vectors in $\mathbb{R}^2$

```
u = {1, 2}; v = {3, 4};

cosine[u_, v_] := (Dot[u, v]) / (Norm[u] Norm[v])

angle[u, v] = N[ArcCos[cosine[u, v]]]

0.179853
```

- The cosine of the angle between two vectors in $\mathbb{R}^3$

```
u = {1, 2, 3}; v = {4, 5, 6};

cosine = (Dot[u, v]) / (Norm[u] Norm[v])
```

$$\frac{16\sqrt{\frac{2}{11}}}{7}$$

```
N[cosine]

0.974632
```

The dot product, norm, and cosine are related by the identity

$$\text{Dot}[x,\ y] == \text{Norm}[x]\ \text{Norm}[y]\ \text{cosine}[x,\ y] \tag{2}$$

- The dot product, Euclidean norm, and cosine identity

```
x = {1, 2}; y = {3, 4};

N[cos[x, y] = (Dot[x, y]) / (Norm[x] Norm[y])]

0.98387
```

```
Dot[x, y] == Norm[x] Norm[y] cos[x, y]
```

True

Covariance

The *covariance* of two vectors in $\mathbb{R}^n$ measures how the vectors vary together. It is based on the dot product of the mean-deviation form of the vectors.

Illustration

- Dot products and the covariance of two data sets

```
x = {2, 6, 3, 1, 8, 9}; y = {6, 2, 0, -3, 9, 2}; mx = Mean[x]; my = Mean[y];
```

$$pcx = \frac{1}{6}\text{Table}[(x[[i]] - mx), \{i, 1, 6\}];$$

$$pcy = \frac{1}{6}\text{Table}[(y[[i]] - my), \{i, 1, 6\}];$$

```
pcv[x, y] = 6 Total[pcx pcy];
```

$$scx = \frac{1}{5}\text{Table}[(x[[i]] - mx), \{i, 1, 6\}];$$

$$scy = \frac{1}{5}\text{Table}[(y[[i]] - my), \{i, 1, 6\}];$$

```
scv[x, y] = 5 Dot[scx, scy]
```

$$\frac{101}{15}$$

```
scv[x, y] == Covariance[x, y]
```

True

Covariance matrix

The (sample) *covariance matrix* of different vectors in $\mathbb{R}^n$ is the matrix whose elements are the pairwise covariances of the vectors. The diagonal elements of the matrix are (sample) variances of the individual sets of data.

Illustration

- Variances and covariances

```
x = {2, 6, 3, 1, 8}; y = {6, 2, 0, -3, 9}; z = {1, 2, 1, 2, -5};
```

```
{Variance[x], Variance[y], Variance[z]}
```

$$\left\{\frac{17}{2}, \frac{227}{10}, \frac{87}{10}\right\}$$

```
lhs = MatrixForm[Covariance[Transpose[{x, y, z}]]]
```

$$\begin{pmatrix} \frac{17}{2} & \frac{37}{4} & -\frac{25}{4} \\ \frac{37}{4} & \frac{227}{10} & -\frac{219}{20} \\ -\frac{25}{4} & -\frac{219}{20} & \frac{87}{10} \end{pmatrix}$$

```
rhs = MatrixForm[
  {{Variance[x], Covariance[x, y], Covariance[x, z]},
   {Covariance[y, x], Variance[y], Covariance[y, z]},
   {Covariance[z, x], Covariance[z, y], Variance[z]}}]
```

$$\begin{pmatrix} \frac{17}{2} & \frac{37}{4} & -\frac{25}{4} \\ \frac{37}{4} & \frac{227}{10} & -\frac{219}{20} \\ -\frac{25}{4} & -\frac{219}{20} & \frac{87}{10} \end{pmatrix}$$

```
lhs == rhs
```

True

Cramer's rule

If $A\mathbf{v} = \mathbf{b}$ is a linear system and A is an n-by-n invertible matrix, then *Cramer's rule* says that for each $1 \leq i \leq n$,

$$v_i = \frac{\text{Det}[A[v_i / b]]}{\text{Det}[A]} \tag{1}$$

where $A[\mathbf{v}_i / \mathbf{b}]$ is the matrix obtained by replacing the ith column of A by the vector $\mathbf{b}$.

Illustration

```
Quit[]
```

- Using Cramer's rule to solve a linear system

```
system = {x + y - z == 4, 2 x - y + 2 z == 18, x - y + z == 5}; variables = {x, y, z};
```

```
Solve[system, variables]
```

$$\left\{\left\{x \to \frac{9}{2},\ y \to 8,\ z \to \frac{17}{2}\right\}\right\}$$

```
MatrixForm[cA = Normal[CoefficientArrays[system, {x, y, z}]][[2]]]
```

$$\begin{pmatrix} 1 & 1 & -1 \\ 2 & -1 & 2 \\ 1 & -1 & 1 \end{pmatrix}$$

```
A[x/b] = {{4, 1, -1}, {18, -1, 2}, {5, -1, 1}};

A[y/b] = {{1, 4, -1}, {2, 18, 2}, {1, 5, 1}};

A[z/b] = {{1, 1, 4}, {2, -1, 18}, {1, -1, 5}};
```

$$\text{solution} = \left\{x = \frac{\text{Det[A[x/b]]}}{\text{Det[cA]}},\ y = \frac{\text{Det[A[y/b]]}}{\text{Det[cA]}},\ z = \frac{\text{Det[A[z/b]]}}{\text{Det[cA]}}\right\}$$

$$\left\{\frac{9}{2},\ 8,\ \frac{17}{2}\right\}$$

This shows that the solution found using Cramer's rule is the same as that found with **Solve**.

Cross product

Although many geometric concepts in $\mathbb{R}^3$ are the obvious extensions of their analogues in $\mathbb{R}^2$, there exist exceptions. One such is the construction of a vector that is orthogonal to two given vectors. Only in $\mathbb{R}^3$ can we define an operation on two vectors known as the *vector cross product* of the two vectors. The built-in *Mathematica* **Cross** function computes vector cross products.

In $\mathbb{R}^3$, any two nonzero linearly independent vectors **u** and **v** can be combined to create a third vector **w** which is orthogonal to both **u** and **v**.

Illustration

- A vector cross product

```
u = {1, 2, 3}; v = {-3, 0, 1};

w = Cross[u, v]

{2, -10, 6}

Dot[u, w] == Dot[v, w] == 0

True
```

- Graphic representation of a vector cross product

```
Graphics3D[{Arrow[{{0, 0, 0}, {1, 2, 3}}],
  Arrow[{{0, 0, 0}, {-3, 0, 1}}], Arrow[{{0, 0, 0}, {2, -10, 6}}]}, Axes → True]
```

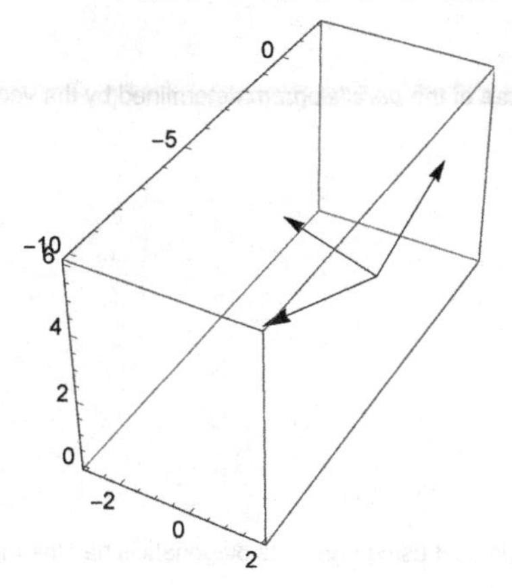

Various geometric constructions are based on the cross product. For a meaningful construction, the two given vectors need to be linearly independent nonzero vectors.

- A vector cross product

Any two nonzero linearly independent vectors **u** and **v** in $\mathbb{R}^3$ determine a vector **w** that is orthogonal to both **u** and **v** in the Euclidean inner product (dot product).

```
u = {1, 2, 3}; v = {4, 5, 6};
```

```
w = Cross[u, v]
```

```
{-3, 6, -3}
```

```
Dot[w, u] == Dot[w, v] == 0
```

```
True
```

The cross product can be calculated using the determinant formula for 1-by-3 vectors, together with the symbolic vectors ***i***, ***j***, and ***k***.

- Using determinants to calculate a vector cross product

```
MatrixForm[A = {{i, j, k}, {1, 2, 3}, {4, 5, 6}}]
```

$$\begin{pmatrix} i & j & k \\ 1 & 2 & 3 \\ 4 & 5 & 6 \end{pmatrix}$$

```
Det[A]
```

```
-3 i + 6 j - 3 k
```

```
i = {1, 0, 0}; j = {0, 1, 0}; k = {0, 0, 1};
```

```
Cross[x, y] == -3 i + 6 j - 3 k
```

```
x × y == {-3, 6, -3}
```

The absolute values of the norm of the cross product represents the area of the *parallelogram* determined by the vectors *x* and *y*.

- Areas of parallelograms and vector cross products

```
u = {1, 2, 3}; v = {4, 5, 6};
```

```
w = Cross[u, v];
```

```
parallelogram = Abs[Norm[w]]
```

$3\sqrt{6}$

The area of one of the *triangles* obtained from the parallelogram by dividing it using one of its diagonals is half the area of the parallelogram.

- Areas of triangles and vector cross products

```
u = {1, -2, 3}; v = {4, 5, 1};
```

```
w = Cross[u, v];
```

$$\text{triangle} = \frac{1}{2}\text{Abs[Norm[w]]}$$

$$\frac{\sqrt{579}}{2}$$

Manipulation

- The vector cross product of two nonzero vectors in $\mathbb{R}^3$

```
Manipulate[Cross[{a, 2, 3}, {4, b, 6}], {a, -3, 3}, {b, -2, 2}]
```

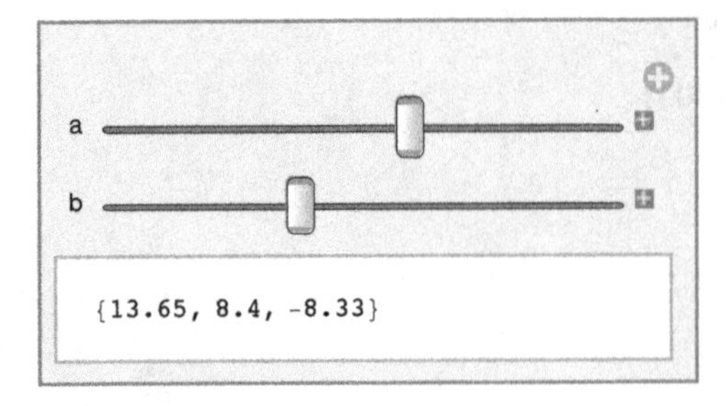

We use **Manipulate** and **Cross** to explore the cross product of two nonzero vectors. As expected, the resulting vector is (approximately) orthogonal to both the vectors {0.6, 2, 3} and {4, - 0.55, 6}:

```
Chop[Dot[{0.6, 2, 3}, {13.65`, 8.399999999999999`, -8.33`}]]
```

0

```
Chop[Dot[{4, -0.55, 6}, {13.65`, 8.399999999999999`, -8.33`}]]
```

0

D

Defective matrix

An *n*-by-*n* matrix is *defective* if it does not have a set of *n* linearly independent eigenvectors. Defective matrices are not diagonalizable.

Illustration

- A 4-by-4 defective matrix

```
MatrixForm[A = Normal[SparseArray[{{2, 3} → 1, {3, 2} → 0}, {4, 4}]]]
```

$$\begin{pmatrix} 0 & 0 & 0 & 0 \\ 0 & 0 & 1 & 0 \\ 0 & 0 & 0 & 0 \\ 0 & 0 & 0 & 0 \end{pmatrix}$$

```
Eigenvectors[A]
```

```
{{0, 0, 0, 1}, {0, 1, 0, 0}, {1, 0, 0, 0}, {0, 0, 0, 0}}
```

The calculation shows that the 4-by-4 matrix *A* has a maximum of three linearly independent eigenvectors. It is therefore a defective matrix.

An *n*-by-*n* real matrix may not have n linearly independent real eigenvectors and may therefore be considered to be defective as a real matrix.

- A real 2-by-2 matrix (defective as a real matrix)

```
MatrixForm[A = {{Cos[π / 3], Sin[π / 3]}, {-Sin[π / 3], Cos[π / 3]}}]
```

$$\begin{pmatrix} \frac{1}{2} & \frac{\sqrt{3}}{2} \\ -\frac{\sqrt{3}}{2} & \frac{1}{2} \end{pmatrix}$$

```
N[Eigenvectors[A]]
```

$\{\{-1.4803\times 10^{-16} - 1.\, i,\ 1.\},\ \{2.22045\times 10^{-16} + 1.\, i,\ 1.\}\}$

The calculations show that the real matrix *A* has no real eigenvectors. Hence it is defective (as a real matrix).

Determinant

The *determinant* of a square matrix is a scalar associated with the matrix. It is defined by induction on the size of the matrix.

Illustration

- The determinant of a 1-by-1 matrix

```
MatrixForm[A = {{a}}]
```

(a)

```
Det[A]
```

a

- The determinant of a 2-by-2 matrix

```
MatrixForm[A = {{a, b}, {c, d}}]
```

$$\begin{pmatrix} a & b \\ c & d \end{pmatrix}$$

```
TraditionalForm[Det[A]]
```

$a d - b c$

- The determinant of a 3-by-3 matrix

```
MatrixForm[A = {{a, b, c}, {d, e, f}, {g, h, i}}]
```

$$\begin{pmatrix} a & b & c \\ d & e & f \\ g & h & i \end{pmatrix}$$

```
Det[A]
```

-c e g + b f g + c d h - a f h - b d i + a e i

Various formulas for calculating determinants exist. Here is an example of the Laplace expansion of the determinant of a 3-by-3 matrix.

- The determinant of a matrix calculated by an expansion along the first row of the matrix

```
MatrixForm[A = {{a, b, c}, {d, e, f}, {g, h, i}}]
```

$$\begin{pmatrix} a & b & c \\ d & e & f \\ g & h & i \end{pmatrix}$$

The determinant of 3-by-3 matrix A can be calculated as a linear combination of the first row of A and the determinants of three 2-by-2 submatrices of A.

```
A11 = {{e, f}, {h, i}};
```

```
A12 = {{d, f}, {g, i}};
```

```
A13 = {{d, e}, {g, h}};

Expand[Det[A] == a Det[A11] - b Det[A12] + c Det[A13]]

True
```

Manipulation

- Determinants of 3-by-3 matrices

```
Manipulate[Det[{{a, b, c}, {1, 2, 3}, {4, 5, 6}}], {a, -2, 2, 1}, {b, -2, 2, 1}, {c, -2, 2, 1}]
```

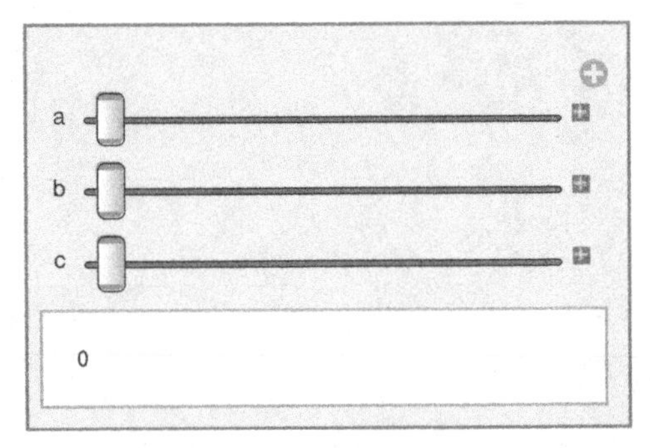

We can combine **Manipulate** and **Det** to explore the determinants of matrices. If we assign the values $a = b = c = -2$, for example, the manipulation shows that the determinant of the resulting matrix is zero. Other assignments to a, b, and c such as a = - 2, b = - 1, and c = - 2 produce a matrix with a nonzero determinant:

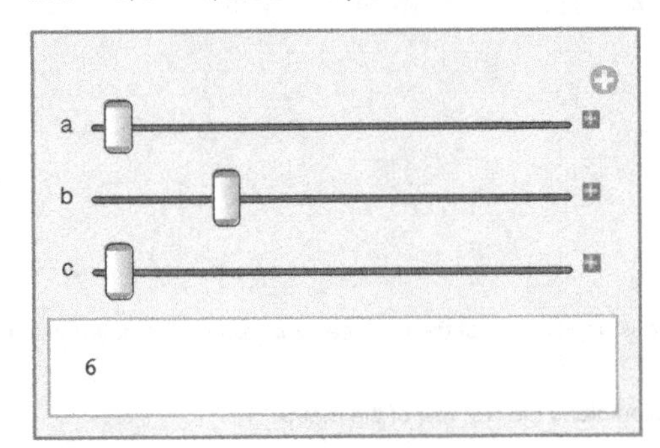

Diagonal

See Diagonal of a matrix, Jordan block, subdiagonal, superdiagonal

Diagonal decomposition

Eigenvalues and eigenvectors are needed for the *diagonal decomposition* of a matrix A into a product of the form

$P \, dM \, P^{-1}$ consisting of an invertible matrix *P* whose columns are eigenvectors of *A* and a diagonal matrix *dM* whose diagonal entries are eigenvalues of *A*. The decomposition of an *n*-by-*n* real matrix requires *n* linearly independent eigenvectors.

Illustration

Eigenvectors and eigenvalues are the building blocks of diagonal decompositions of real matrices. Suppose we would like to rewrite a matrix *A* as a product

```
P.DiagonalMatrix[Eigenvalues[A]].Inverse[P]
```
(1)

then the diagonal matrix **DiagonalMatrix[Eigenvalues[A]]** must consist of the eigenvalues of *A* and the columns of *P* must be associated eigenvectors.

- A diagonal decomposition of a 3-by-3 real matrix

```
MatrixForm[A = {{2, 1, 1}, {4, 1, 7}, {5, 3, 0}}]
```

$$\begin{pmatrix} 2 & 1 & 1 \\ 4 & 1 & 7 \\ 5 & 3 & 0 \end{pmatrix}$$

```
evalues = Eigenvalues[A]
```

```
{7, -4, 0}
```

```
MatrixForm[dM = DiagonalMatrix[evalues]]
```

$$\begin{pmatrix} 7 & 0 & 0 \\ 0 & -4 & 0 \\ 0 & 0 & 0 \end{pmatrix}$$

```
MatrixForm[evectors = Eigenvectors[A]]
```

$$\begin{pmatrix} 1 & 3 & 2 \\ 1 & -19 & 13 \\ -3 & 5 & 1 \end{pmatrix}$$

Mathematica outputs the eigenvectors of *A* as row vectors. In order to form a matrix whose columns are eigenvectors, we must transpose them.

```
MatrixForm[A == Transpose[evectors].dM.Inverse[Transpose[evectors]]]
```

```
True
```

The matrices

```
MatrixForm[P = Transpose[evectors]]
```

and

```
MatrixForm[dM = DiagonalMatrix[evalues]]
```

yield a diagonal decomposition of the matrix A. The matrix P is not unique. Different choices of eigenvectors produce different decompositions.

Manipulation

All real 3-by-3 matrices have at least one real eigenvalue since their characteristic polynomials are real polynomials of odd degree and real polynomials of odd degree cut the x-axis at least once. The point of intersection of the graph of the polynomial and of the x-axis corresponds to a real eigenvalue of the matrix.

- Using **Manipulate** to explore eigenvalues

The matrix

```
MatrixForm[A = {{1, 0, 5}, {6, 2, 0}, {1, 0, 3}}]
```

$$\begin{pmatrix} 1 & 0 & 5 \\ 6 & 2 & 0 \\ 1 & 0 & 3 \end{pmatrix}$$

has three real eigenvalues.

```
Eigenvalues[A]
```

$$\{2 + \sqrt{6}, 2, 2 - \sqrt{6}\}$$

We can use the **Plot** function to visualize these eigenvalues as the roots of the characteristic polynomials of A.

```
cpA = CharacteristicPolynomial[A, t]
```

$$-4 - 6 t + 6 t^2 - t^3$$

```
Plot[cpA, {t, -5, 5}]
```

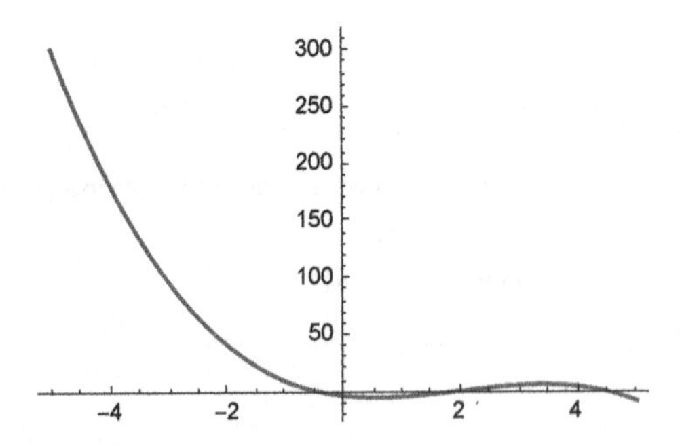

What happens if we replace the third row of A by {a, 0, 3}, with a ranging over a wider interval of scalars?

```
A = {{1, 0, 5}, {6, 2, 0}, {a, 0, 3}}
```

```
{{1, 0, 5}, {6, 2, 0}, {a, 0, 3}}
```

```
Manipulate[Eigenvalues[{{1, 0, 5}, {6, 2, 0}, {a, 0, 3}}], {a, -5, 5}]
```

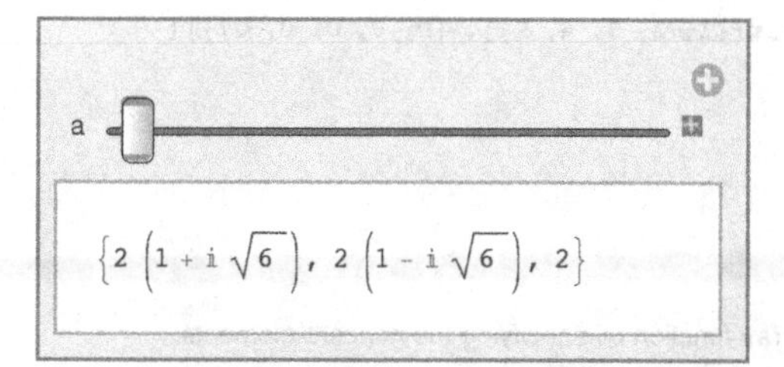

We can combine **Manipulate** and **Eigenvalues** to explore the nature of the eigenvalues of matrices. The manipulation shows that for negative values of *a*, some of the eigenvalues of the real matrix *A* are not real. Therefore the resulting matrix is not diagonalizable, although it has three distinct (complex) eigenvalues.

Diagonal matrix

A *diagonal matrix A* is a square array whose elements $A_{[[i,j]]}$ in the ith row and j^{th} column are zero if $i \neq j$. For some applications it is convenient to extend this definition to rectangular matrices. In that case, the matrices are padded with either zero rows and/or zero columns and are sometimes called *generalized* diagonal matrices.

Illustration

- A 5-by-5 diagonal matrix

```
MatrixForm[DiagonalMatrix[{1, 2, 3, 4, 5}]]
```

$$\begin{pmatrix} 1 & 0 & 0 & 0 & 0 \\ 0 & 2 & 0 & 0 & 0 \\ 0 & 0 & 3 & 0 & 0 \\ 0 & 0 & 0 & 4 & 0 \\ 0 & 0 & 0 & 0 & 5 \end{pmatrix}$$

- A generalized diagonal matrix obtained by appending a row of zeros

```
MatrixForm[Append[DiagonalMatrix[{1, 2, 3, 4, 5}], {0, 0, 0, 0, 0}]]
```

$$\begin{pmatrix} 1 & 0 & 0 & 0 & 0 \\ 0 & 2 & 0 & 0 & 0 \\ 0 & 0 & 3 & 0 & 0 \\ 0 & 0 & 0 & 4 & 0 \\ 0 & 0 & 0 & 0 & 5 \\ 0 & 0 & 0 & 0 & 0 \end{pmatrix}$$

- A generalized diagonal matrix obtained by appending a column of zeros

```
MatrixForm[Transpose[Append[DiagonalMatrix[{1, 2, 3, 4, 5}], {0, 0, 0, 0, 0}]]]
```

$$\begin{pmatrix} 1 & 0 & 0 & 0 & 0 & 0 \\ 0 & 2 & 0 & 0 & 0 & 0 \\ 0 & 0 & 3 & 0 & 0 & 0 \\ 0 & 0 & 0 & 4 & 0 & 0 \\ 0 & 0 & 0 & 0 & 5 & 0 \end{pmatrix}$$

Diagonal matrices can be created using the **SparseArray** function by specifying the nonzero elements.

- A 4-by-4 diagonal matrix

```
MatrixForm[Normal[SparseArray[{{1, 1} → 5, {2, 2} → 2, {3, 3} → 5}, {4, 4}]]]
```

$$\begin{pmatrix} 5 & 0 & 0 & 0 \\ 0 & 2 & 0 & 0 \\ 0 & 0 & 5 & 0 \\ 0 & 0 & 0 & 0 \end{pmatrix}$$

- A 4-by-5 diagonal matrix

```
MatrixForm[Normal[SparseArray[{{1, 1} → 5, {2, 2} → 2, {3, 3} → 5, {4, 4} → 6}, {4, 5}]]]
```

$$\begin{pmatrix} 5 & 0 & 0 & 0 & 0 \\ 0 & 2 & 0 & 0 & 0 \\ 0 & 0 & 5 & 0 & 0 \\ 0 & 0 & 0 & 6 & 0 \end{pmatrix}$$

Diagonal of a matrix

The *diagonal* of an *m*-by-*n* matrix *A* is the list of all elements $A_{[[i,i]]}$ of *A* for *i* from 1 to *m*.

Illustration

- Diagonal of a 4-by-6 matrix

```
A = RandomInteger[{0, 9}, {4, 6}];
```

$$A = \begin{pmatrix} 2 & 8 & 0 & 0 & 8 & 7 \\ 0 & 3 & 2 & 4 & 4 & 1 \\ 1 & 0 & 7 & 8 & 2 & 0 \\ 5 & 8 & 7 & 0 & 9 & 1 \end{pmatrix};$$

```
Diagonal[A]
```

```
{2, 3, 7, 0}
```

- Diagonal of a 4-by-4 matrix

$$A = \begin{pmatrix} 8 & 8 & 9 & 3 \\ 4 & 0 & 6 & 7 \\ 8 & 3 & 8 & 5 \\ 8 & 6 & 4 & 5 \end{pmatrix};$$

```
diagonalA = {A[[1, 1]], A[[2, 2]], A[[3, 3]], A[[4, 4]]}
{8, 0, 8, 5}

diagonalA == Diagonal[A]
True
```

- Diagonal of a 4-by-5 matrix

```
A = RandomInteger[{0, 9}, {4, 5}];
```

$$A = \begin{pmatrix} 1 & 4 & 1 & 1 & 8 \\ 9 & 1 & 9 & 9 & 3 \\ 2 & 2 & 7 & 2 & 6 \\ 0 & 5 & 6 & 1 & 9 \end{pmatrix};$$

```
diagonalA = {A[[1, 1]], A[[2, 2]], A[[3, 3]], A[[4, 4]]}
{1, 1, 7, 1}
```

The superdiagonal of an *m*-by-*n* matrix *A* is the list of all elements $A_{[[i,i+1]]}$ for *i* from 1 to $(m - 1)$.

- The superdiagonal of a 4-by-6 matrix

$$A = \begin{pmatrix} 2 & 8 & 0 & 0 & 8 & 7 \\ 0 & 3 & 2 & 4 & 4 & 1 \\ 1 & 0 & 7 & 8 & 2 & 0 \\ 5 & 8 & 7 & 0 & 9 & 1 \end{pmatrix};$$

```
Diagonal[A, 1]
{8, 2, 8, 9}
```

The subdiagonal of an *m*-by-*n* matrix *A* is the list of all elements $A_{[[i+1,i]]}$ for *i* from 2 to *m*.

- The subdiagonal of a 4-by-6 matrix

$$A = \begin{pmatrix} 2 & 8 & 0 & 0 & 8 & 7 \\ 0 & 3 & 2 & 4 & 4 & 1 \\ 1 & 0 & 7 & 8 & 2 & 0 \\ 5 & 8 & 7 & 0 & 9 & 1 \end{pmatrix};$$

```
Diagonal[A, -1]
{0, 0, 7}
```

Difference equation

If the vectors in a list $\{\mathbf{v}_0, \mathbf{v}_1, \mathbf{v}_2, \ldots, \mathbf{v}_n, \ldots\}$ are connected by a matrix *A* for which $\mathbf{v}_{n+1} = A\mathbf{v}_n$ for $n = 0, 1, 2, \ldots$, then the equation $\mathbf{v}_{n+1} = A\mathbf{v}_n$ is called a linear *difference equation.*

Illustration

- A difference equation based on a 2-by-2 matrix

```
MatrixForm[A = {{0.75, 0.5}, {0.25, 0.5}}]
```

$$\begin{pmatrix} 0.75 & 0.5 \\ 0.25 & 0.5 \end{pmatrix}$$

```
v0 = {100 000, 200 000}
```

```
{100 000, 200 000}
```

The list consisting of the first three elements of the list $\{\mathbf{v}_0, \mathbf{v}_1, \mathbf{v}_2, \ldots, \mathbf{v}_n, \ldots\}$ is

```
{v0, v1 = A.v0, v2 = A.v1}
```

```
{{100 000, 200 000}, {175 000., 125 000.}, {193 750., 106 250.}}
```

Dimension of a vector space

A vector space is *finite-dimensional* if it has a basis consisting of a finite number of basis vectors. Since all bases of a finite-dimensional vector space have the same number of elements, this number is defined to be the *dimension* of the space.

Illustration

- A two-dimensional vector space

The space $\mathbb{R}^2$ of all pairs of real numbers $\{a, b\}$ is a two-dimensional vector space. The sets

```
B1 = {e1 = {1, 0}, e2 = {0, 1}}
```

```
B2 = {b1 = {3, -4}, b2 = {1, 1}}
```

are two bases for the same space.

- A one-dimensional vector space

The space $\mathbb{C}$ of all complex numbers is a one-dimensional complex vector space. The set

```
C = {1}
```

```
{1}
```

is a basis for $\mathbb{C}$ since every complex number *z* is a multiple of 1.

- A four-dimensional vector space

The space ℝ[t,3] of real polynomials of degree 3 or less is a four-dimensional vector space since the set

$$B = \{1, t, t^2, t^3\}$$

is a basis for the space.

- A four-dimensional vector space

```
A = RandomInteger[{0, 9}, {4, 5}];
```

$$A = \begin{pmatrix} 7 & 1 & 9 & 7 & 5 \\ 2 & 9 & 8 & 2 & 9 \\ 0 & 6 & 4 & 0 & 6 \\ 4 & 7 & 7 & 7 & 5 \end{pmatrix};$$

```
B = RowReduce[A];
```

$$B = \begin{pmatrix} 1 & 0 & 0 & 0 & -4 \\ 0 & 1 & 0 & 0 & -1 \\ 0 & 0 & 1 & 0 & 3 \\ 0 & 0 & 0 & 1 & 1 \end{pmatrix};$$

The first four columns of the matrix *B* are the pivot columns of the matrix *A*. They therefore form a basis for the column space of *A*. We can use the **Length** function to calculate its dimension.

```
Length[B]
```

4

Dimensions of a matrix

The numbers of rows and columns of a matrix, in that order, are called the *dimensions* of the matrix.

Illustration

- A matrix of dimensions {3, 4}

```
A = RandomInteger[{0, 9}, {3, 4}];
```

$$A = \begin{pmatrix} 0 & 7 & 2 & 2 \\ 5 & 7 & 8 & 5 \\ 9 & 4 & 7 & 1 \end{pmatrix};$$

```
Dimensions[A]
```

{3, 4}

- A matrix of dimensions {4, 3}

```
A = RandomInteger[{0, 9}, {4, 3}];
```

$$A = \begin{pmatrix} 4 & 3 & 0 \\ 3 & 4 & 7 \\ 6 & 1 & 5 \\ 3 & 5 & 2 \end{pmatrix};$$

```
Dimensions[A]
```

```
{4, 3}
```

- Dimensions of a square matrix

```
A = RandomInteger[{0, 9}, {4, 4}];
```

$$A = \begin{pmatrix} 3 & 4 & 5 & 6 \\ 0 & 4 & 3 & 1 \\ 3 & 0 & 5 & 8 \\ 9 & 4 & 2 & 9 \end{pmatrix};$$

```
Dimensions[A]
```

```
{4, 4}
```

Dirac matrix

The *Dirac* matrices are 4-by-4 matrices arising in quantum electrodynamics. They are Hermitian and unitary.

Illustration

- The 4-by-4 *Dirac matrices*

```
MatrixForm[I4 = {{1, 0, 0, 0}, {0, 1, 0, 0}, {0, 0, 1, 0}, {0, 0, 0, 1}}]
```

$$\begin{pmatrix} 1 & 0 & 0 & 0 \\ 0 & 1 & 0 & 0 \\ 0 & 0 & 1 & 0 \\ 0 & 0 & 0 & 1 \end{pmatrix}$$

```
MatrixForm[σ1 = {{0, 1, 0, 0}, {1, 0, 0, 0}, {0, 0, 0, 1}, {0, 0, 1, 0}}]
```

$$\begin{pmatrix} 0 & 1 & 0 & 0 \\ 1 & 0 & 0 & 0 \\ 0 & 0 & 0 & 1 \\ 0 & 0 & 1 & 0 \end{pmatrix}$$

```
MatrixForm[σ2 = {{0, -i, 0, 0}, {i, 0, 0, 0}, {0, 0, 0, -i}, {0, 0, i, 0}}]
```

$$\begin{pmatrix} 0 & -i & 0 & 0 \\ i & 0 & 0 & 0 \\ 0 & 0 & 0 & -i \\ 0 & 0 & i & 0 \end{pmatrix}$$

```
MatrixForm[σ3 = {{1, 0, 0, 0}, {0, -1, 0, 0}, {0, 0, 1, 0}, {0, 0, 0, -1}}]
```

$$\begin{pmatrix} 1 & 0 & 0 & 0 \\ 0 & -1 & 0 & 0 \\ 0 & 0 & 1 & 0 \\ 0 & 0 & 0 & -1 \end{pmatrix}$$

```
MatrixForm[ρ1 = {{0, 0, 1, 0}, {0, 0, 0, 1}, {1, 0, 0, 0}, {0, 1, 0, 0}}]
```

$$\begin{pmatrix} 0 & 0 & 1 & 0 \\ 0 & 0 & 0 & 1 \\ 1 & 0 & 0 & 0 \\ 0 & 1 & 0 & 0 \end{pmatrix}$$

```
MatrixForm[ρ2 = {{0, 0, -i, 0}, {0, 0, 0, -i}, {i, 0, 0, 0}, {0, i, 0, 0}}]
```

$$\begin{pmatrix} 0 & 0 & -i & 0 \\ 0 & 0 & 0 & -i \\ i & 0 & 0 & 0 \\ 0 & i & 0 & 0 \end{pmatrix}$$

```
MatrixForm[ρ3 = {{1, 0, 0, 0}, {0, 1, 0, 0}, {0, 0, -1, 0}, {0, 0, 0, -1}}]
```

$$\begin{pmatrix} 1 & 0 & 0 & 0 \\ 0 & 1 & 0 & 0 \\ 0 & 0 & -1 & 0 \\ 0 & 0 & 0 & -1 \end{pmatrix}$$

```
{HermitianMatrixQ[σ1], HermitianMatrixQ[ρ3]}

{True, True}

{UnitaryMatrixQ[σ1], UnitaryMatrixQ[ρ3]}

{True, True}
```

Direct sum of vector spaces

The zero subspaces are useful for the definition of direct sums of subspaces. If two subspaces U and V of a vector space W are *disjoint*, in other words, if they share only the zero vector of the space, and if B_U is a basis for U and B_V is a basis for V, then every vector **w** in W can be written as a unique sum **u** + **v**, with **u** in B_U and **v** in B_V. The union of U and V, in that order, is called the direct sum of U and V and is written as $U \oplus V$.

The direct sum symbol ⊕ is produced by typing Esc c+ Esc.

Illustration

- A direct sum of two subspaces of $\mathbb{R}^4$

If B_1 and B_2 are the two bases

```
B1 = {{1, 0, 0, 0}, {0, 1, 0, 0}};
```

```
B2 = {{0, 0, 1, 0}};
```

of subspaces of $\mathbb{R}^4$ and V is the subspace of all vectors of the form $\{a, b, c, 0\}$, then $W = \text{span}[B_1] \oplus \text{span}[B_2]$:

```
w = {a, b, c, 0} == a {1, 0, 0, 0} + b {0, 1, 0, 0} + c {0, 0, 1, 0}
```

```
True
```

- The direct sums of four vector spaces generated by the 3-by-5 matrix

```
A = {{3, 1, 0, 2, 4}, {1, 1, 0, 0, 2}, {5, 2, 0, 3, 7}};
```

```
Dimensions[A]
```

```
{3, 5}
```

- The coordinate space $\mathbb{R}^5$ as a direct sum of the null space and the row space of a matrix A

```
nsA = NullSpace[A]
```

```
{{-1, -1, 0, 0, 1}, {-1, 1, 0, 1, 0}, {0, 0, 1, 0, 0}}
```

```
rsA = RowReduce[A]
```

```
{{1, 0, 0, 1, 1}, {0, 1, 0, -1, 1}, {0, 0, 0, 0, 0}}
```

The null space and the row space are subspaces of $\mathbb{R}^5$ with dimensions 3 and 2. The spaces are disjoint and the sum of their dimensions is therefore 5.

```
Solve[nsA[[1]] == a rsA[[1]] + b rsA[[2]], {a, b}]
```

```
{}
```

```
Solve[nsA[[2]] == a rsA[[1]] + b rsA[[2]], {a, b}]
```

```
{}
```

```
Solve[nsA[[3]] == a rsA[[1]] + b rsA[[2]], {a, b}]
```

```
{}
```

```
Solve[rsA[[1]] == a nsA[[1]] + b nsA[[2]] + c nsA[[3]], {a, b, c}]
```

```
{}
```

```
Solve[rsA[[2]] == a nsA[[1]] + b nsA[[2]] + c nsA[[3]], {a, b, c}]
```

```
{}
```

The union of *nsA* and *rsA* forms a basis for $\mathbb{R}^5$. This is expressed by saying that $\mathbb{R}^5$ is a direct sum of the null and row spaces.

The notation `NullSpace[A] ⊕RowSpace[A] == ℝ^5` expresses the fact that the direct sum of the two disjoint subspaces is all of $\mathbb{R}^5$.

- The coordinate space $\mathbb{R}^3$ as a direct sum of the left null space and the column space of A

```
lnsA = NullSpace[Transpose[A]]
```

```
{{-3, -1, 2}}
```

```
csA = RowReduce[Transpose[A]]
```

$$\left\{\left\{1, 0, \frac{3}{2}\right\}, \left\{0, 1, \frac{1}{2}\right\}, \{0, 0, 0\}, \{0, 0, 0\}, \{0, 0, 0\}\right\}$$

The left null space and the column space are subspaces of $\mathbb{R}^3$ with dimensions 1 and 2. The spaces are disjoint and the sum of their dimensions 3.

```
Solve[lnsA[[1]] == a csA[[1]] + b csA[[2]], {a, b}]
```

```
{}
```

```
Solve[csA[[1]] == a lnsA[[1]], a]
```

```
{}
```

```
Solve[csA[[2]] == a lnsA[[1]], a]
```

```
{}
```

Hence we can form the direct sum `leftnullSpace[A]⊕ columnSpace[A]` of the left null space and the column space to build $\mathbb{R}^3$.

It may happen that some of the four subspaces of a matrix are zero spaces. In that case, their bases are empty and their dimensions therefore are zero.

- Direct sums involving zero subspaces

```
MatrixForm[A = {{1, 2}, {3, 4}}]
```

$$\begin{pmatrix} 1 & 2 \\ 3 & 4 \end{pmatrix}$$

```
NullSpace[A]
```

```
{}
```

```
MatrixForm[RowReduce[A]]
```

$$\begin{pmatrix} 1 & 0 \\ 0 & 1 \end{pmatrix}$$

This shows that the null space of A is the zero subspace $Z = \{\{0, 0\}\}$ of $\mathbb{R}^2$ and therefore

```
Z ⊕ RowSpace[A] = RowSpace[A] = ℝ²
```

Similarly,

```
NullSpace[Transpose[A]]
```

```
{}
```

```
MatrixForm[RowReduce[Transpose[A]]]
```

$$\begin{pmatrix} 1 & 0 \\ 0 & 1 \end{pmatrix}$$

The left null space of A is therefore also the zero subspace $Z = \{\{0, 0\}\}$ of $\mathbb{R}^2$. Hence

```
Z ⊕ ColumnSpace[A] = ColumnSpace[A] = ℝ²
```

Discrete Fourier transform

The *discrete Fourier transform* converts a list of data into a list of Fourier series coefficients. The *Mathematica* **Fourier** function and its inverse, the **InverseFourier** function, are the built-in tools for the conversion. The **Fourier** function can also be defined explicitly in terms of matrix multiplication using Fourier matrices.

Illustration

- A Fourier transform and its matrix equivalent

```
data = {-1, -1, -1, -1, 1, 1, 1, 1};
```

```
dftdata = Fourier[data]
```

```
{0. + 0. i, -0.707107 - 1.70711 i, 0. + 0. i, -0.707107 - 0.292893 i,
 0. + 0. i, -0.707107 + 0.292893 i, 0. + 0. i, -0.707107 + 1.70711 i}
```

```
sfmdata = N[Simplify[FourierMatrix[8].data]]
```

```
{0., -0.707107 - 1.70711 i, 0., -0.707107 - 0.292893 i,
 0., -0.707107 + 0.292893 i, 0., -0.707107 + 1.70711 i}
```

```
dftdata == sfmdata
```

```
True
```

- An inverse Fourier transform and its matrix equivalent

```
fcdata = {0. + 0. I, -0.707107 - 1.70711 I, 0. + 0. I, -0.707107 - 0.292893 I,
   0. + 0. I, -0.707107 + 0.292893 I, 0. + 0. I, -0.707107 + 1.70711 I};
```

```
ifdata = InverseFourier[fcdata]

{-1., -1., -1., -1., 1., 1., 1., 1.}

ifF = Simplify[Inverse[FourierMatrix[8]]];

sfmdata = Chop[ifF.fcdata]

{-1., -1., -1., -1., 1., 1., 1., 1.}
```

Discriminant of a Hessian matrix

See Hessian matrix

Disjoint subspaces

Two subspaces *U* and *V* of a vector space *W* are *disjoint* if they only have the zero vector in common.

Illustration

- Two disjoint proper subspaces

```
Quit[]

W = ℝ^5; S = {{1, 0, 0, 0, 0}, {0, 1, 0, 0, 0}}; T = {{0, 0, 1, 0, 0}, {0, 0, 0, 1, 0}};

U = {a S[[1]] + b S[[2]]}

V = {c T[[1]] + d T[[2]]}
```

By construction, the subspaces *U* and *V* are disjoint:

```
Solve[a S[[1]] + b S[[2]] == c T[[1]] + d T[[2]], {a, b, c, d}]
```

Distance between a point and a plane

The Euclidean distance *d[point,plane]* between a point {*p*, *q*, *r*} and a plane *ax* + *by* + *cz* + *d* = 0 in the space $\mathbb{E}^3$ is

```
Abs[a x + b y + c z + d /. {x → p, y → q, z → r}] / Sqrt[a^2 + b^2 + c^2]     (1)
```

Illustration

- The Euclidean distance between a point and a plane

```
{p, q, r} = {1, 2, 3};
```

```
plane = 3 x - y + 4 z - 9 == 0;
```

```
numerator = Abs[3 x - y + 4 z - 9 /. {x → 1, y → 2, z → 3}]
```

$$\frac{61}{2}$$

```
denominator = Sqrt[3^2 + (-1)^2 + 4^2]
```

$$\sqrt{26}$$

```
distance = numerator / denominator
```

$$\frac{61}{2\sqrt{26}}$$

- Using projections and normals to compute the Euclidean distance between a point and a plane

The Euclidean distance between an external point $P\{p, q, r\}$ and the point $Q\{x_0, y_0, z_0\}$ in the plane $ax + by + cz + d = 0$ is also equal to the Euclidean norm of the orthogonal projection of the vector $(Q - P) = \{x_0 - p, y_0 - q, z_0 - r\}$ onto the normal $\{a, b, c\}$ of the given plane.

```
Clear[x, y, z, p, q, r]
```

```
plane = 3 x - y + 4 z - 9 == 0;
```

```
externalpoint = {p, q, r} = {1, 2, 3};
```

```
normal = {3, -1, 4};
```

```
Reduce[plane, {x, y, z}]
```

$$z == \frac{9}{4} - \frac{3x}{4} + \frac{y}{4}$$

```
planarpoint = {x0, y0, z0} = {0, 0, 9/4};
```

```
Projection[externalpoint - planarpoint, normal]
```

$$\left\{\frac{6}{13}, -\frac{2}{13}, \frac{8}{13}\right\}$$

```
Norm[%]
```

$$2\sqrt{\frac{2}{13}}$$

Distance function

A *distance function* on a vector space *V* is a function that assigns a nonnegative real number *d*(**u**, **v**) to every pair of vectors {**u**, **v**} in *V* and has the following properties:

Properties of distance functions

$d[\mathbf{u}, \mathbf{v}] \geq 0$ (1)

$d[\mathbf{u}, \mathbf{v}] = d[\mathbf{v}, \mathbf{u}]$ (2)

$d[\mathbf{u}, \mathbf{v}] \leq d[\mathbf{u}, \mathbf{w}] + d[\mathbf{w}, \mathbf{v}]$ (3)

$d[\mathbf{u}, \mathbf{v}] = 0$ if and only if $\mathbf{u} = \mathbf{v}$ (4)

Illustration

- The distance between two vectors determined by a norm

Like all other norms, the Euclidean norm and the p-norms define distance functions. But there are others. The function

$$d[\mathbf{u}, \mathbf{v}] = \begin{cases} 1 & \text{if } \mathbf{u} \neq \mathbf{v} \\ 0 & \text{otherwise} \end{cases} \quad (5)$$

is a distance function.

In topology and other fields, distance functions are called metrics and spaces equipped with metrics are called metric spaces.

- A distance function for $\mathbb{R}^2$

```
d[u_, v_] := If[u ≠ v, 1, 0]

{d[{1, 2}, {2, 3}], d[{a, b}, {a, b}]}

{1, 0}
```

- The Euclidean distance function for $\mathbb{R}^2$

```
d[{u_, v_}, {r_, s_}] := Sqrt[(u - r)^2 + (v - s)^2]

d[{1, 2}, {4, -5}]
```

$\sqrt{58}$

```
d[{1, 2}, {4, -5}] == Norm[{1 - 4, 2 + 5}, 2]

True
```

Domain of a linear transformation

The *domain* of a linear transformation *T* is the vector space on which *T* acts. The notation $T : A \longrightarrow B$ identifies the vector space *A* as the domain of *T* and the vector space *B* as its codomain.

Illustration

- The domain, codomain, and range of a linear transformation *T* from $\mathbb{R}^2$ to $\mathbb{R}^3$

```
T[{x_, y_}] := {x, y, 0}
```

```
T[{1, 1}]
```

```
{1, 1, 0}
```

The domain of *T* is $\mathbb{R}^2$, the codomain of *T* is $\mathbb{R}^3$, and the range of *T* is the subspace of $\mathbb{R}^3$ consisting of all vectors of the form {x, y, 0}.

Dot product

The *dot product* of two real vectors is the sum of the componentwise products of the vectors. In spite of its name, *Mathematica* does not use a dot (.) to represent this function. It must be written in the **Dot** notation. The period (the dot) is used to designate *matrix multiplication*.

Properties of dot products

```
Dot[u, v] = Dot[v, u]                              (1)

Dot[u, v + w] = Dot[u, v] + Dot[u, w]              (2)

Dot[u, r v + w] = r Dot[u, v] + Dot[u, w]          (3)

Dot[r u, s v] = (r s) Dot[u, v]                    (4)
```

Illustration

- The dot product of two vectors in $\mathbb{R}^2$

```
Clear[a, b, v, w]
```

```
v = {1, 2}; w = {a, b};
```

```
Dot[v, w]
```

```
a + 2 b
```

- The dot product of two vectors in $\mathbb{R}^3$

```
Clear[x, a, b, c]
```

```
x = {1, 2, 3}; y = {a, b, c};

Dot[x, y]

a + 2 b + 3 c
```

- The dot product of two vectors in $\mathbb{R}^5$

```
Clear[a, b, c, d, e, r, s]

r = {1, 2, 3, 4, 5}; s = {a, b, c, d, e};

Dot[r, s]

a + 2 b + 3 c + 4 d + 5 e
```

- The dot product and the standard deviation

```
data = Range[10];

average = Mean[data];

dot = Dot[(data - average), (data - average)];

sample = Length[data] - 1;
```

$$\text{stdevdata} = \text{Sqrt}\left[\frac{1}{\text{sample}}\,\text{dot}\right]$$

$$\sqrt{\frac{55}{6}}$$

```
stdevdata == StandardDeviation[data]

True
```

Manipulation

- Exploring the dot product

```
Manipulate[Dot[{a, b}, {-5, 2}], {a, -3, 3, 1}, {b, -4, 4, 1}]
```

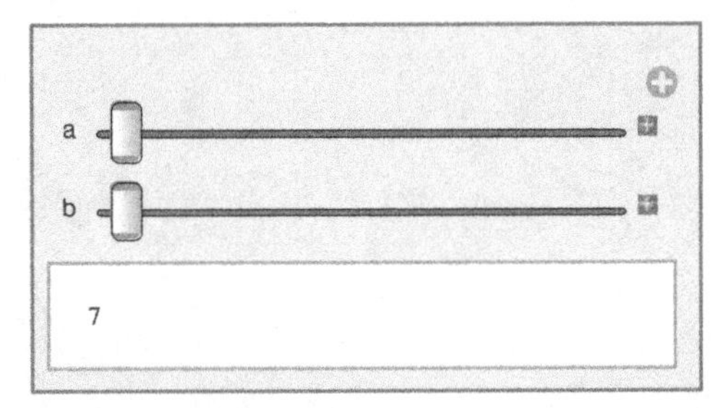

We use **Manipulate** and **Dot** to explore the dot product. If we let $a = -3$ and $b = -4$, then the manipulation shows, for example, that the dot product of the generated vectors is 7.

- Sample standard deviations

```
Manipulate[data = Range[n]; average = Mean[data];
 dot = Dot[(data - average), (data - average)]; sample = Length[data] - 1;
 stdevdata = Sqrt[1/sample dot], {n, 2, 100, 1}]
```

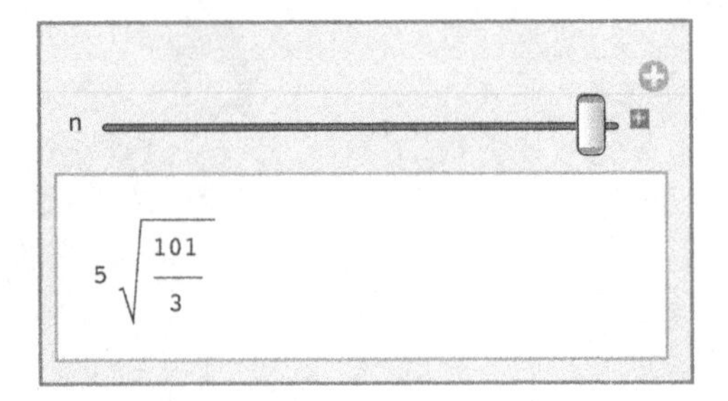

We use **Manipulate**, **Range**, **Mean**, **Dot**, **Length**, and **Sqrt** to explore sample standard deviations.

```
StandardDeviation[Range[100]]
```

$$5\sqrt{\frac{101}{3}}$$

For $n = 100$, the manipulation shows that the sample standard deviation of the list {1, 2, ..., 100} is $5\sqrt{\frac{101}{3}}$.

Dual space

Consider the two-dimensional coordinate space $V = \mathbb{R}^2$. A linear functional f is a function $\mathbb{R}^2 \to \mathbb{R}$ preserving linear combinations. Each dual space has a basis consisting of linear functionals. It's called a *dual basis* and defined on the *dual space* of V.

Illustration

- A dual space of $\mathbb{R}^3$

Let V be the real vector space $\mathbb{R}^3$ and consider the following linear functionals on V:

```
f1[{x_, y_, z_}] := 3 x + y;
```

```
f2[{x_, y_, z_}] := y - z;
```

```
f3[{x_, y_, z_}] := x + y + 2 z;
```

We show that set $\{f_1, f_2, f_3\}$ is a basis for V^*:

```
Clear[x, y, z, a, b, c];
```

```
Expand[a (3 x + y) + b (y - z) + c (x + y + 2 z)]
```

```
3 a x + c x + a y + b y + c y - b z + 2 c z
```

- The set of linear functionals $\{f_1, f_2, f_3\}$ spans V^*.

```
Solve[{d, e, f} == {3 a + c, a + b + c, -b + 2 c}, {a, b, c}]
```

$$\left\{\left\{a \to \frac{1}{8}(3 d - e - f),\ b \to \frac{1}{4}(-d + 3 e - f),\ c \to \frac{1}{8}(-d + 3 e + 3 f)\right\}\right\}$$

This shows that every linear combination of linear functionals on V can be written uniquely as a linear combination of the linear functionals f_1, f_2, and f_3.

- The set of linear functionals $\{f_1, f_2, f_3\}$ is also linearly independent.

```
Solve[{3 a + c, a + b + c, -b + 2 c} == {0, 0, 0}, {a, b, c}]
```

```
{{a → 0, b → 0, c → 0}}
```

This shows that the zero linear functional can only be written as the trivial linear combination of the linear functionals f_1, f_2, and f_3.

- Construction of a basis for V for which $\{f_1, f_2, f_3\}$ is a dual basis.

To show that $\{f_1, f_2, f_3\}$ is a dual basis, there must exist a basis $\{e_1, e_2, e_3\}$ for V for which $f_i(e_j) = 1$ if $i = j$ and 0 if $i \neq j$. Let

```
B = {e1 = {x1, y1, z1}, e2 = {x2, y2, z2}, e3 = {x3, y3, z3}};
```

be the required basis. Then $\{f_1, f_2, f_3\}$ is a dual basis, provided that e_1, e_2, and e_3 are the following vectors:

```
solution1 = Flatten[Solve[{3 x1 + y1 == 1, y1 - z1 == 0, x1 + y1 + 2 z1 == 0}, {x1, y1, z1}]]
```

$$\left\{x_1 \to \frac{3}{8},\ y_1 \to -\frac{1}{8},\ z_1 \to -\frac{1}{8}\right\}$$

```
solution2 = Flatten[Solve[{3 x2 + y2 == 0, y2 - z2 == 1, x2 + y2 + 2 z2 == 0}, {x2, y2, z2}]]
```

$\left\{x_2 \to -\frac{1}{4},\ y_2 \to \frac{3}{4},\ z_2 \to -\frac{1}{4}\right\}$

```
solution3 = Flatten[Solve[{3 x3 + y3 == 0, y3 - z3 == 0, x3 + y3 + 2 z3 == 1}, {x3, y3, z3}]]
```

$\left\{x_3 \to -\frac{1}{8},\ y_3 \to \frac{3}{8},\ z_3 \to \frac{3}{8}\right\}$

```
e1 = {x1, y1, z1} /. solution1
```

$\left\{\frac{3}{8}, -\frac{1}{8}, -\frac{1}{8}\right\}$

```
e2 = {x2, y2, z2} /. solution2
```

$\left\{-\frac{1}{4}, \frac{3}{4}, -\frac{1}{4}\right\}$

```
e3 = {x3, y3, z3} /. solution3
```

$\left\{-\frac{1}{8}, \frac{3}{8}, \frac{3}{8}\right\}$

To show that $B = \{e_1, e_2, e_3\}$ is a basis for V, it suffices to show that the matrix B is invertible.

```
B = {e1, e2, e3}
```

$\left\{\left\{\frac{3}{8}, -\frac{1}{8}, -\frac{1}{8}\right\}, \left\{-\frac{1}{4}, \frac{3}{4}, -\frac{1}{4}\right\}, \left\{-\frac{1}{8}, \frac{3}{8}, \frac{3}{8}\right\}\right\}$

```
Det[B]
```

$\frac{1}{8}$

The following calculations show that $\{f_1, f_2, f_3\}$ is a dual basis for V with respect to the basis $\{e_1, e_2, e_3\}$:

```
{f1[e1], f1[e2], f1[e3]}
```

{1, 0, 0}

```
{f2[e1], f2[e2], f2[e3]}
```

{0, 1, 0}

```
{f3[e1], f3[e2], f3[e3]}
```

{0, 0, 1}

E

Echelon form

See Row echelon matrix

Eigenspace

The span of the set of all eigenvectors associated with an eigenvalue λ is the *eigenspace* of λ.

Illustration

- The eigenspaces of a 4-by-4 matrix

```
Clear[A, B, t, u, v, w]
```

```
MatrixForm[A = {{4, 7, 2, 0}, {7, 7, 9, 2}, {2, 0, 1, 9}, {8, 8, 2, 4}}]
```

$$\begin{pmatrix} 4 & 7 & 2 & 0 \\ 7 & 7 & 9 & 2 \\ 2 & 0 & 1 & 9 \\ 8 & 8 & 2 & 4 \end{pmatrix}$$

```
MatrixForm[B = UpperTriangularize[{{4, 7, 2, 0}, {7, 7, 9, 2}, {2, 0, 1, 9}, {8, 8, 2, 4}}]]
```

$$\begin{pmatrix} 4 & 7 & 2 & 0 \\ 0 & 7 & 9 & 2 \\ 0 & 0 & 1 & 9 \\ 0 & 0 & 0 & 4 \end{pmatrix}$$

```
Eigensystem[B]
```

```
{{7, 4, 4, 1}, {{7, 3, 0, 0}, {1, 0, 0, 0}, {0, 0, 0, 0}, {17, -9, 6, 0}}}
```

- The eigenspace of the eigenvalue 7

```
Reduce[B.{t, u, v, w} == 7 {t, u, v, w}, {t, u, v, w}]
```

$$u == \frac{3\,t}{7}\ \&\&\ v == 0\ \&\&\ w == 0$$

Therefore the eigenspace of the eigenvalue 7 consists of all vectors of the form

```
{t, u, v, w} /. {u → 3 t/7, v → 0, w → 0}
```

$$\left\{t, \frac{3t}{7}, 0, 0\right\}$$

In particular, the vector

```
{t, u, v, w} /. {t → 7, u → 3, v → 0, w → 0}
```

{7, 3, 0, 0}

computed with the **Eigensystem** function yields a basis for the eigenspace of the eigenvalue 7.

- The eigenspace of the eigenvalue 4

```
Reduce[B.{t, u, v, w} == 4 {t, u, v, w}, {t, u, v, w}]
```

u == 0 && v == 0 && w == 0

Therefore the eigenspace of the eigenvalue 4 consists of all vectors of the form

```
{t, u, v, w} /. {u → 0, v → 0, w → 0}
```

{t, 0, 0, 0}

In particular, the vector

```
{t, u, v, w} /. {t → 1, u → 0, v → 0, w → 0}
```

{1, 0, 0, 0}

computed with the **Eigensystem** function, forms a basis for the eigenspace of the eigenvalue 4.

- The eigenspace of the eigenvalue 1

```
Reduce[B.{t, u, v, w} == {t, u, v, w}, {t, u, v, w}]
```

$$u == -\frac{9t}{17} \;\&\&\; v == \frac{6t}{17} \;\&\&\; w == 0$$

Therefore the eigenspace of the eigenvalue 1 consists of all vectors of the form

```
{t, u, v, w} /. {u → -9 t/17, v → 6 t/17, w → 0}
```

$$\left\{t, -\frac{9t}{17}, \frac{6t}{17}, 0\right\}$$

The vector

```
{t, u, v, w} /. {t → 17, u → -9, v → 6, w → 0}
```

```
{17, -9, 6, 0}
```

forms a basis for the eigenspace of the eigenvalue 1.

Manipulation

- Eigenvalues and eigenvectors

```
Manipulate[Eigensystem[A = {{4, a, a, a}, {0, a, b, 2}, {0, 0, 1, 9}, {0, 0, 0, 4}}],
 {a, -1, 1, 1}, {b, -2, 2, 1}]
```

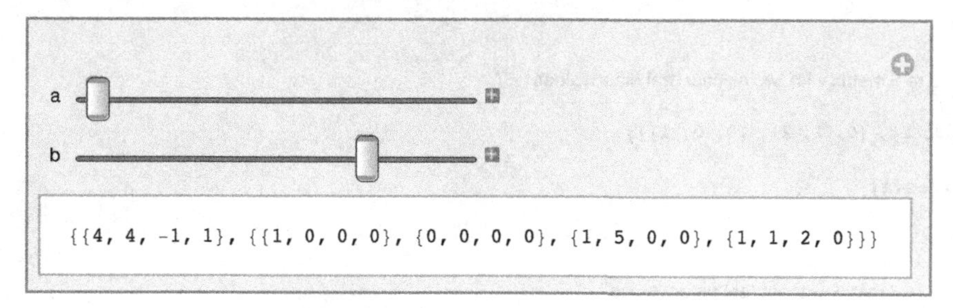

We use **Manipulate** and **Eigensystem** to explore eigenvalues and eigenvectors. The example shows, for example, that if $a = -1$ and $b = 1$, the input matrix

```
A = {{4, -1, -1, -1}, {0, -1, 1, 2}, {0, 0, 1, 9}, {0, 0, 0, 4}};
```

has the eigenvalues

```
valuesA = {4, 4, -1, 1};
```

and associated eigenvectors

```
vectorsA = {{1, 0, 0, 0}, {0, 0, 0, 0}, {1, 5, 0, 0}, {1, 1, 2, 0}};
```

The following table confirms that this is correct:

```
Table[A.vectorsA[[i]] == valuesA[[i]] vectorsA[[i]], {i, 1, 4}]
```

```
{True, True, True, True}
```

The **Eigensystem** function outputs both zero and nonzero vectors of eigenspaces.

Eigenvalue

An *eigenvalue* of a square matrix A is a scalar λ for which there exists a nonzero vector $\mathbf{v}$ with the property that $A\mathbf{v} = \lambda\mathbf{v}$. The eigenvalues of a real square matrix may be all real, both real and complex, or all complex. All n-by-n triangular real matrices have n real eigenvalues.

Illustration

- A 3-by-3 real matrix with three distinct real eigenvalues

```
A = {{2, 3, 1}, {0, 3, 2}, {0, 0, 4}};
```

```
Eigenvalues[A]
```

```
{4, 3, 2}
```

- A 3-by-3 real matrix with two distinct real eigenvalues

```
A = {{2, 3, 1}, {0, 1, 2}, {0, 0, 1}};
```

```
Eigenvalues[A]
```

```
{2, 1, 1}
```

- A 2-by-2 real matrix without real eigenvalues

Although every n-by-n numerical matrix has n (not necessarily distinct) eigenvalues, some real matrices may only have complex eigenvalues.

```
A = {{0, -1}, {1, 0}};
```

```
cpA = CharacteristicPolynomial[A, t]
```

$1 + t^2$

```
Eigenvalues[A]
```

```
{i, -i}
```

$\text{Solve}[t^2 + 1 == 0, t]$

```
{{t → -i}, {t → i}}
```

It is easy to explain geometrically why the matrix A has no real eigenvalues. The matrix represents a rotation transformation and rotations do not map nonzero vectors to multiples of themselves.

- A 2-by-2 family of real matrices without real eigenvalues

If **Sin**[x] is not equal to zero then the following matrix has no real eigenvalues.

```
A = {{Cos[x], Sin[x]}, {-Sin[x], Cos[x]}};
```

```
Eigenvalues[A]
```

```
{Cos[x] - i Sin[x], Cos[x] + i Sin[x]}
```

- A 4-by-4 real matrix with two real and two complex eigenvalues

$$A = \begin{pmatrix} 0 & 1 & 1 & -1 \\ -1 & -1 & -3 & 3 \\ -3 & 0 & 3 & -2 \\ 0 & 1 & 1 & -3 \end{pmatrix};$$

```
cpA = CharacteristicPolynomial[A, t]
```

$-18 - 5t - 6t^2 + t^3 + t^4$

```
Plot[cpA, {t, -5, 5}]
```

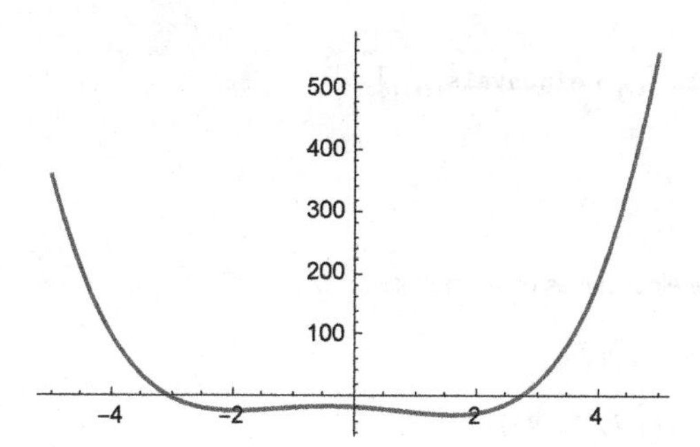


```
N[Eigenvalues[A]]
```

{-3.05739, 2.73717, -0.339888 + 1.42667 i, -0.339888 - 1.42667 i}

The determinant of a matrix is equal to the product of the eigenvalues of the matrix.

- Eigenvalues and determinants

```
A = RandomInteger[{0, 9}, {4, 4}];

MatrixForm[A = {{8, 6, 3, 0}, {1, 0, 5, 5}, {9, 8, 8, 7}, {4, 1, 5, 8}}]
```

$$\begin{pmatrix} 8 & 6 & 3 & 0 \\ 1 & 0 & 5 & 5 \\ 9 & 8 & 8 & 7 \\ 4 & 1 & 5 & 8 \end{pmatrix}$$

```
eigenvals = N[Eigenvalues[A]]
```

{19.9177, 3.06525 + 0.922228 i, 3.06525 - 0.922228 i, -2.04819}

```
Chop[eigenvals[[1]] eigenvals[[2]] eigenvals[[3]] eigenvals[[4]]]
```

-418.

```
Det[A]
```

-418

The trace of a matrix is equal to the sum of the eigenvalues of the matrix.

- Eigenvalues and traces

```
A = RandomInteger[{0, 9}, {4, 4}];

A = {{8, 6, 3, 0}, {1, 0, 5, 5}, {9, 8, 8, 7}, {4, 1, 5, 8}};

Tr[A]

24

Chop[eigenvals[[1]] + eigenvals[[2]] + eigenvals[[3]] + eigenvals[[4]]]

24.
```

- Eigenvalues and determinants

The determinant of a matrix is equal to the product of the eigenvalues of the matrix.

```
A = RandomInteger[{0, 9}, {4, 4}];

A = {{8, 6, 3, 0}, {1, 0, 5, 5}, {9, 8, 8, 7}, {4, 1, 5, 8}};

Det[A]

-418

eigenvals = N[Eigenvalues[A]]

{19.9177, 3.06525 + 0.922228 i, 3.06525 - 0.922228 i, -2.04819}

Chop[eigenvals[[1]] eigenvals[[2]] eigenvals[[3]] eigenvals[[4]]]

-418.
```

Manipulation

- Eigenvalues of a 4-by-4 matrix

```
Clear[A]

MatrixForm[A = {{1 + a, 6, 3, 0}, {1, 2 - b, 5, 5}, {9, 8, 8, 3 + c}, {4, 1, 5, 8}}]
```

$$\begin{pmatrix} 1+a & 6 & 3 & 0 \\ 1 & 2-b & 5 & 5 \\ 9 & 8 & 8 & 3+c \\ 4 & 1 & 5 & 8 \end{pmatrix}$$

```
Manipulate[Evaluate[Eigenvalues[A]], {a, -4, 4, 1}, {b, -5, 5, 1}, {c, -3, 3, 1}]
```

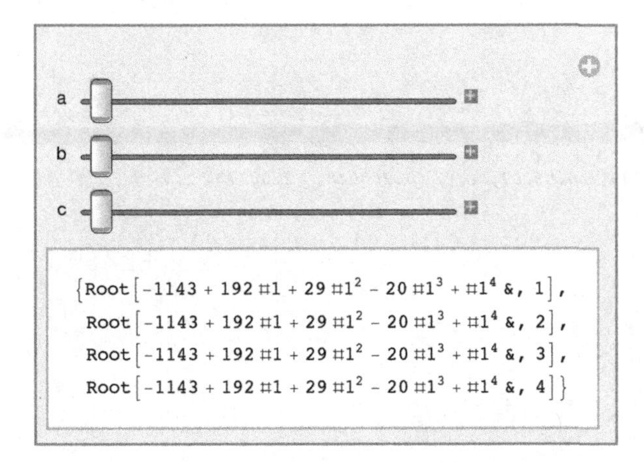

We combine **Manipulate, Evaluate,** and **Eigenvalues** to explore the eigenvalues of different matrices by varying the values of *a*, *b*, and *c*. By letting *a* = - 4, *b* = - 5, and *c* = - 3, for example, the manipulation displays the eigenvalues of the generated matrix as the roots of the polynomial

```
p[t_] := -1143 + 192 t + 29 t^2 - 20 t^3 + t^4

N[{Root[-1143 + 192 #1 + 29 #1^2 - 20 #1^3 + #1^4 &, 1], Root[-1143 + 192 #1 + 29 #1^2 - 20 #1^3 + #1^4 &, 2],
  Root[-1143 + 192 #1 + 29 #1^2 - 20 #1^3 + #1^4 &, 3],
  Root[-1143 + 192 #1 + 29 #1^2 - 20 #1^3 + #1^4 &, 4]}]
{-3.92763, 17.9912, 2.96821 - 2.71388 i, 2.96821 + 2.71388 i}
```

The **NSolve** command confirms the results:

```
NSolve[p[t] == 0, t]

{{t → -3.92763}, {t → 2.96821 - 2.71388 i}, {t → 2.96821 + 2.71388 i}, {t → 17.9912}}
```

Eigenvector

An *eigenvector* of a square real matrix *A* is a nonzero vector **v** for which there exists an eigenvalue λ for which $A\mathbf{v} = \lambda\mathbf{v}$.

Illustration

- An approximate eigenvector of a real matrix

```
A = RandomInteger[{0, 9}, {3, 3}];
```

```
MatrixForm[A = {{7, 1, 1}, {7, 8, 9}, {8, 3, 7}}]
```

$$\begin{pmatrix} 7 & 1 & 1 \\ 7 & 8 & 9 \\ 8 & 3 & 7 \end{pmatrix}$$

```
{v1, v2, v3} = N[Eigenvectors[A]]

{{0.343958, 1.67981, 1.}, {2.23599, -6.83207, 1.}, {0.260048, -2.08773, 1.}}

Reduce[A.v1 == λ v1]

λ == 14.7911

N[Eigenvalues[A]]

{14.7911, 4.39173, 2.81718}
```

- Eigenvectors of a diagonal matrix

```
A = DiagonalMatrix[{1, 2, 2, 4}]

{{1, 0, 0, 0}, {0, 2, 0, 0}, {0, 0, 2, 0}, {0, 0, 0, 4}}

Eigenvectors[A]
{{0, 0, 0, -1}, {0, 0, 1, 0}, {0, 1, 0, 0}, {1, 0, 0, 0}}

Eigenvalues[A]

{4, 2, 2, 1}

A.{0, 0, 0, -1} == 4 {0, 0, 0, -1}

True

A.{0, 0, 1, 0} == 2 {0, 0, 1, 0}

True

A.{0, 1, 0, 0} == 2 {0, 1, 0, 0}

True

A.{1, 0, 0, 0} == {1, 0, 0, 0}

True
```

- Eigenvectors of a triangular matrix

```
A = RandomInteger[{0, 9}, {3, 3}];
```

```
MatrixForm[A = {{5, 0, 8}, {2, 2, 9}, {3, 1, 0}}]
```

$$\begin{pmatrix} 5 & 0 & 8 \\ 2 & 2 & 9 \\ 3 & 1 & 0 \end{pmatrix}$$

```
tA = LowerTriangularize[A]
```

```
{{5, 0, 0}, {2, 2, 0}, {3, 1, 0}}
```

```
Eigenvectors[tA]
```

```
{{15, 10, 11}, {0, 2, 1}, {0, 0, 1}}
```

```
Eigenvalues[tA]
```

```
{5, 2, 0}
```

```
tA.{15, 10, 11} == 5 {15, 10, 11}
```

```
True
```

```
tA.{0, 2, 1} == 2 {0, 2, 1}
```

```
True
```

```
tA.{0, 0, 1} == 0 {0, 0, 1}
```

```
True
```

- Eigenvectors as bases for null spaces

```
MatrixForm[A = {{5, 0, 0}, {2, 2, 0}, {3, 1, 0}}]
```

$$\begin{pmatrix} 5 & 0 & 0 \\ 2 & 2 & 0 \\ 3 & 1 & 0 \end{pmatrix}$$

```
Eigensystem[A]
```

```
{{5, 2, 0}, {{15, 10, 11}, {0, 2, 1}, {0, 0, 1}}}
```

```
NullSpace[(A - 5 IdentityMatrix[3])]
```

```
{{15, 10, 11}}
```

```
NullSpace[(A - 2 IdentityMatrix[3])]
```

```
{{0, 2, 1}}
```

```
NullSpace[(A - 0 IdentityMatrix[3])]
```

```
{{0, 0, 1}}
```

This shows that the eigenvectors of A associated with each λ are the basis vectors for the null spaces of each of the matrices (A - λ *IdentityMatrix*).

- Eigenvectors and diagonalization of matrices

If a square matrix A is equal to a product $(P\,dM\,P^{-1})$ of an invertible matrix P and a diagonal matrix dM, then the columns of P are eigenvectors of A.

```
MatrixForm[A = {{5, 0, 0}, {2, 2, 0}, {3, 1, 4}}]
```

$$\begin{pmatrix} 5 & 0 & 0 \\ 2 & 2 & 0 \\ 3 & 1 & 4 \end{pmatrix}$$

```
{eigenvalues, eigenvectors} = Eigensystem[A]
```

```
{{5, 4, 2}, {{3, 2, 11}, {0, 0, 1}, {0, -2, 1}}}
```

```
Transpose[eigenvectors].DiagonalMatrix[eigenvalues].Inverse[Transpose[eigenvectors]]
```

```
{{5, 0, 0}, {2, 2, 0}, {3, 1, 4}}
```

The resulting matrix product is equal to the original matrix A.

Elementary matrix

An *elementary matrix* is a matrix obtained from an identity matrix by applying one elementary row operation to the identity matrix.

Illustration

- An elementary matrix resulting from the interchange of two rows

```
A = IdentityMatrix[3]
```

```
{{1, 0, 0}, {0, 1, 0}, {0, 0, 1}}
```

```
A = {A[[1]], A[[3]], A[[2]]}
```

```
{{1, 0, 0}, {0, 0, 1}, {0, 1, 0}}
```

- An elementary matrix resulting from multiplication of a row by a nonzero constant

```
A = IdentityMatrix[3]
```

```
{{1, 0, 0}, {0, 1, 0}, {0, 0, 1}}
```

```
A[[2]] = 3 A[[2]];
A
```

```
{{1, 0, 0}, {0, 3, 0}, {0, 0, 1}}
```

- An elementary matrix resulting from the addition of a multiple of a row to another row

```
A = IdentityMatrix[3]
```

```
{{1, 0, 0}, {0, 1, 0}, {0, 0, 1}}
```

```
A[[3]] = A[[3]] - 4 A[[1]];
A
```

```
{{1, 0, 0}, {0, 1, 0}, {-4, 0, 1}}
```

Elementary row operation

The *elementary row operations* on a matrix are the interchange of two rows, the multiplication of a row by a nonzero constant, and the addition of a multiple of a row to another row.

Illustration

- Interchange of two rows

$$A = \begin{pmatrix} 2 & 6 & 9 & 0 & 1 \\ 7 & 6 & 7 & 9 & 2 \\ 1 & 0 & 7 & 2 & 9 \end{pmatrix};$$

```
MatrixForm[B = {A[[3]], A[[2]], A[[1]]}]
```

$$\begin{pmatrix} 1 & 0 & 7 & 2 & 9 \\ 7 & 6 & 7 & 9 & 2 \\ 2 & 6 & 9 & 0 & 1 \end{pmatrix}$$

- Multiplication of a row by a nonzero constant

$$A = \begin{pmatrix} 2 & 6 & 9 & 0 & 1 \\ 7 & 6 & 7 & 9 & 2 \\ 1 & 0 & 7 & 2 & 9 \end{pmatrix};$$

```
A[[2]] = 5 A[[2]]; MatrixForm[A]
```

$$\begin{pmatrix} 2 & 6 & 9 & 0 & 1 \\ 35 & 30 & 35 & 45 & 10 \\ 1 & 0 & 7 & 2 & 9 \end{pmatrix}$$

- Addition of a multiple of one row to another row

$$A = \begin{pmatrix} 2 & 6 & 9 & 0 & 1 \\ 7 & 6 & 7 & 9 & 2 \\ 1 & 0 & 7 & 2 & 9 \end{pmatrix};$$

```
A[[3]] = A[[3]] - 4 A[[2]]; MatrixForm[A]
```

$$\begin{pmatrix} 2 & 6 & 9 & 0 & 1 \\ 7 & 6 & 7 & 9 & 2 \\ -27 & -24 & -21 & -34 & 1 \end{pmatrix}$$

Euclidean distance

The *Euclidean distance* between two vectors **u** and **v** in the space $\mathbb{R}^n$ is the two-norm of the difference vector (**u** - **v**).

Illustration

- The distance between two vectors in $\mathbb{R}^2$

```
x = {1, -2}; y = {-3, 4};

EuclideanDistance[x, y]
```

$2\sqrt{13}$

- The distance between two vectors in $\mathbb{R}^2$

```
x = {1, -2}; y = {-3, 4};

distance[x_, y_] := Norm[x - y, 2]

distance[x, y]
```

$2\sqrt{13}$

- The distance between two vectors in $\mathbb{R}^3$

Since **Norm[v]** = **Norm[v**, 2] by default, we can omit the 2-option when using the **Norm** function for calculating Euclidean distances.

```
x = {1, 2, 3}; y = {4, 5, 6};

distance[x_, y_] := Norm[x - y]

distance[x, y]
```

$3\sqrt{3}$

- The distance between two vectors in $\mathbb{R}^4$

```
distance[x_List, y_List] /; Length[x] == Length[y] := Sqrt[Total[(x - y)^2]]
```

```
vector1 = {1, 2, 3, 4}; vector2 = {5, 6, 7, 8};

distance[vector1, vector2]

8
```

Euclidean norm

The Euclidean norm **Norm[v,** *2*] or simply **Norm[v]** = ||**v**|| function on a coordinate space $\mathbb{R}^n$ is the square root of the sum of the squares of the coordinates of **v**.

Properties of Euclidean norms

$\|\mathbf{v}\| > 0$ when $\mathbf{v} \neq \mathbf{0}$ (1)

$\|\mathbf{v}\| = 0$ if and only if $\mathbf{v} = \mathbf{0}$ (2)

$\|k\,\mathbf{v}\| = |k|\,\|\mathbf{v}\|$ for all scalars *k* and all vectors v (3)

$\|\mathbf{v} + \mathbf{w}\| \leq \|\mathbf{v}\| + \|\mathbf{w}\|$ (4)

(where $|k|$ denotes the absolute value of the scalar *k*)

Illustration

- Euclidean norm (length) of a vector in $\mathbb{R}^2$

```
vector = {a, b};

||vector|| = Norm[vector]
```

$\sqrt{\text{Abs}[a]^2 + \text{Abs}[b]^2}$

- A non-Euclidean norm (length) of a vector in $\mathbb{R}^2$

```
vector = {a, b};

||vector||₃ = Norm[vector, 3]
```

$(\text{Abs}[a]^3 + \text{Abs}[b]^3)^{1/3}$

For any real number $p \geq 1$, we can define a non-Euclidean "*p*-norm:"

```
||vector||p = Norm[vector, p]
```

$(\text{Abs}[a]^p + \text{Abs}[b]^p)^{\frac{1}{p}}$

Manipulation

- A Euclidean norm

```
Manipulate[Norm[Range[n]], {n, 1, 9, 1}]
```

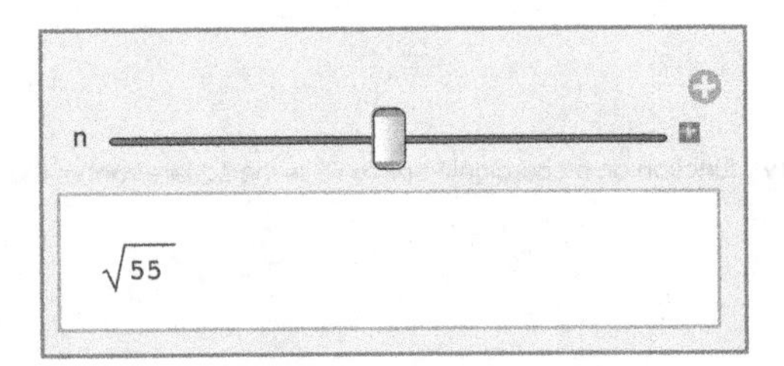


```
Norm[{1, 2, 3, 4, 5}]
```

$\sqrt{55}$

We use **Manipulate**, **Norm**, and **Range** to explore the Euclidean norm of real vectors. If we let $n = 5$, for example, the manipulation displays the Euclidean norm of the vector {1, 2, 3, 4, 5}.

Euclidean space

The coordinate spaces $\mathbb{R}^n$, equipped with the dot products as inner products, are usually called Euclidean spaces and are often denoted by $\mathbb{E}^n$. The dot product serves as the basis for defining the standard geometric concepts in these spaces.

Illustration

- Length of a vector in $\mathbb{E}^3$

```
v = {1, 2, 3};
```

```
lengthv = Sqrt[Dot[v, v]]
```

$\sqrt{14}$

```
Norm[v]
```

$\sqrt{14}$

- Orthogonality of two vectors in $\mathbb{E}^3$

The vectors $u = \{1, 2, 3\}$ and $v = \{-2, 1, 0\}$ are orthogonal in $\mathbb{E}^3$:

```
Dot[{1, 2, 3}, {-2, 1, 0}]
```

0

We can represent them graphically by two perpendicular arrows in $\mathbb{E}^3$:

```
Graphics3D[{Arrow[{{0, 0, 0}, {1, 2, 3}}], Arrow[{{0, 0, 0}, {-2, 1, 0}}]}, Axes → True]
```

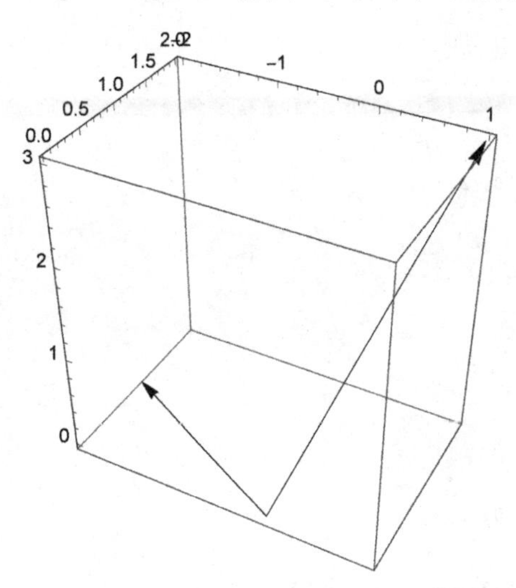

■ Cosine of the angle between two vectors in $\mathbb{E}^2$

```
v = {1, 0}; w = {1, 1};

           Dot[v, w]
cosvw = ─────────────────
        Norm[v] Norm[w]
```

$\frac{1}{\sqrt{2}}$

■ Angle between two vectors in $\mathbb{E}^2$

```
v = {1, 0}; w = {1, 1};

           Dot[v, w]
cosvw = ─────────────────;
        Norm[v] Norm[w]

angle = ArcCos[cosvw]
```

$\frac{\pi}{4}$

The standard basis of $\mathbb{E}^4$ are the columns (or rows) of the 4-by-4 identity matrix.

■ Standard basis of $\mathbb{E}^4$

```
MatrixForm[A = IdentityMatrix[4]]
```

$$\begin{pmatrix} 1 & 0 & 0 & 0 \\ 0 & 1 & 0 & 0 \\ 0 & 0 & 1 & 0 \\ 0 & 0 & 0 & 1 \end{pmatrix}$$

```
stBasis = {A[[All,1]], A[[All,2]], A[[All,3]], A[[All,4]]}
```

```
{{1, 0, 0, 0}, {0, 1, 0, 0}, {0, 0, 1, 0}, {0, 0, 0, 1}}
```

- Orthogonal matrix with columns and rows in $\mathbb{E}^3$

```
A = RandomInteger[{0, 9}, {3, 3}];
```

$$A = \begin{pmatrix} 3 & 4 & 2 \\ 8 & 7 & 2 \\ 8 & 5 & 2 \end{pmatrix};$$

```
B = Orthogonalize[A]
```

```
{{1, 0, 0, 0}, {0, 1, 0, 0}, {0, 0, 1, 0}, {0, 0, 0, 1}}
```

```
{Dot[B[[1]], B[[2]]] == 0, Dot[B[[1]], B[[3]]] == 0, Dot[B[[2]], B[[3]]] == 0}
```

```
{True, True, True}
```

```
Transpose[B] == Inverse[B]
```

```
True
```

Exact solution

See Linear system

Expansion along a coordinate axis

For any real number $s > 1$, a left-multiplication of a vector **v** in $\mathbb{R}^2$ by the matrix

```
MatrixForm[A = {{s, 0}, {0, 1}}]
```

$$\begin{pmatrix} s & 0 \\ 0 & 1 \end{pmatrix}$$

is an expansion along the *x*-axis. A left-multiplication by the matrix

```
MatrixForm[B = {{1, 0}, {0, s}}]
```

$$\begin{pmatrix} 1 & 0 \\ 0 & s \end{pmatrix}$$

is an expansion along the *y*-axis.

Illustration

- Expansions along the coordinate axes

```
Clear[A, B, s]
```

```
MatrixForm[A = {{s, 0}, {0, 1}}]
```

$$\begin{pmatrix} s & 0 \\ 0 & 1 \end{pmatrix}$$

```
MatrixForm[B = {{1, 0}, {0, s}}]
```

$$\begin{pmatrix} 1 & 0 \\ 0 & s \end{pmatrix}$$

```
v = {3, 5}; s = 2;
```

```
{A.v, B.v}
```

```
{{6, 5}, {3, 10}}
```

This shows that the vector {3, 5} is expanded to {6, 5} by left-multiplication by *A*, and expanded to {3,10} by left-multiplication by *B*.

Exponential form of complex numbers

The exponential form $e^{i\theta}$ of a complex number (x + *i* y) is based on Euler's formula

```
Exp[i θ] == Cos[θ] + i Sin[θ]        (1)
```

Illustration

- Exponential form of the complex number {x, y} = $\left\{\frac{1}{\sqrt{2}}, \frac{1}{\sqrt{2}}\right\}$

```
{1/Sqrt[2], 1/Sqrt[2]} == {Re[Exp[I π/4]], Im[Exp[I π/4]]}
```

```
True
```

```
ComplexExpand[Exp[I π / 4]] == 1/√2 + 1/√2 i
```

True

- Euler's formula for the complex number $\{x, y\} = \left\{\frac{1}{\sqrt{2}}, \frac{1}{\sqrt{2}}\right\}$

```
ComplexExpand[Exp[I π / 4]] == ComplexExpand[Cos[π / 4] + I Sin[π / 4]]
```

True

F

Finite-dimensional vector space

A *finite-dimensional vector space* is a vector space that has a finite basis. Every finite-dimensional real or complex vector space is isomorphic, as a vector space, to a coordinate space $\mathbb{R}^n$ or $\mathbb{C}^n$. The number of elements n of any basis of a space is called the *dimension* of the space.

Illustration

- The five-dimensional real coordinate space

The space $\mathbb{R}^5$ consists of all lists of the form {*a*, *b*, *c*, *d*, *e*}, where *a*, *b*, *c*, *d*, and *e* are real numbers. The addition and scalar multiplications are defined as follows:

Vector addition

$$\{a_1, a_2, a_3, a_4, a_5\} + \{b_1, b_2, b_3, b_4, b_5\} == \{a_1 + b_1, a_2 + b_2, a_3 + b_3, a_4 + b_4, a_5 + b_5\} \tag{1}$$

```
True
```

Scalar multiplication

$$a\,\{a_1, a_2, a_3, a_4, a_5\} == \{a\,a_1, a\,a_2, a\,a_3, a\,a_4, a\,a_5\} \tag{2}$$

```
True
```

The dimension of the space is 5.

In linear algebra, the space $\mathbb{R}^5$ is usually defined in terms of columns instead of rows of real numbers. The definition using rows in this guide is designed to match the way *Mathematica* calculates with vectors and matrices.

If column vectors are required, they can be defined as lists of singleton lists. For example, a column vector of height 5 can be defined as {{*a*}, {*b*}, {*c*}, {*d*}, {*e*}}. The **MatrixForm** and **TraditionalForm** functions can then be used to display the vectors in column format.

```
MatrixForm[vector1 = {{a}, {b}, {c}, {d}, {e}}]
```

$$\begin{pmatrix} a \\ b \\ c \\ d \\ e \end{pmatrix}$$

```
vector1
```

```
{{a}, {b}, {c}, {d}, {e}}
```

```
TraditionalForm[vector2 = {{1}, {2}, {3}}]
```

$$\begin{pmatrix} 1 \\ 2 \\ 3 \end{pmatrix}$$

```
vector2
```

```
{{1}, {2}, {3}}
```

- The five-dimensional real polynomial space ℝ[t,4]

The vectors of this space are real polynomials of degree 4 or less and the scalars are real numbers.

Vector addition

$$\text{Collect}\left[\left(a_0 + a_1 t + a_2 t^2 + a_3 t^3 + a_4 t^4\right) + \left(b_0 + b_1 t + b_2 t^2 + b_3 t^3 + b_4 t^4\right), t\right] \quad (3)$$

$$a_0 + b_0 + t\,(a_1 + b_1) + t^2\,(a_2 + b_2) + t^3\,(a_3 + b_3) + t^4\,(a_4 + b_4)$$

Scalar multiplication

$$\text{Expand}\left[a\left(a_0 + a_1 t + a_2 t^2 + a_3 t^3 + a_4 t^4\right)\right] \quad (4)$$

$$a\,a_0 + a\,t\,a_1 + a\,t^2\,a_2 + a\,t^3\,a_3 + a\,t^4\,a_4$$

Forward substitution

A linear system $A\mathbf{v} = \mathbf{b}$ can sometimes be solved by decomposing the coefficient matrix A into a product LU, where L is a lower-triangular matrix and U is an upper-triangular matrix. The system $A\mathbf{v} = \mathbf{b}$ can then be solved by solving the systems $L\mathbf{w} = \mathbf{b}$ and $U\mathbf{v} = \mathbf{w}$. Since L is lower-triangular, the system $L\mathbf{w} = \mathbf{b}$ can then be solved by *forward substitution*. (The associated system $U\mathbf{v} = \mathbf{w}$ is then solved by *back substitution*.)

Illustration

- Solving a linear system by forward substitution

```
Clear[x, y, z]
```

```
system = {eq1, eq2, eq3} = {x == 2, x - 3 y == 9, 2 x + 4 y - 6 z == 8};
```

The matrix of coefficients representing this system is lower-triangular. Each variable can be solved by forward substitution of the previous variable's value.

```
solx = Reduce[eq1]
```

```
x == 2
```

```
soly = Reduce[eq2 /. {x → 2}]
```

$y == -\frac{7}{3}$

```
solz = Reduce[eq3 /. {x → 2, y → -7 / 3}]
```

$z == -\frac{20}{9}$

```
Flatten[Reduce[system, {x, y, z}]] == (solx && soly && solz)
```

```
True
```

The last equation shows that the solutions found by forward substitution are the same as those found by solving the system with the **Reduce** command.

Fourier matrix

A *Fourier matrix* is a scalar multiple of the *n*-by-*n* Vandermonde matrix for the roots of unity $\omega = e^{-(2\pi i)/n}$. The scalar $\frac{1}{\sqrt{n}}$ is a normalization factor which makes the associated Vandermonde matrix unitary. The *Mathematica* **FourierMatrix** function does not use a scaling factor.

$$1/\text{Sqrt}[n]\ \{\{\{1, 1, 1, \ldots, 1\}, \{1, \omega, \omega^2, \ldots, \omega^{n-1}\}, \{1, \omega^2, \ldots, \omega^{2(n-1)}\}, \ldots, \{1, \omega^{n-1}, \ldots, \omega^{(n-1)(n-1)}\}\} \quad (1)$$

where $\omega = e^{-(2\pi i)/n}$ is an nth root of unity in which $i = \sqrt{-1}$.

Illustration

- A 2-by-2 Fourier matrix

```
ω = Exp[- (2 π I) / 2]
```

```
-1
```

```
MatrixForm[F2 = (1 / Sqrt[2]) {{1, 1}, {1, ω}}]
```

$$\begin{pmatrix} \frac{1}{\sqrt{2}} & \frac{1}{\sqrt{2}} \\ \frac{1}{\sqrt{2}} & -\frac{1}{\sqrt{2}} \end{pmatrix}$$

```
UnitaryMatrixQ[F2]
```

```
True
```

The built-in *Mathematica* function produces the same result.

```
MatrixForm[FourierMatrix[2]]
```

$$\begin{pmatrix} \frac{1}{\sqrt{2}} & \frac{1}{\sqrt{2}} \\ \frac{1}{\sqrt{2}} & -\frac{1}{\sqrt{2}} \end{pmatrix}$$

- A 4-by-4 Fourier matrix

```
ω = Exp[-(2 π I) / 4]
```

```
-i
```

```
MatrixForm[F4 = (1/Sqrt[4]) {{1, 1, 1, 1}, {1, ω, ω^2, ω^3}, {1, ω^2, ω^4, ω^6}, {1, ω^3, ω^6, ω^9}}]
```

$$\begin{pmatrix} \frac{1}{2} & \frac{1}{2} & \frac{1}{2} & \frac{1}{2} \\ \frac{1}{2} & -\frac{i}{2} & -\frac{1}{2} & \frac{i}{2} \\ \frac{1}{2} & -\frac{1}{2} & \frac{1}{2} & -\frac{1}{2} \\ \frac{1}{2} & \frac{i}{2} & -\frac{1}{2} & -\frac{i}{2} \end{pmatrix}$$

```
UnitaryMatrixQ[F4]
```

```
True
```

The *Mathematica* **FourierMatrix** function uses $\mathbf{Exp}\left(\frac{2\pi i}{4}\right)$ instead of $\mathbf{Exp}\left(-\frac{2\pi i}{4}\right)$:

```
ρ = Exp[(2 π I) / 4]
```

```
i
```

```
MatrixForm[F = FourierMatrix[4]]
```

$$\begin{pmatrix} \frac{1}{2} & \frac{1}{2} & \frac{1}{2} & \frac{1}{2} \\ \frac{1}{2} & \frac{i}{2} & -\frac{1}{2} & -\frac{i}{2} \\ \frac{1}{2} & -\frac{1}{2} & \frac{1}{2} & -\frac{1}{2} \\ \frac{1}{2} & -\frac{i}{2} & -\frac{1}{2} & \frac{i}{2} \end{pmatrix}$$

```
MatrixForm[ρF = (1/Sqrt[4]) {{1, 1, 1, 1}, {1, ρ, ρ^2, ρ^3}, {1, ρ^2, ρ^4, ρ^6}, {1, ρ^3, ρ^6, ρ^9}}]
```

$$\begin{pmatrix} \frac{1}{2} & \frac{1}{2} & \frac{1}{2} & \frac{1}{2} \\ \frac{1}{2} & \frac{i}{2} & -\frac{1}{2} & -\frac{i}{2} \\ \frac{1}{2} & -\frac{1}{2} & \frac{1}{2} & -\frac{1}{2} \\ \frac{1}{2} & -\frac{i}{2} & -\frac{1}{2} & \frac{i}{2} \end{pmatrix}$$

```
F == ρF
```

True

Fourier transform

See Discrete Fourier transform

Fredholm's theorem

Fredholm's theorem states that if A is an m-by-n matrix, then the orthogonal complement of the row space of A is the null space of A and the orthogonal complement of the column space of A is the left null space of A.

Illustration

- Comparing the null space of A and the orthogonal complement of the row space of A

```
Clear[a, b, c, d, e]
```

```
MatrixForm[A = {{3, 1, 0, 2, 4}, {1, 1, 0, 0, 2}, {5, 2, 0, 3, 7}}]
```

$$\begin{pmatrix} 3 & 1 & 0 & 2 & 4 \\ 1 & 1 & 0 & 0 & 2 \\ 5 & 2 & 0 & 3 & 7 \end{pmatrix}$$

```
nsA = NullSpace[A]
```

{{-1, -1, 0, 0, 1}, {-1, 1, 0, 1, 0}, {0, 0, 1, 0, 0}}

```
rrA = RowReduce[A]
```

{{1, 0, 0, 1, 1}, {0, 1, 0, -1, 1}, {0, 0, 0, 0, 0}}

```
ovrrA1 = Reduce[Dot[{a, b, c, d, e}, {1, 0, 0, 1, 1}] == 0, {a, b, c, d, e}]
```

e == -a - d

```
ovrrA2 = Reduce[Dot[{a, b, c, d, e}, {0, 1, 0, -1, 1}] == 0, {a, b, c, d, e}]
```

```
e == -b + d
```

```
Reduce[-a - d == d - b, {a, b, d}]
```

```
       a   b
d == - — + —
       2   2
```

With this last result, e becomes:

```
e = -a - d = -a - (b / 2 - a / 2) = -a + a / 2 - b / 2 = -a / 2 - b / 2
```

Hence all vectors *v* in orthogonal complement of the row space of *A* are of the form

```
v = {a, b, c, b / 2 - a / 2, -a / 2 - b / 2}
```

```
             a   b    a   b
{a, b, c, - — + —, - — - —}
             2   2    2   2
```

These vectors also defined the null space of *A*:

```
Simplify[A.v]
```

```
{0, 0, 0}
```

Free variable of a linear system

The *free variables* of a linear system $A\mathbf{v} = \mathbf{b}$ are the variables that are not determined by the pivot columns of the coefficient matrix *A*.

Illustration

- Free variables of a linear system

```
Clear[x, y, z]
```

```
system = {3 x + 4 y - z - w == 9, x + y + 4 z == 1};
```

```
A = {{3, 4, -1, -1}, {1, 1, 4, 0}};
```

```
MatrixForm[RowReduce[A]]
```

$$\begin{pmatrix} 1 & 0 & 17 & 1 \\ 0 & 1 & -13 & -1 \end{pmatrix}$$

Since the first and second columns of *A* are the pivot columns of the linear system, the variables *x* and *y* are basic variables of the system and the variable *z* and *w* are the free variables of the system.

Frobenius companion matrix

See Companion matrix

Frobenius norm

The Frobenius norm $\|A\|_F$ of an m-by-n real matrix A is the square root of the trace of the matrix $A^T A$.

Illustration

- The Frobenius norm of a general 2-by-3 real matrix

```
A = {{a11, a12, a13}, {a21, a22, a23}};
```

```
Norm[A, Frobenius]
```

$$\sqrt{\left(\text{Abs}[a_{11}]^2 + \text{Abs}[a_{12}]^2 + \text{Abs}[a_{13}]^2 + \text{Abs}[a_{21}]^2 + \text{Abs}[a_{22}]^2 + \text{Abs}[a_{23}]^2\right)}$$

```
Sqrt[Tr[Transpose[A].A]]
```

$$\sqrt{a_{11}^2 + a_{12}^2 + a_{13}^2 + a_{21}^2 + a_{22}^2 + a_{23}^2}$$

- The Frobenius norm of a numerical 3-by-2 real matrix

```
A = {{1, 2}, {3, 4}, {5, 6}};
```

```
Norm[A, Frobenius]
```

$$\sqrt{91}$$

The definition of the Frobenius norm, using the trace, produces the same result:

```
Norm[A, Frobenius] == Sqrt[Tr[Transpose[A].A]]
```

```
True
```

Manipulation

- The Frobenius norm of a 4-by-5 real matrix

$$A = \begin{pmatrix} 2 & 1 & 4 & 2 & 5 & 7 \\ 3 & 9a & 8 & 2 & 2 & 9 \\ 2 & 8 & 7 & 5b & 6 & 3 \\ 9 & 3 & 6 & 5 & 6 & 5 \end{pmatrix};$$

```
Manipulate[Evaluate[Sqrt[Tr[Transpose[A].A]]], {a, -2, 2, 1}, {b, -3, 3, 1}]
```

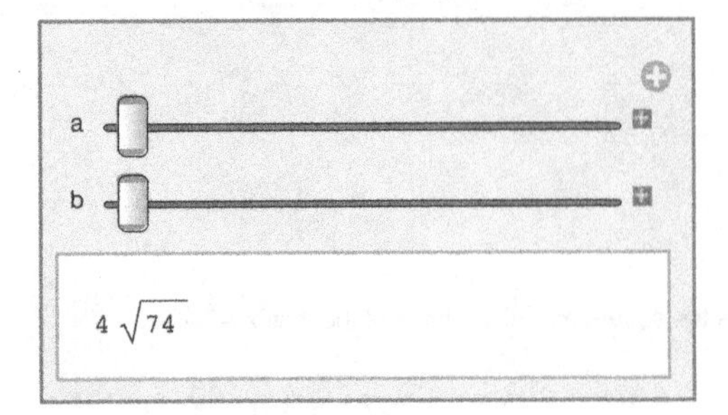

We use **Manipulate**, **Sqrt**, **Tr**, and **Transpose** to explore the Frobenius norm of real matrices. For $a = -2$ and $b = -3$, for example, the manipulation displays the Frobenius norm of the generated matrix. By varying the values of a and b, we can calculate the Frobenius norms of other matrices.

Full rank of a matrix

An n-by-n matrix is of *full rank* if its rank is n. A square matrix that is not of full rank is said to be rank deficient.

Illustration

- A real matrix that is of full rank

```
A = RandomReal[{0, 9}, {3, 3}];
```

```
A = (1.2672497552861213`  7.260324652231585`  6.404178280507901`
     6.373852525598297`   4.371460236056876`  5.717126196739134`
     3.2916555250409`     3.608127689047624`  1.369108253600869`);
```

```
MatrixRank[A]
```

```
4
```

This shows that the matrix A is of full rank.

- A complex 4-by-4 matrix of full rank

```
MatrixForm[A = {{1, 2 I, 3 - I, 4}, {4 I, 3, 2, 2 I}, {0, 8 - 4 I, 0, 0}, {1, 1, 1, 1}}]
```

$$\begin{pmatrix} 1 & 2 i & 3 - i & 4 \\ 4 i & 3 & 2 & 2 i \\ 0 & 8 - 4 i & 0 & 0 \\ 1 & 1 & 1 & 1 \end{pmatrix}$$

```
MatrixRank[A]
```

```
4
```

- A real 4-by-4 matrix that is rank deficient (not of full rank)

```
MatrixForm[A = {{1, 0, 0, 0}, {-1, 0, 0, 0}, {0, 1, 0, 0}, {0, 0, 0, 1}}]
```

$$\begin{pmatrix} 1 & 0 & 0 & 0 \\ -1 & 0 & 0 & 0 \\ 0 & 1 & 0 & 0 \\ 0 & 0 & 0 & 1 \end{pmatrix}$$

```
MatrixRank[A]
```

```
3
```

Fundamental subspace

See Column space, left null space, matrix-based subspace, null space, row space

Fundamental theorem of algebra

The *fundamental theorem of algebra* says that every polynomial $a_0 + a_1 x + \cdots + a_n x^n$ with complex coefficients $a_0, a_1, \ldots, a_n$ factors into linear factors over the field $\mathbb{C}$ of complex numbers. The theorem is used in linear algebra to guarantee the existence of eigenvalues of real and complex square matrices.

A theorem usually referred to *the unsolvability of the quintic* establishes that there is no general algorithm for finding the linear factors of arbitrary real or complex polynomials of degree *5* or higher, even if their coefficients are integers. In general, numerical techniques must therefore be used to calculate eigenvalues.

Illustration

- Linear factors of a polynomial of degree 2 with rational roots

```
p = 3 + 2 x - 5 x^2;
```

```
Factor[p]
```

```
{0, -13, -36}
```

- Linear factors of a polynomial of degree 4 with complex roots

```
p = 1 + x^4;
```

```
Factor[p, Extension -> {Sqrt[2], I}]
```

```
{2, 17, 82}
```

- Linear factors of a polynomial of degree **p = 3 + 2 x - 5 x²;**

```
Clear[p, x]
```

```
p = 1 + x^5;
```

```
Roots[p == 0, x]
```

$x == -1 \;||\; x == -(-1)^{2/5} \;||\; x == -(-1)^{4/5} \;||\; x == (-1)^{1/5} \;||\; x == (-1)^{3/5}$

```
p == Simplify[ComplexExpand[(x + 1) (x + (-1)^(2/5)) (x + (-1)^(4/5)) (x - (-1)^(1/5)) (x - (-1)^(3/5))]]
```

```
True
```

G

Gaussian elimination

Gaussian elimination is a procedure for converting a matrix to row echelon form using elementary row operations. Neither the resulting row echelon form nor the steps of the process is unique.

The difference between Gaussian and Gauss–Jordan elimination is that the former produces a matrix in row echelon form, while the latter produces a matrix in unique reduced row echelon form.

Illustration

- Conversion of a 4-by-4 matrix to row echelon form

```
MatrixForm[A = {{0, 2, 3, 4}, {0, 0, 0, 0}, {6, 7, 0, 8}, {0, 4, 1, 8}}]
```

$$\begin{pmatrix} 0 & 2 & 3 & 4 \\ 0 & 0 & 0 & 0 \\ 6 & 7 & 0 & 8 \\ 0 & 4 & 1 & 8 \end{pmatrix}$$

Interchange rows 2 and 4:

```
A = {A[[1]], A[[4]], A[[3]], A[[2]]}
```

```
{{0, 2, 3, 4}, {0, 4, 1, 8}, {6, 7, 0, 8}, {0, 0, 0, 0}}
```

Interchange rows 1 and 3 :

```
A = {A[[3]], A[[2]], A[[1]], A[[4]]}
```

```
{{6, 7, 0, 8}, {0, 4, 1, 8}, {0, 2, 3, 4}, {0, 0, 0, 0}}
```

Add - *1/2* row 2 to row 3 :

$$\mathbf{MatrixForm}\left[\mathbf{A} = \left\{\mathbf{A}_{[[1]]}, \mathbf{A}_{[[2]]}, \mathbf{A}_{[[3]]} - \frac{1}{2}\mathbf{A}_{[[2]]}, \mathbf{A}_{[[4]]}\right\}\right]$$

$$\begin{pmatrix} 6 & 7 & 0 & 8 \\ 0 & 4 & 1 & 8 \\ 0 & 0 & \frac{5}{2} & 0 \\ 0 & 0 & 0 & 0 \end{pmatrix}$$

The final matrix is in row echelon form.

Gauss–Jordan elimination

Gauss–Jordan elimination is a procedure for converting a matrix to reduced row echelon form using elementary row operations. It is a refinement of Gaussian elimination. The reduced row echelon form of a matrix is unique, but the steps of the procedure are not.

Two linear systems are equivalent (have the same solutions) if and only if the reduced row echelon forms of the augmented matrices of the two systems obtained by Gauss–Jordan elimination are identical. This property is known as the Church–Rosser property of the underlying equivalence relation.

Illustration

- Gauss–Jordan elimination applied to a 3-by-5 real matrix

```
A = {{6, 6, 0, 3, 5}, {4, 0, 5, 3, 8}, {3, 7, 3, 4, 4}};
```

```
RowReduce[A]
```

$$\left\{\left\{1, 0, 0, \frac{13}{64}, \frac{199}{192}\right\}, \left\{0, 1, 0, \frac{19}{64}, -\frac{13}{64}\right\}, \left\{0, 0, 1, \frac{7}{16}, \frac{37}{48}\right\}\right\}$$

- Gauss–Jordan elimination applied to a 5-by-3 real matrix

```
A = {{4, 8, 3}, {3, 1, 7}, {0, 0, 1}, {3, 2, 5}, {3, 6, 9}};
```

```
RowReduce[A]
```

```
{{1, 0, 0}, {0, 1, 0}, {0, 0, 1}, {0, 0, 0}, {0, 0, 0}}
```

- Gauss–Jordan elimination applied to a 4-by-4 real matrix

$$A = \begin{pmatrix} 0 & 2 & 3 & 4 \\ 0 & 0 & 0 & 0 \\ 6 & 7 & 0 & 8 \\ 0 & 4 & 1 & 8 \end{pmatrix}; B = \begin{pmatrix} 6 & 7 & 0 & 8 \\ 0 & 4 & 1 & 8 \\ 0 & 0 & \frac{5}{2} & 0 \\ 0 & 0 & 0 & 0 \end{pmatrix};$$

```
MatrixForm[RowReduce[B]]
```

$$\begin{pmatrix} 1 & 0 & 0 & -1 \\ 0 & 1 & 0 & 2 \\ 0 & 0 & 1 & 0 \\ 0 & 0 & 0 & 0 \end{pmatrix}$$

```
RowReduce[B] == RowReduce[A]
```

```
True
```

This last result shows that two different matrices can have the same reduced row echelon form.

General solution of a linear system

The *general solution* of a linear system is a linear system that expresses the basic variables of the system in terms of the free variables of the system.

Illustration

- The general solution of a linear system of three equations in four variables

```
eq1 = -5 x2 + 15 x3 + 4 x4 == 7; eq2 = x1 - 2 x2 - 4 x3 + 3 x4 == 6; eq3 = 2 x1 + 4 x3 + 3 x4 == 1;
```

The augmented matrix of the system in the variables x_1, x_2, x_3, x_4, in that order, is

$$A = \begin{pmatrix} 0 & -5 & 15 & 4 & 7 \\ 1 & -2 & -4 & 3 & 6 \\ 2 & 0 & 4 & 3 & 1 \end{pmatrix};$$

The reduced row echelon form of the matrix A is

```
MatrixForm[B = RowReduce[A]]
```

$$\begin{pmatrix} 1 & 0 & 0 & \frac{89}{60} & \frac{19}{20} \\ 0 & 1 & 0 & -\frac{31}{40} & -\frac{83}{40} \\ 0 & 0 & 1 & \frac{1}{120} & -\frac{9}{40} \end{pmatrix}$$

The pivot columns of the matrix A are its first three columns, so that the basic variables of the system are x_1, x_2, and x_3, and the free variable of the system is x_4. The matrix B is the augmented matrix of the linear system

$$\left\{x_1 + \frac{89}{60} x_4 == \frac{19}{20},\ x_2 - \frac{31}{40} x_4 == -\frac{83}{40},\ x_3 + \frac{1}{120} x_4 == -\frac{9}{40}\right\};$$

This list of equations is the *general solution* of the original system since it provides explicit expressions of the basic variables x_1, x_2, and x_3 of the system in terms of the free variable x_4 of the system.

If we assign the value 0 to the free variable x_4, for example, we get the following values for x_1, x_2, and x_3:

$$\left\{x_1 + \frac{89}{60} x_4 == \frac{19}{20},\ x_2 - \frac{31}{40} x_4 == -\frac{83}{40},\ x_3 + \frac{1}{120} x_4 == -\frac{9}{40}\right\} /. \{x_4 \to 0\}$$

$$\left\{x_1 == \frac{19}{20},\ x_2 == -\frac{83}{40},\ x_3 == -\frac{9}{40}\right\}$$

The four calculated values form a particular solution of the original system:

$$\{eq1, eq2, eq3\} /. \left\{x_1 \to \frac{19}{20},\ x_2 \to -\frac{83}{40},\ x_3 \to -\frac{9}{40},\ x_4 \to 0\right\}$$

```
{True, True, True}
```

Geometric multiplicity of an eigenvalue

The *geometric multiplicity* of an eigenvalue of a matrix is the dimension of the eigenspace associated with the eigenvalue. The geometric multiplicity of an eigenvalue is always less than or equal to the algebraic multiplicity of the eigenvalue.

A matrix is diagonalizable if and only if all geometric and algebraic multiplicities of the eigenvalues of the matrix are equal.

Illustration

```
Clear[A, p, t]
```

```
A = UpperTriangularize[RandomInteger[{0, 3}, {5, 5}]];
```

$$A = \begin{pmatrix} 1 & 2 & 0 & 3 & 1 \\ 0 & 0 & 2 & 0 & 0 \\ 0 & 0 & 0 & 2 & 1 \\ 0 & 0 & 0 & 3 & 3 \\ 0 & 0 & 0 & 0 & 3 \end{pmatrix};$$

```
p[A, t] = CharacteristicPolynomial[A, t]
```

$9\,t^2 - 15\,t^3 + 7\,t^4 - t^5$

```
Factor[p[A, t]]
```

$-(-3 + t)^2\,(-1 + t)\,t^2$

```
Eigenvalues[A]
```

```
{3, 3, 1, 0, 0}
```

The algebraic multiplicity of eigenvalue 1 is 1, and that of the eigenvalues 0 and 3 is 2.

```
Eigensystem[A]
```

```
{{3, 3, 1, 0, 0},
 {{35, 8, 12, 18, 0}, {0, 0, 0, 0, 0}, {1, 0, 0, 0, 0}, {-2, 1, 0, 0, 0}, {0, 0, 0, 0, 0}}}
```

Mathematica displays the vectors {{35, 8, 12, 18, 0} and {0, 0, 0, 0, 0}} as two candidates for eigenvectors associated with the eigenvalue 3. But since the zero vector is not an eigenvector, the eigenspace of the eigenvalue 3 is only the span of the single vector {35, 8, 12, 18, 0} and has dimension 1. For the same reason, the dimension of the eigenvalue 0 is also 1. This shows that the matrix *A* is not diagonalizable since the sum of the dimensions of the eigenspaces is less than the dimension of the matrix.

Geometric transformation

Mathematica comes with a variety of geometric transforms and *geometric matrix transformations* corresponding to operations in transformational geometry. Here are some examples. The full list of transforms can be explored in the *Mathematica* Help file: *Help > Documentation Center > Geometric Transforms.*

Illustration

- Rotation of a rectangle

```
Graphics[Rotate[Rectangle[], 30 Degree]]
```

- Translation of a circle in $\mathbb{R}^2$

```
Graphics[Translate[Circle[], {1, 0}], Axes → True]
```

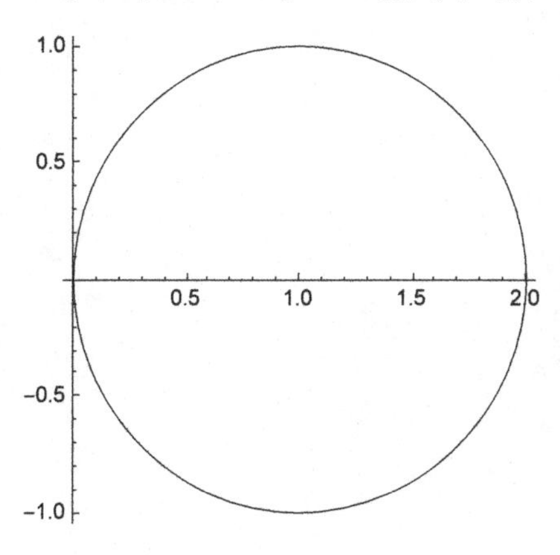

- Translation of a sphere in $\mathbb{R}^3$

```
Graphics3D[Translate[Sphere[], {1, 0, 0}], Axes → True]
```

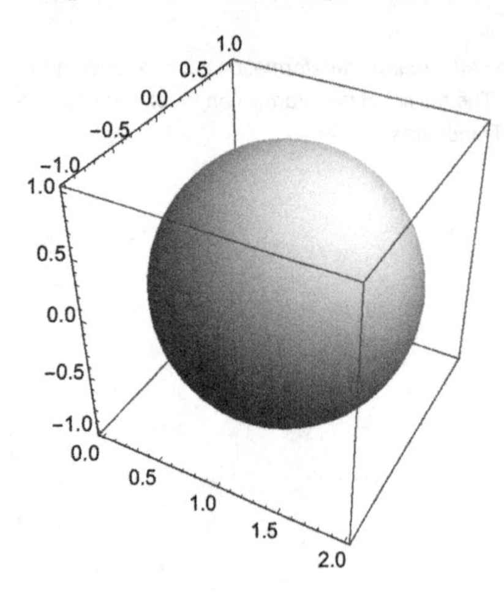

- Scaling of a disk in $\mathbb{R}^2$

```
Graphics[{Pink, Scale[Disk[], {1, 1/2}]}]
```

- Scaling of a cylinder in $\mathbb{R}^3$

```
Graphics3D[Scale[Cylinder[], {1/2, 1, 1}, {0, 0, 0}], Boxed → False]
```

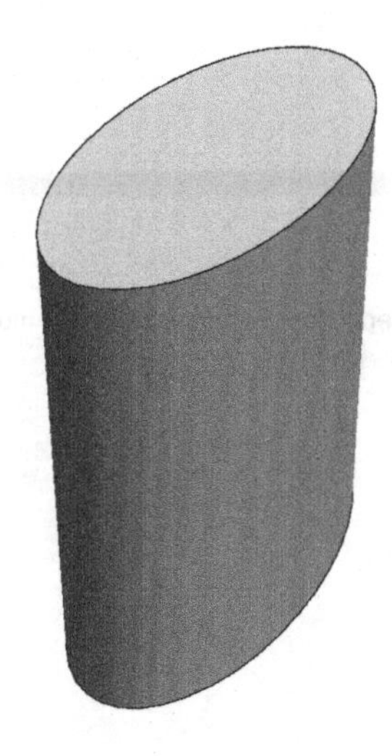

- Shearing of a rectangle in $\mathbb{R}^2$

```
ShearingTransform[θ, {1, 0}, {0, 1}];
```

```
Graphics[GeometricTransformation[Rectangle[],
  ShearingTransform[30 Degree, {1, 0}, {0, 1}]], Frame → True]
```

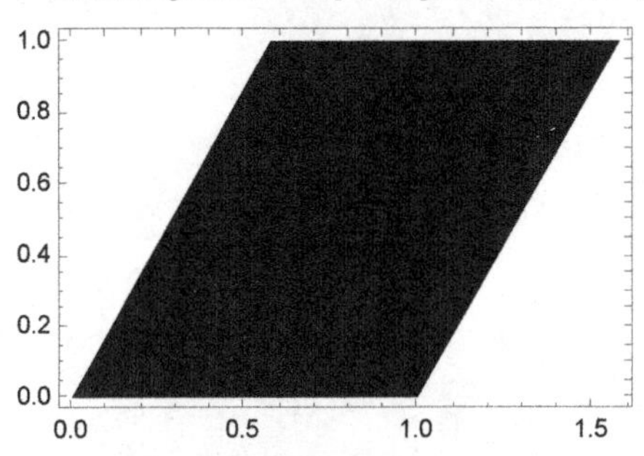

- Reflection of a point in $\mathbb{R}^2$ about the line $y = x$

```
rt = ReflectionTransform[{1, -1}];
```

```
rt[{x, y}]
```

```
{y, x}
```

- Reflection of a point of $\mathbb{R}^3$ in the z-plane

```
rt = ReflectionTransform[{0, 0, 1}];

rt[{x, y, z}]

{x, y, -z}
```

Gram–Schmidt process

The *Gram–Schmidt process* is an algorithm for converting a list of linearly independent vectors into a list of mutually orthogonal vectors. It is based on the orthogonal decomposition of given vectors.

Illustration

- Orthogonalization of two vectors in $\mathbb{R}^3$

```
vectors = {{5, 2, 3}, {8, 5, 4}}

{{5, 2, 3}, {8, 5, 4}}

orthovectors = Orthogonalize[vectors]
```

$$\left\{\left\{\frac{5}{\sqrt{38}}, \sqrt{\frac{2}{19}}, \frac{3}{\sqrt{38}}\right\}, \left\{-\frac{3}{\sqrt{1387}}, \frac{33}{\sqrt{1387}}, -\frac{17}{\sqrt{1387}}\right\}\right\}$$

```
Dot[orthovectors[[1]], orthovectors[[2]]]

0
```

- The Gram–Schmidt process applied to three vectors

```
Clear[u, v, w]

u = {1, 0, 1}; v = {1, 1, 1}; w = {0, 1, 1};

ov2 = Orthogonalize[{u, v}]
```

$$\left\{\left\{\frac{1}{\sqrt{2}}, 0, \frac{1}{\sqrt{2}}\right\}, \{0, 1, 0\}\right\}$$

```
Dot[ov2[[1]], ov2[[2]]]

0
```

```
MatrixForm[ov3 = Orthogonalize[{u, v, w}]]
```

$$\begin{pmatrix} \frac{1}{\sqrt{2}} & 0 & \frac{1}{\sqrt{2}} \\ 0 & 1 & 0 \\ -\frac{1}{\sqrt{2}} & 0 & \frac{1}{\sqrt{2}} \end{pmatrix}$$

The resulting set of vectors is orthogonal.

```
{Dot[ov3[[1]], ov3[[2]]], Dot[ov3[[1]], ov3[[3]]], Dot[ov3[[2]], ov3[[3]]]}
```

```
{0, 0, 0}
```

- Another illustration of the Gram–Schmidt process

```
Proj[u_, v_] := (Dot[u, v] / Dot[u, u]) u
```

The defined function *Proj*[u,v] will produce the projection of the vector **v** onto the vector **u.**

```
v1 = {1, 0, 1}; v2 = {1, 1, 1}; v3 = {0, 1, 1};
```

Next we use the projections of **v2** onto **v1**, and **v3** onto **v1** and **v2,** to create orthogonal vectors.

```
u1 = v1; u2 = v2 - Proj[u1, v2]; u3 = v3 - Proj[u1, v3] - Proj[u2, v3];
```

```
MatrixForm[ov4 = {u1, u2, u3}]
```

$$\begin{pmatrix} 1 & 0 & 1 \\ 0 & 1 & 0 \\ -\frac{1}{2} & 0 & \frac{1}{2} \end{pmatrix}$$

The resulting set of vectors is orthogonal

```
{Dot[ov4[[1]], ov4[[2]]], Dot[ov4[[1]], ov4[[3]]], Dot[ov4[[2]], ov4[[3]]]}
```

```
{0, 0, 0}
```

since the pairwise dot products of its vectors are all zero.

Manipulation

- Orthogonalization of two linearly independent vectors in $\mathbb{R}^2$

```
u = {1, 2}; v = {-2, 2};
```

```
Manipulate[Orthogonalize[Evaluate[{{1 + a, 2}, {-2, 2 + b}}]], {a, -3, 3, 1}, {b, -3, 3, 1}]
```

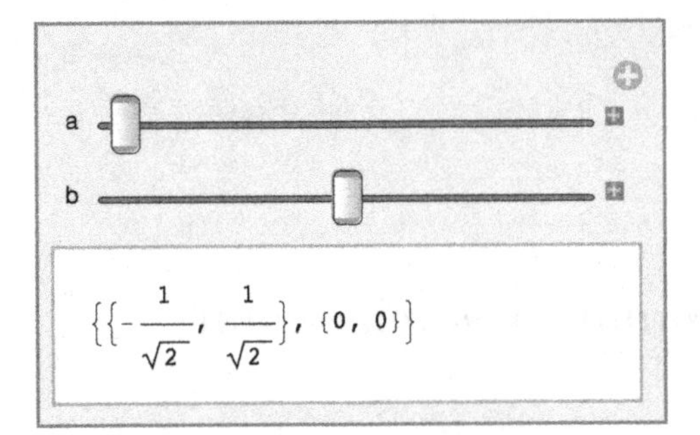

We use **Manipulate, Orthogonalize,** and **Evaluate** to explore the effect of the Gram–Schmidt process on two linearly independent vectors in $\mathbb{R}^2$. If a = - 3 and b = 0, then the two vectors are identical and cannot be orthogonalized. *Mathematica* normalizes the first vector and converts the second vector to the zero vector in $\mathbb{R}^2$. If a = - 3 and b = 1, for example, the manipulation produces two normal orthogonal vectors.

- Orthogonalization of three linearly independent vectors in $\mathbb{R}^3$

```
u = {1, 0, 0}; v = {0, 1, 0}; w = {0, 0, 1};
```

```
Manipulate[Orthogonalize[Evaluate[{{1 + a, 0, 0}, {0, 1 + b, 0}, {0, 0, 1 + c}}]],
 {a, -3, 3, 1}, {b, -3, 3, 1}, {c, -3, 3, 1}]
```

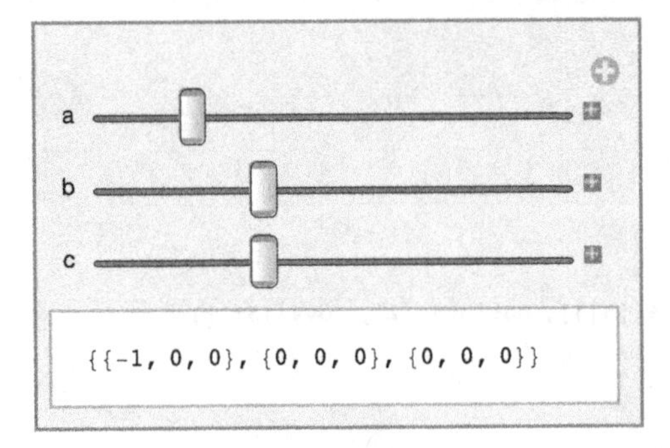

We use **Manipulate, Orthogonalize,** and **Evaluate** to explore the effect of the Gram–Schmidt process on three linearly independent vectors in $\mathbb{R}^3$. If a = - 2 and b = c = - 1, for example, then the manipulation shows that the three resulting vectors cannot be orthogonalized. On the other hand, if a = - 3, b = 2, and c = - 2, for example, the manipulation produces three normal orthogonal vectors.

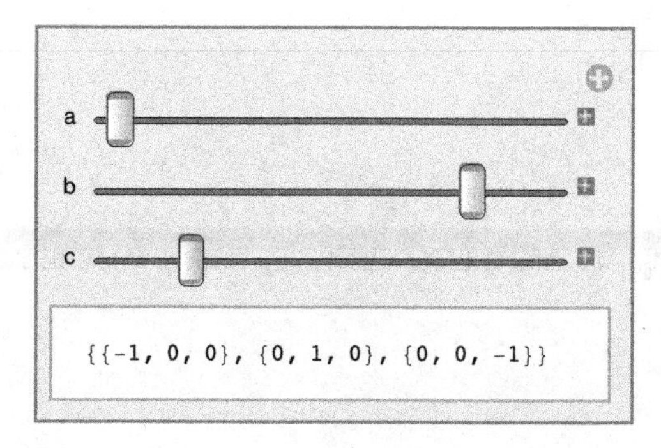
a
b
c
{{-1, 0, 0}, {0, 1, 0}, {0, 0, -1}}

H

Hankel matrix

A *Hankel matrix* is a square matrix A whose elements $A_{[[i,j]]}$ are equal to its elements $A_{[[i-1,j+1]]}$. In other words, the skew-diagonals of a Hankel matrix are constant.

Illustration

- A 4-by-4 Hankel matrix

```
MatrixForm[H = HankelMatrix[4]]
```

$$\begin{pmatrix} 1 & 2 & 3 & 4 \\ 2 & 3 & 4 & 0 \\ 3 & 4 & 0 & 0 \\ 4 & 0 & 0 & 0 \end{pmatrix}$$

The skew-diagonals of H are {1}, {2, 2}, {3, 3, 3}, {4, 4, 4, 4}, {0, 0, 0}, {0, 0}, and {0}.

- A general 4-by-4 Hankel matrix

```
MatrixForm[H = HankelMatrix[{a, b, c, d}]]
```

$$\begin{pmatrix} a & b & c & d \\ b & c & d & 0 \\ c & d & 0 & 0 \\ d & 0 & 0 & 0 \end{pmatrix}$$

- A complex Hankel matrix

```
MatrixForm[H = HankelMatrix[{1, 1 + 2 I, 3 + 4 I}]]
```

$$\begin{pmatrix} 1 & 1+2i & 3+4i \\ 1+2i & 3+4i & 0 \\ 3+4i & 0 & 0 \end{pmatrix}$$

- A more general complex Hankel Matrix

```
MatrixForm[H = HankelMatrix[{1, 1 + 2 I, 3 + 4 I}, {3 + 4 I, 5 - I, 2 I}]]
```

$$\begin{pmatrix} 1 & 1+2i & 3+4i \\ 1+2i & 3+4i & 5-i \\ 3+4i & 5-i & 2i \end{pmatrix}$$

Manipulation

- Exploring 4-by-4 Hankel matrices

```
Manipulate[MatrixForm[H = HankelMatrix[{a, 2, 3, 4}, {4, b, c, 5}]],
 {a, -2, 2, 1}, {b, -3, 3, 1}, {c, 0, 4, 1}]
```

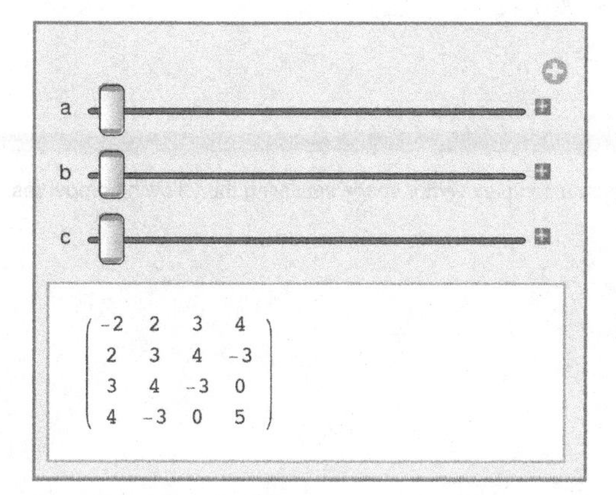

We use **Manipulate, MatrixForm,** and **HankelMatrix** to explore Hankel matrices. If we let $a = -2$, $b = -3$, and $c = 0$, for example, the manipulation displays a Hankel matrix. Other choices of values for *a*, *b*, and *c* produce other Hankel matrices.

Height of a column vector

The *height* of a column vector is the number of rows of the vector.

Illustration

- The height of a column vector with three rows

```
MatrixForm[v = RandomInteger[{0, 9}, {3, 1}]]
```

$$v = \begin{pmatrix} 7 \\ 6 \\ 5 \end{pmatrix};$$

```
Dimensions[v]
```

```
{3, 1}
```

The height of the vector **v** is 3, the first component of the **Dimensions** output.

- The height of a complex column vector with two rows

```
MatrixForm[v = {3 - I, 6 + 8 I}]
```

$$\begin{pmatrix} 3 - i \\ 6 + 8 i \end{pmatrix}$$

```
Dimensions[v]
```

{2}

Hermitian inner product

A *Hermitian inner product* is a bilinear functional **<u, v>** on a complex vector space satisfying the following properties:

Properties of Hermitian inner products

⟨u, v⟩ = Conjugate[⟨v, u⟩] (1)

⟨u + v, w⟩ = ⟨u, w⟩ + ⟨v, w⟩ (2)

⟨u, v + w⟩ = ⟨u, v⟩ + ⟨u, w⟩ (3)

⟨a u, v⟩ = a ⟨u, v⟩ (4)

⟨u, a v⟩ = Conjugate[a] ⟨u, v⟩ (5)

⟨u, u⟩ ≥ 0 (6)

⟨u, u⟩ = 0 if and only if **u** = **0** (7)

Illustration

- A Hermitian inner product

```
⟨u_, v_⟩ := u.A.Conjugate[v]
```

where A is a Hermitian positive-definite matrix.

In pencil-and-paper linear algebra, the vectors **u** and **v** are assumed to be column vectors. Therefore the vector **v** must be transposed in the definition and the inner product is defined as the product of a column vector **u** times a Hermitian positive-definite matrix A times the conjugate transpose of the column vector **v**. In *Mathematica* transposition is not required since the appropriate orientation of the vectors is understood from the context.

- A complex inner product determined by an identity matrix

```
MatrixForm[A = IdentityMatrix[2]]
```

$$\begin{pmatrix} 1 & 0 \\ 0 & 1 \end{pmatrix}$$

```
u = {3 + I, -I}; v = {4, 2 + I};
```

```
u.A.Conjugate[v]
```

11 + 2 i

- A Hermitian inner product determined by a diagonal matrix

```
MatrixForm[A = {{3, 0}, {0, 5}}]
```

$$\begin{pmatrix} 3 & 0 \\ 0 & 5 \end{pmatrix}$$

```
PositiveDefiniteMatrixQ[A]
```

```
True
```

```
u = {3 + I, -I}; v = {4, 2 + I};
```

```
u.A.Conjugate[v]
```

```
31 + 2 i
```

- The Hermitian inner product of two complex vectors in $\mathbb{C}^2$ computed using the dot product

```
u = {a + b I, c + d I}; v = {e + f I, g + h I};
```

```
hDot[u_, v_] := Dot[u, Conjugate[v]]
```

```
hDot[u, v]
```

```
(a + i b) (Conjugate[e] - i Conjugate[f]) + (c + i d) (Conjugate[g] - i Conjugate[h])
```

```
u = {1 + I, 2 - I}; v = {3 - 2 I, 1 + I};
```

```
hDot[u, v]
```

```
2 + 2 i
```

- Verification of the inner product calculation of two complex vectors in $\mathbb{C}^2$ using matrix multiplication

```
u.IdentityMatrix[2].Conjugate[v] == hDot[u, v]
```

```
True
```

Hermitian matrix

A *Hermitian matrix* is a square matrix with complex entries that is equal to its own conjugate transpose. A real matrix is Hermitian if it is symmetric.

Illustration

- A 2-by-2 Hermitian matrix *A*

```
MatrixForm[A = {{3 + I, -I}, {-I, 6}}]
```

$$\begin{pmatrix} 3+i & -i \\ -i & 6 \end{pmatrix}$$

The matrix *A* is not Hermitian.

```
HermitianMatrixQ[A]
```

```
False
```

However, its conjugate transpose is Hermitian.

```
MatrixForm[ConjugateTranspose[A]]
```

$$\begin{pmatrix} 3-i & i \\ i & 6 \end{pmatrix}$$

```
HermitianMatrixQ[{{1, 3 + 4 I}, {3 - 4 I, 2}}]
```

```
True
```

- A 3-by-3 Hermitian matrix *A*

```
MatrixForm[A = {{2, 2 + I, 4}, {2 - I, 3, I}, {4, -I, 1}}]
```

$$\begin{pmatrix} 2 & 2+i & 4 \\ 2-i & 3 & i \\ 4 & -i & 1 \end{pmatrix}$$

```
HermitianMatrixQ[A]
```

```
True
```

- Another 3-by-3 Hermitian matrix *H*

The next matrix and its transpose are both Hermitian and so is its conjugate transpose.

```
MatrixForm[H = {{1, 2 I, 3 + 4 I}, {-2 I, 5, 6 - 7 I}, {3 - 4 I, 6 + 7 I, 8}}]
```

$$\begin{pmatrix} 1 & 2i & 3+4i \\ -2i & 5 & 6-7i \\ 3-4i & 6+7i & 8 \end{pmatrix}$$

```
HermitianMatrixQ[H]
```

```
True
```

```
MatrixForm[Transpose[H]]
```

$$\begin{pmatrix} 1 & -2\,i & 3-4\,i \\ 2\,i & 5 & 6+7\,i \\ 3+4\,i & 6-7\,i & 8 \end{pmatrix}$$

```
HermitianMatrixQ[Transpose[H]]
```

```
True
```

```
MatrixForm[ConjugateTranspose[H]]
```

$$\begin{pmatrix} 1 & 2\,i & 3+4\,i \\ -2\,i & 5 & 6-7\,i \\ 3-4\,i & 6+7\,i & 8 \end{pmatrix}$$

```
H == ConjugateTranspose[H]
```

```
True
```

```
HermitianMatrixQ[{{1, 2 I, 3 + 4 I}, {-2 I, 5, 6 - 7 I}, {3 - 4 I, 6 + 7 I, 8}}]
```

```
True
```

- A real Hermitian matrix

```
MatrixForm[A = RandomInteger[{0, 9}, {2, 3}]]
```

$$\begin{pmatrix} 3 & 9 & 4 \\ 6 & 0 & 1 \end{pmatrix}$$

```
MatrixForm[H = Transpose[A].A]
```

$$\begin{pmatrix} 45 & 27 & 18 \\ 27 & 81 & 36 \\ 18 & 36 & 17 \end{pmatrix}$$

```
HermitianMatrixQ[H]
```

```
True
```

Hessenberg matrix

A *Hessenberg matrix* is a square matrix that is close to being triangular. An upper Hessenberg matrix has zeros below the first subdiagonal, and a lower Hessenberg matrix has zeros above the first superdiagonal.

A *Hessenberg decomposition* of a matrix *A* is a matrix decomposition of a matrix $PHP^H = A$ into a unitary matrix *P* and a Hessenberg matrix *H* and the conjugate transpose P^H of *P*.

Illustration

- The Hessenberg decomposition of a 4-by-4 matrix

```
{p, h} = HessenbergDecomposition[(1.  2.  3.  4.
                                  5.  6.  7.  8.
                                  9.  10. 11. 12.
                                  13. 14. 15. 16.)];
```

The matrix h is an upper Hessenberg matrix:

```
MatrixForm[h]
```

$$\begin{pmatrix} 1. & -5.3669 & 0.443129 & -4.50542\times10^{-16} \\ -16.5831 & 33.0873 & -9.55746 & -7.76819\times10^{-15} \\ 0. & -2.20899 & -0.0872727 & -6.32932\times10^{-16} \\ 0. & 0. & -3.11293\times10^{-15} & -5.99553\times10^{-16} \end{pmatrix}$$

Hessian matrix

A Hessian matrix is a square matrix whose elements are second-order partial derivatives of a given function.

Illustration

Determinants can be used to classify critical points of differentiable functions. For example, if $f: \mathbb{R}^2 \longrightarrow \mathbb{R}$ is a function with continuous second partial derivatives f_{xx}, f_{xy}, f_{yx}, and f_{yy}, then the matrix

```
MatrixForm[H_f = {{f_xx, f_xy}, {f_yx, f_yy}}]
```

$$\begin{pmatrix} f_{xx} & f_{xy} \\ f_{yx} & f_{yy} \end{pmatrix}$$

is a Hessian matrix. Its determinant $Det[H_f]$ is called the discriminant of *f*. We know from calculus that if **v** = {*x*, *y*} is a critical point of *f*, in other words, if $f_x[\mathbf{v}] = f_y[\mathbf{v}] = 0$ and $Det[H_f] > 0$, then *f*[**v**] is a local minimum of f if $f_{xx}[\mathbf{v}] > 0$ and a local maximum if $f_{xx}[\mathbf{v}] < 0$. If $Det[H_f] < 0$, then **v** is a saddle point of *f*. If $Det[H_f] = 0$, the test fails.

- Hessian matrix

```
f[x_, y_] := x^2 + y^2 - x + 4 y + 3;
```

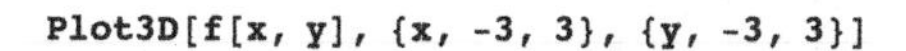

```
Plot3D[f[x, y], {x, -3, 3}, {y, -3, 3}]
```

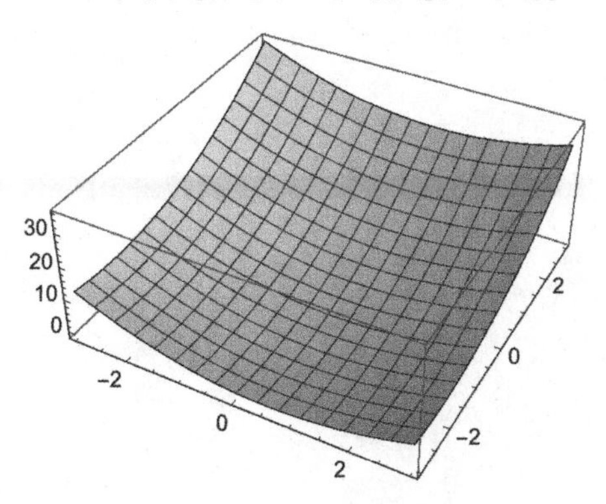


```
fxx = D[f[x, y], x, x]; fxy = D[f[x, y], x, y];
```

```
fyx = D[f[x, y], y, x]; fyy = D[f[x, y], y, y];
```

```
Hf = {{fxx, fxy}, {fyx, fyy}}
```

```
{{2, 0}, {0, 2}}
```

```
Det[Hf]
```

```
4
```

```
{Reduce[D[f[x, y], x] == 0, {x, y}], Reduce[D[f[x, y], y] == 0, {x, y}]}
```

$$\left\{x == \frac{1}{2},\ y == -2\right\}$$

Since the determinant of $H_f > 0$, f has a local minimum at {x, y} = {1/2, -2}.

Hilbert matrix

A *Hilbert matrix* is a square matrix H whose ijth element $H_{[[i,j]]}$ is $(i + j - 1)^{-1}$. Hilbert matrices are canonical examples of ill-conditioned matrices that are difficult to use in numerical computations. Their ill-conditioned nature is measured in terms of their condition numbers. The larger the condition numbers, the more ill-conditioned the matrix.

Illustration

The Hilbert matrices of dimensions 2, 3, and 4 are the following:

- The 2-by-2 Hilbert matrix

```
MatrixForm[HilbertMatrix[2]]
```

$$\begin{pmatrix} 1 & \frac{1}{2} \\ \frac{1}{2} & \frac{1}{3} \end{pmatrix}$$

- The 3-by-3 Hilbert matrix

```
MatrixForm[HilbertMatrix[3]]
```

$$\begin{pmatrix} 1 & \frac{1}{2} & \frac{1}{3} \\ \frac{1}{2} & \frac{1}{3} & \frac{1}{4} \\ \frac{1}{3} & \frac{1}{4} & \frac{1}{5} \end{pmatrix}$$

- The 4-by-4 Hilbert matrix

```
MatrixForm[HilbertMatrix[4]]
```

$$\begin{pmatrix} 1 & \frac{1}{2} & \frac{1}{3} & \frac{1}{4} \\ \frac{1}{2} & \frac{1}{3} & \frac{1}{4} & \frac{1}{5} \\ \frac{1}{3} & \frac{1}{4} & \frac{1}{5} & \frac{1}{6} \\ \frac{1}{4} & \frac{1}{5} & \frac{1}{6} & \frac{1}{7} \end{pmatrix}$$

Manipulation

- Exploring *n*-by-*n* Hilbert matrices

```
Manipulate[MatrixForm[HilbertMatrix[n]], {n, 2, 6, 1}]
```

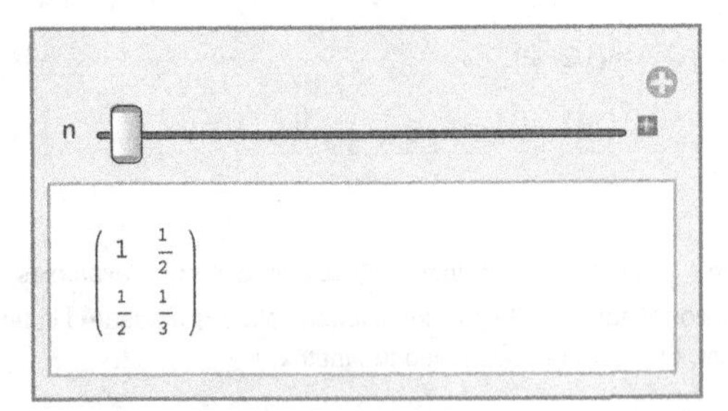

We use **Manipulate, MatrixForm,** and **HilbertMatrix** to generate Hilbert matrices. If n = 2, for example, the manipulation displays the Hilbert matrix of dimension 2.

Homogeneous coordinate

For any point {*x*, *y*} in $\mathbb{R}^2$, the coordinates of the point {*x*, *y*, 1} in $\mathbb{R}^3$ are *homogeneous coordinates* since $\frac{x}{1} = x$ and $\frac{y}{1} = y$. These coordinates are used in affine geometry to express affine transformations by matrix multiplication.

Illustration

- Point in $\mathbb{R}^2$ with homogeneous coordinates in $\mathbb{R}^3$

```
hcoordinates = {1, 2, 3}
```

{1, 2, 3}

```
point = {1 / 3, 2 / 3}
```

$\left\{\frac{1}{3}, \frac{2}{3}\right\}$

- Point in $\mathbb{R}^2$ with homogeneous coordinates in $\mathbb{R}^3$

```
hcoordinates = {6, 9, 3}
```

{6, 9, 3}

```
point = {2, 3}
```

{2, 3}

- Homogeneous coordinates in $\mathbb{R}^3$ associated with a point in $\mathbb{R}^2$

```
point = {3, 5}
```

{3, 5}

```
hcoordinates = {3, 5, 1}
```

{3, 5, 1}

- Homogeneous coordinates in $\mathbb{R}^3$ associated with a point in $\mathbb{R}^2$

```
point = {3, 2}
```

{3, 2}

```
hcoordinates = {9, 6, 3}
```

{9, 6, 3}

- Homogeneous coordinates of a linear equation

Using the homogeneous coordinates {x_1, x_2, x_2} for {*x*, *y*}, we can rewrite the linear equation

$a_1 x + a_2 y + a_3 == 0$

as

```
a1 x1 + a2 x2 + a3 x3 == 0
```

Homogeneous linear system

A *homogeneous linear system* is a linear system of the form $A\mathbf{v} = \mathbf{0}$. The solution space of the system can be found in various ways.

Illustration

- Solving a homogeneous linear system with infinitely many solutions

```
system = {x + 2 y + z == 0, x - y - 2 z == 0, 2 x + y - z == 0};
```

```
Solve[system]
```

```
{{y → -x, z → x}}
```

We assign a parameter *t* to replace the variable *z*

```
{t, -t, t}
```

```
{t, -t, t}
```

Next we verify that all values of the parameter *t* satisfy the given system.

```
system /. {x → t, y → -t, z → t}
```

```
{True, True, True}
```

Next we identify a basis for the solutions space:

```
{t, -t, t} == t {1, -1, 1}
```

```
True
```

Hence the set {{1, -1, 1}} is a basis for the solution space of the given system.

The **NullSpace** function of *Mathematica* can be used to solve a homogeneous linear system by applying it to the "coefficient matrix" of the linear system.

- Using the **NullSpace** function to solve a homogeneous linear system

```
system = {x + 2 y + z == 0, x - y - 2 z == 0, 2 x + y - z == 0};
```

```
A = {{1, 2, 1}, {1, -1, -2}, {2, 1, -1}};
```

```
NullSpace[A]
```

```
{{1, -1, 1}}
```

- A homogeneous linear system with two equations in three variables

```
equations = {3 x + 4 y - z == 0, x - 2 y + 5 z == 0};
```

```
solution = Solve[equations, {x, y, z}]
```

Solve::svars : Equations may not give solutions for all "solve" variables. ≫

$$\left\{\left\{y \to -\frac{8 x}{9}, z \to -\frac{5 x}{9}\right\}\right\}$$

```
equations /. solution
```

```
{{True, True}}
```

```
coefficientmatrix = {{3, 4, -1}, {1, -2, 5}};
```

The **LinearSolve** command can also be used to find a solution of a homogeneous linear system. However, it fails to produce all solutions.

```
LinearSolve[coefficientmatrix, {0, 0}]
```

```
{0, 0, 0}
```

```
coefficientmatrix.{0, 0, 0}
```

```
{0, 0}
```

Manipulation

- Manipulating a homogeneous linear system in three equations and three variables

```
equations = {3 x + 4 y - z == 0, x - 2 y + 5 z == 0, x + y + z == 0};
```

```
coefficients = {{3, 4, -1}, {1, -2, 5}, {1, 1, 1}};
```

```
f = LinearSolve[coefficients]
```

LinearSolveFunction[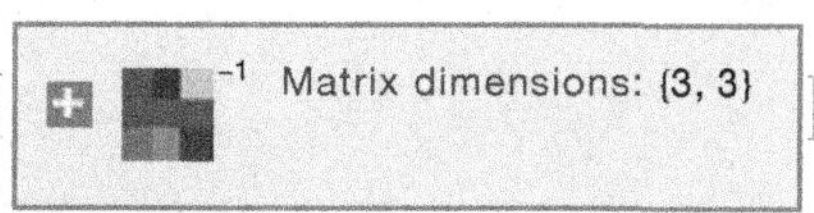]


```
f[{0, 0, 0}]
```

```
{0, 0, 0}
```

```
Manipulate[Evaluate[f[{a, b, c}]], {a, -2, 2, 1}, {b, -2, 2, 1}, {c, -2, 2, 1}]
```

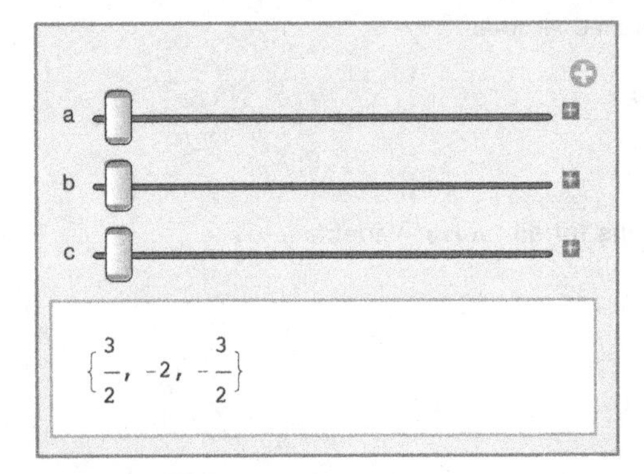

We use **Manipulate** and **Evaluate** to explore the solutions of linear systems. If $a = b = c = -2$, for example, the manipulation displays the solution {3/2, - 2, - 3/2} of the associated linear system.

- A family of vectors generated by using the linear combinations determined by a linear system

```
Manipulate[x {1, 1, 2} + y {2, -1, 1} + z {1, -2, -1}, {x, -5, 5}, {y, -6, 6}, {z, -8, 8}]
```

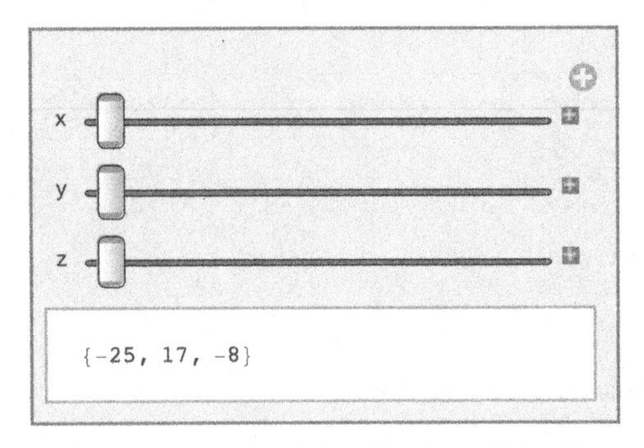

We use **Manipulate** to calculate the vector obtained by assigning numerical values to the coefficients *x*, *y*, and *z*. If we let $x = -5$, $y = -6$, and $z = -8$, for example, the manipulation produces the vector {- 25, 17, - 8}.

The substitution operator /. confirms this solution.

```
x {1, 1, 2} + y {2, -1, 1} + z {1, -2, -1} /. {x → -5, y → -6, z → -8}
```

```
{-25, 17, -8}
```

Householder matrix

A real square matrix is a *Householder matrix* if it is of the form

$$\text{IdentityMatrix} - \frac{2}{\text{Dot}[v, v]}\, \text{u.Transpose}[u] \tag{1}$$

determined by a nonzero column vector **u** and the row vector **v** associated with **u**.

Every Householder matrix is orthogonal. Householder matrices describe reflections about a plane or hyperplane containing the origin. Householder matrices are used in numerical linear algebra to perform *QR* decompositions.

Illustration

```
u = {{1}, {2}, {3}};
```

```
u.Transpose[u]
```

```
{{1, 2, 3}, {2, 4, 6}, {3, 6, 9}}
```

```
v = Flatten[u]
```

```
{1, 2, 3}
```

$$\text{MatrixForm}\left[h = \text{IdentityMatrix}[3] - \frac{2}{\text{Dot}[v, v]}(u.\text{Transpose}[u])\right]$$

$$\begin{pmatrix} \frac{6}{7} & -\frac{2}{7} & -\frac{3}{7} \\ -\frac{2}{7} & \frac{3}{7} & -\frac{6}{7} \\ -\frac{3}{7} & -\frac{6}{7} & -\frac{2}{7} \end{pmatrix}$$

```
OrthogonalMatrixQ[h]
```

```
True
```

Identity matrix

An *identity matrix* is a square matrix with 1s on the main diagonal and 0s elsewhere. It has the property that for any *n*-by-*n* matrix,

```
A.IdentityMatrix[n] == IdentityMatrix[n].A
```
(1)

The *Mathematica* function **IdentityMatrix[n]** computes the *n*-by-*n* identity matrix.

Illustration

- The 3-by-3 identity matrix

```
MatrixForm[IdentityMatrix[3]]
```

$$\begin{pmatrix} 1 & 0 & 0 \\ 0 & 1 & 0 \\ 0 & 0 & 1 \end{pmatrix}$$

Identity matrices are identities with respect to vector multiplication in matrix spaces.

```
MatrixForm[A = {{9, 1, 6}, {9, 4, 1}, {3, 4, 0}}]
```

$$\begin{pmatrix} 9 & 1 & 6 \\ 9 & 4 & 1 \\ 3 & 4 & 0 \end{pmatrix}$$

```
MatrixForm[A.IdentityMatrix[3] == A]
```

```
True
```

Identity matrices can also be formed using the **SparseArray** function by specifying the values of the nonzero elements.

- The 2-by-2 identity matrix

```
MatrixForm[Normal[s = SparseArray[{{1, 1} → 1, {2, 2} → 1, {3, 3} → 1}]]]
```

$$\begin{pmatrix} 1 & 0 & 0 \\ 0 & 1 & 0 \\ 0 & 0 & 1 \end{pmatrix}$$

Ill-conditioned matrix

A matrix is *ill-conditioned* if a small change in its elements produces significant errors in the solutions of associated linear systems.

```
MatrixForm[A = {{1, 2}, {2, 3.999}}]
```

$$\begin{pmatrix} 1 & 2 \\ 2 & 3.999 \end{pmatrix}$$

```
b = {4, 7.999};

LinearSolve[A, b]

{2., 1.}

c = {4.001, 7.998};

LinearSolve[A, c]

{-3.999, 4.}

N[LinearAlgebra`MatrixConditionNumber[A]]

35 988.

A = {{1, 1}, {1, 0.999}}

{{1, 1}, {1, 0.999}}

N[LinearAlgebra`MatrixConditionNumber[A]]

4000.

b = {1, 0.999}; c = {1.0001, 0.998};

LinearSolve[A, b]

{0., 1.}

LinearSolve[A, c]

{-1.0999, 2.1}
```

Illustration

- An ill-conditioned 2-by-2 matrix

```
MatrixForm[A = {{1, 1}, {1, 0.999}}]
```

$$\begin{pmatrix} 1 & 1 \\ 1 & 0.999 \end{pmatrix}$$

```
N[LinearAlgebra`MatrixConditionNumber[A]]
```

4000.

```
b = {1, 0.999}; c = {1.0001, 0.998};
```

```
LinearSolve[A, b]
```

{0., 1.}

```
LinearSolve[A, c]
```

{-1.0999, 2.1}

This calculation shows that by making small changes in the constant vector *b*, the solutions of the associated linear systems change significantly.

- An ill-conditioned 2-by-2 matrix

```
MatrixForm[A1 = {{1, -1}, {-1, 1}}]
```

$$\begin{pmatrix} 1 & -1 \\ -1 & 1 \end{pmatrix}$$

```
N[LinearAlgebra`MatrixConditionNumber[A1]]
```

∞

```
MatrixForm[A2 = {{1, -1}, {-1, 1.00001}}]
```

$$\begin{pmatrix} 1 & -1 \\ -1 & 1.00001 \end{pmatrix}$$

```
N[LinearAlgebra`MatrixConditionNumber[A2]]
```

400 004.

The singularity of the matrix A_1 is also affected by the small change in one of its elements:

```
{Det[A1], Det[A2]}
```

{0, 0.00001}

This suggests that whereas the matrix A_1 is singular, the matrix A_2 is nonsingular since its determinant is nonzero.

- An ill-conditioned matrix in terms of its singular values

```
MatrixForm[A = {{1.001, 2.001}, {2.001, 3.001}}]
```

$$\begin{pmatrix} 1.001 & 2.001 \\ 2.001 & 3.001 \end{pmatrix}$$

```
s = SingularValueList[A]
```

```
{4.23796, 0.235962}
```

```
conditionnumberA = s[[1]] / s[[2]]
```

```
17.9603
```

Image of a linear transformation

The *image* of a linear transformation $T : V \longrightarrow W$ is the set of all vectors $T[\mathbf{v}]$ in W. The image of T is a subspace of W. If T is a matrix transformation, then the image of T is the column space of T.

Illustration

- The trace of a linear transformation $T : \mathbb{R}^{2\times2} \longrightarrow \mathbb{R}$ as an image

```
Clear[A, B, s, T, a, b, c, d, e, f, g, h]
```

```
T[{{a_, b_}, {c_, d_}}] := Tr[{{a, b}, {c, d}}]
```

```
A = {{a, b}, {c, d}}; B = {{e, f}, {g, h}};
```

```
Expand[s (T[A] + T[B])]
```

```
a s + d s + e s + h s
```

```
Expand[s T[A] + s T[B]]
```

```
a s + d s + e s + h s
```

- The image of a matrix transformation $T : \mathbb{R}^3 \longrightarrow \mathbb{R}^2$

```
A = {{1, 2, 3}, {4, 5, 6}};
T[{x_, y_, z_}] := A.{x, y, z}
```

The 2-by-3 matrix takes vectors from $\mathbb{R}^3$ to vectors in $\mathbb{R}^2$.

```
T[{x, y, z}]
```

```
{x + 2 y + 3 z, 4 x + 5 y + 6 z}
```

On the other hand, the 3-by-2 matrix

```
A = {{1, 2}, {3, 4}, {5, 6}};
T[{x_, y_}] := A.{x, y}
```

takes vectors from $\mathbb{R}^2$ to vectors in $\mathbb{R}^3$.

```
T[{x, y}]
```

```
{x + 2 y, 3 x + 4 y, 5 x + 6 y}
```

Incidence matrix

The *incidence matrix A* of an *undirected* graph has a row for each vertex and a column for each edge of the graph. The element $A_{[[i,j]]}$ of A is 1 if the ith vertex is a vertex of the jth edge and 0 otherwise.

The *incidence matrix A* of a *directed* graph has a row for each vertex and a column for each edge of the graph. The element $A_{[[i,j]]}$ of A is -1 if the ith vertex is an initial vertex of the jth edge, 1 if the ith vertex is a terminal vertex, and 0 otherwise.

Illustration

- The incidence matrix of an undirected graph

```
CycleGraph[4]
```

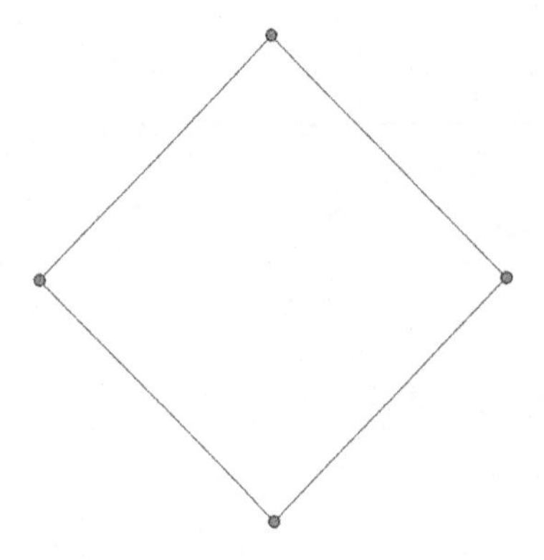

```
MatrixForm[IncidenceMatrix[%]]
```

$$\begin{pmatrix} 1 & 1 & 0 & 0 \\ 1 & 0 & 1 & 0 \\ 0 & 0 & 1 & 1 \\ 0 & 1 & 0 & 1 \end{pmatrix}$$

- The incidence matrix of a directed graph

```
Graph[{1 ↔ 2, 2 ↔ 3, 3 ↔ 4, 4 ↔ 1}]
```

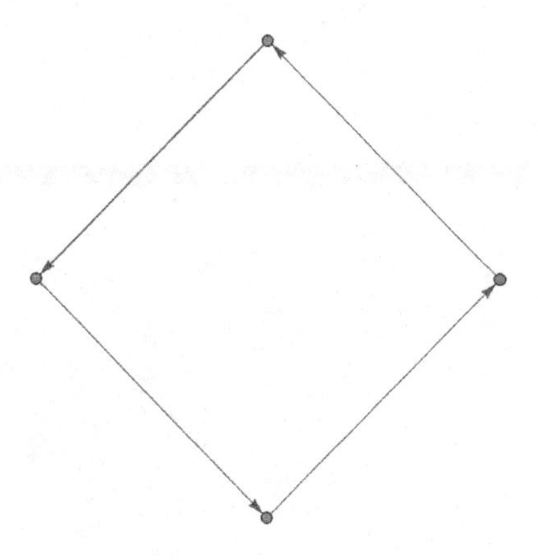

```
MatrixForm[IncidenceMatrix[%]]
```

$$\begin{pmatrix} -1 & 0 & 0 & 1 \\ 1 & -1 & 0 & 0 \\ 0 & 1 & -1 & 0 \\ 0 & 0 & 1 & -1 \end{pmatrix}$$

- The incidence matrix of an undirected graph has no negative entries

```
g = Graph[{1 ↔ 2, 2 ↔ 3, 3 ↔ 1}]
```

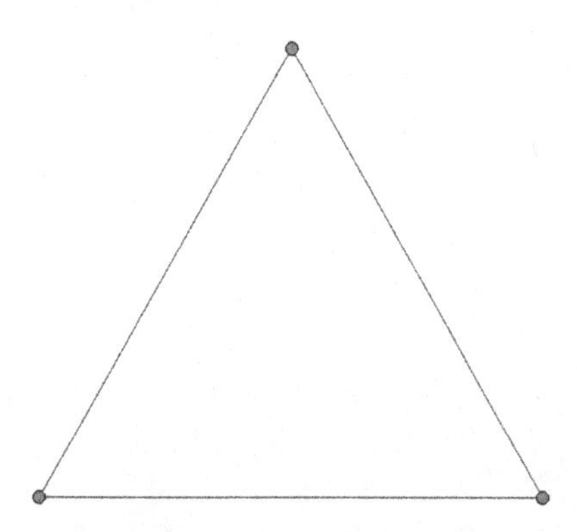

```
MatrixForm[IncidenceMatrix[g]]
```

$$\begin{pmatrix} 1 & 0 & 1 \\ 1 & 1 & 0 \\ 0 & 1 & 1 \end{pmatrix}$$

The sum of the elements in any column of incidence matrix of an undirected graph is always 2.

- The incidence matrix of a directed graph has some negative entries

```
g = Graph[{1 ↦ 2, 2 ↦ 3, 3 ↦ 1}]
```

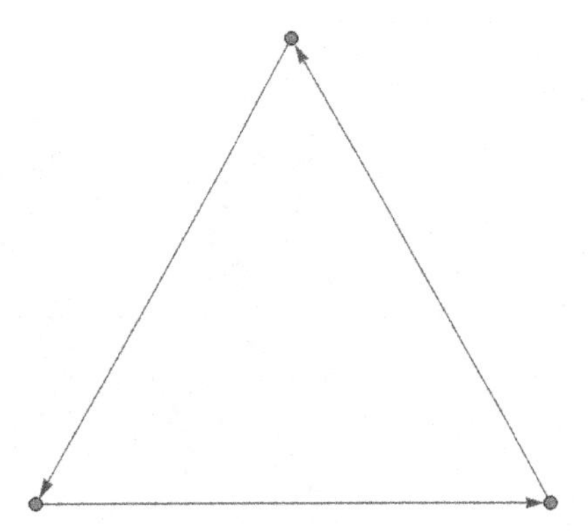

If a directed graph has no self-loops, the sum of the elements of its incidence matrix is always 0.

- The incidence matrix of a graph with self-loops has entries equal to 2.

```
g = Graph[{1 ↔ 2, 2 ↔ 3, 3 ↔ 1, 3 ↔ 3}]
```

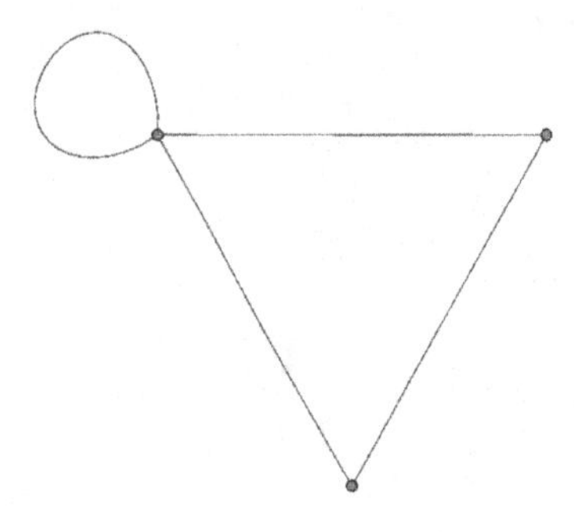

```
MatrixForm[IncidenceMatrix[g]]
```

$$\begin{pmatrix} 1 & 0 & 1 & 0 \\ 1 & 1 & 0 & 0 \\ 0 & 1 & 1 & 2 \end{pmatrix}$$

- A matrix plot of a large incidence matrix

```
Graph[Table[i → Mod[i^2, 10^3], {i, 0, 10^3 - 1}]];
```

```
Timing[A = IncidenceMatrix[%]]
```

```
{0.000160, SparseArray[ Specified elements: 1996
                        Dimensions: {1000, 1000} ]}
```

```
MatrixPlot[A]
```

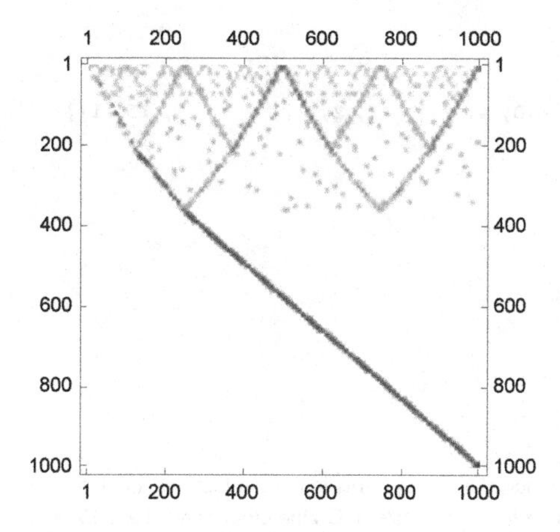

Inconsistent linear system

An *inconsistent* linear system is a linear system that has no solution.

Illustration

- An inconsistent linear system in four equations and three unknowns

```
Clear[A, b, x, y, z]
```

$$A = \begin{pmatrix} 1 & 3 & 5 \\ 9 & 1 & 8 \\ 2 & 2 & 3 \\ 0 & 6 & 3 \end{pmatrix}; b = \begin{pmatrix} 1 \\ 2 \\ 3 \\ 4 \end{pmatrix};$$

```
system = Thread[A.{x, y, z} == Flatten[b], Alignment → Right]
```

```
{x + 3 y + 5 z, 9 x + y + 8 z, 2 x + 2 y + 3 z, 6 y + 3 z} == {1, 2, 3, 4}
```

```
Solve[system, {x, y, z}]
```

```
{}
```

This shows that there are no values of *x*, *y*, and *z* for which the system has a solution. The same result can be obtained by showing that the associated matrix equation has no solution.

```
LinearSolve[A, b];
```

LinearSolve::nosol : Linear equation encountered that has no solution. ≫

Manipulation

- Exploring the consistency of a linear system

```
Manipulate[LinearSolve[{{1, 3, 5}, {9, 1, 8}, {10, 4, a}}, {1, 2, 2}], {a, 10, 15, 1}]
```

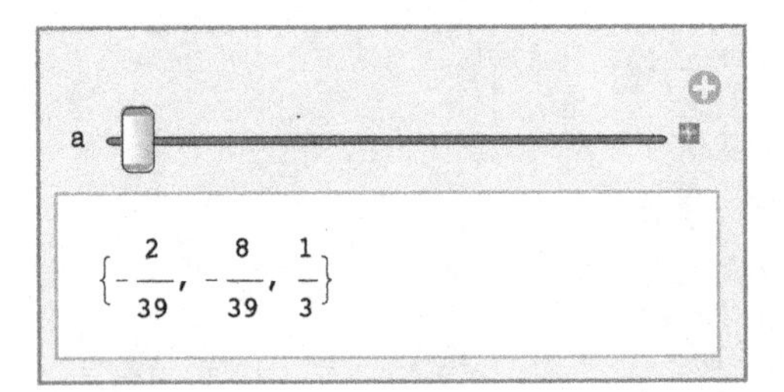

We use **Manipulate** and **LinearSolve** to explore the consistency of linear systems in three equations and three variables. The manipulation shows that if we let *a* = 10, the generated system is consistent. On the other hand, if a = 13, an inconsistent system results.

```
LinearSolve[{{1, 3, 5}, {9, 1, 8}, {10, 4, 13}}, {1, 2, 2}]
```

LinearSolve::nosol : Linear equation encountered that has no solution. ≫

```
Manipulate[{{1, 3, 5}, {9, 1, 8}, {10, 4, a}}.{x, y, z} == {1, 2, 2},
 {a, 10, 16, 1}, {x, -5, 5, 1}, {y, -5, 5, 1}, {z, -5, 5, 1}]
```

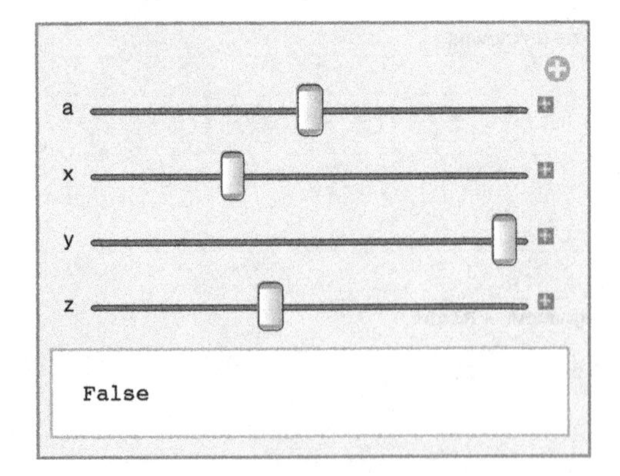

We use **Manipulate** to show that the linear system

```
{x + 3 y + 5 z, 9 x + y + 8 z, 10 x + 4 y + 13 z} == {1, 2, 2}
```

is inconsistent.

```
LinearSolve[{{1, 3, 5}, {9, 1, 8}, {10, 4, 13}}, {1, 2, 2}]
```

LinearSolve::nosol : Linear equation encountered that has no solution. ≫

```
LinearSolve[{{1, 3, 5}, {9, 1, 8}, {10, 4, 13}}, {1, 2, 2}]
```

Injective linear transformation

A linear transformation $T : V \longrightarrow W$ from a vector space V to a vector space W is *injective* (one-to-one) if T[**u**] = T[**v**] only if **u** = **v** for all vectors **u** and **v** in V.

Illustration

- An injective linear transformation T : ℝ[t,2] ⟶ $\mathbb{R}^2$

```
Clear[T, a, b, c, t]

T[a_ + b_ t_ + c_ t_^2] := {a, b}

T[a + b t + c t^2] == T[c + d t + e t^2]

{a, b} == {c, d}
```

- An injective linear transformation $T : \mathbb{R}^{2\times 3} \longrightarrow \mathbb{R}^6$

```
Clear[T, a, b, c, d, e, f, A, B]

T[{{a_, b_}, {c_, d_}, {e_, f_}}] := {a, b, c, d, e, f}

MatrixForm[A = {{a, b}, {c, d}, {e, f}}]
```

$$\begin{pmatrix} a & b \\ c & d \\ e & f \end{pmatrix}$$

```
T[A]

{a, b, c, d, e, f}

Clear[g, h, i, j, k, l]

MatrixForm[B = {{g, h}, {i, j}, {k, l}}]
```

$$\begin{pmatrix} g & h \\ i & j \\ k & l \end{pmatrix}$$

```
T[B]
```

```
{g, h, i, j, k, l}
```

Hence

```
T[A] == T[B]
```

```
{a, b, c, d, e, f} == {g, h, i, j, k, l}
```

if and only if $A = B$.

Inner product

An *inner product* < **u**, **v** > on a vector space V is a bilinear functional in the variables **u** and **v** on the vector space V satisfying the following equations:

Properties of inner products

$\langle u, v \rangle = \langle v, u \rangle$ (1)

$\langle a\, u, v \rangle = a \langle u, v \rangle$ (2)

$\langle u + v, w \rangle = \langle u, w \rangle + \langle v, w \rangle$ (3)

$\langle u, u \rangle \geq 0$ (4)

$\langle u, u \rangle = 0$ if and only if $u = \mathbf{0}$. (5)

Illustration

- The inner product on the Euclidean vector space $\mathbb{E}^2$

By definition, a Euclidean n-space is the coordinate space $\mathbb{R}^n$, equipped with the *dot product* inner product.

```
⟨u_, v_⟩ := Sqrt[Dot[u, v]]
```

```
u = {1, 2}; v = {3, 4};
```

```
⟨u, v⟩
```

$\sqrt{11}$

```
Norm[{1, 2}] == ⟨{1, 2}, {1, 2}⟩
```

```
True
```

Any positive-definite symmetric n-by-n matrix A can be used to define an inner product. If A is an identity matrix, the inner product defined by A is the Euclidean inner product.

- A nonstandard inner product on the coordinate vector space $\mathbb{R}^2$

```
Clear[A]
```

```
MatrixForm[A = DiagonalMatrix[{2, 3}]]
```

$$\begin{pmatrix} 2 & 0 \\ 0 & 3 \end{pmatrix}$$

```
PositiveDefiniteMatrixQ[A]
```

```
True
```

The following sample calculations show that the function

```
⟨u_, v_⟩ := u.A.v
```

defines an inner product.

```
u = {1, 2}; v = {3, 4};
```

```
⟨u, v⟩
```

```
30
```

```
⟨u, v⟩ == ⟨v, u⟩
```

```
True
```

```
⟨{1, 2}, {3, 4}⟩ == ⟨{3, 4}, {1, 2}⟩
```

```
True
```

```
⟨a u, v⟩ == a ⟨u, v⟩
```

```
True
```

```
⟨{a, 2 a}, {3, 4}⟩ == a ⟨{1, 2}, {3, 4}⟩
```

```
True
```

```
w = {5, 6};
```

```
⟨u + w, v⟩ == ⟨u, v⟩ + ⟨w, v⟩
```

```
True
```

```
⟨{6, 8}, {3, 4}⟩ == ⟨{1, 2}, {3, 4}⟩ + ⟨{5, 6}, {3, 4}⟩
```

```
True
```

```
⟨u, u⟩ ≥ 0
```

True

```
⟨{1, 2}, {1, 2}⟩ ≥ 0
```

True

```
Clear[x, y]
```

```
Reduce[⟨{x, y}, {x, y}⟩ == 0, {x, y}, Reals]
```

x == 0 && y == 0

The function ⟨u,v⟩ satisfies the definition of an inner product:

```
⟨u, v⟩ == ⟨v, u⟩
```

True

```
⟨{1, 2}, {3, 4}⟩ == ⟨{3, 4}, {1, 2}⟩
```

True

```
{1, 2}.A.{3, 4}
```

30

```
{3, 4}.A.{1, 2}
```

30

```
⟨v, u⟩
```

30

```
⟨{3, 4}, {1, 2}⟩
```

30

Manipulation

- Exploring inner products on $\mathbb{R}^3$

```
Manipulate[{1, 2, 3}.DiagonalMatrix[{a, b, 1}].{4, 5, 6}, {a, -5, 5, 1}, {b, -5, 5, 1}]
```

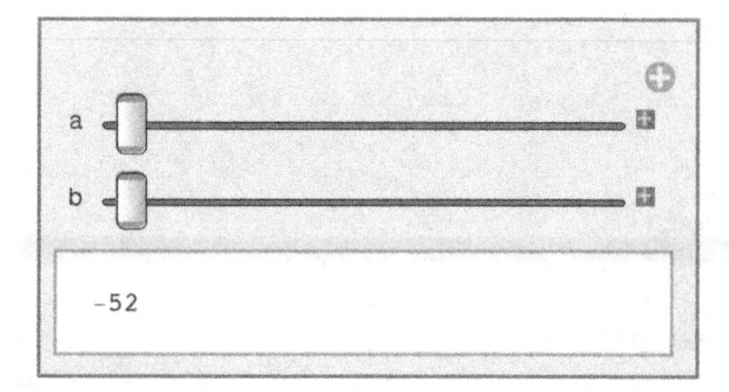

We use **Manipulate** and **DiagonalMatrix** to explore inner products on $\mathbb{R}^3$. The manipulation displays the value of the inner product if for $a = b = -5$.

Inner product norm

An *inner product norm* is a vector norm $\|\mathbf{w}\| = \sqrt{<\mathbf{w}, \mathbf{w}>}$ determined by an inner product <**u**, **v**>.

Illustration

- The space $\mathbb{R}^3$ as an inner product space with a nonstandard inner product norm

```
Clear[A]
```

```
MatrixForm[A = DiagonalMatrix[{1, 2, 3}]]
```

$$\begin{pmatrix} 1 & 0 & 0 \\ 0 & 2 & 0 \\ 0 & 0 & 3 \end{pmatrix}$$

```
⟨u_, v_⟩ := u.A.v
```

```
u = {2, -1, 3}; v = {1, 1, 1};
```

```
⟨u, v⟩
```

9

The associated vector norm is

```
‖w_‖ := Sqrt[⟨w, w⟩]
```

```
‖u‖
```

$\sqrt{33}$

and the normalized vector **n** corresponding to vector **u** is

$$n = \frac{1}{\|u\|} u$$

$$\left\{\frac{2}{\sqrt{33}}, -\frac{1}{\sqrt{33}}, \sqrt{\frac{3}{11}}\right\}$$

$\|n\|$

1

Manipulation

- Exploring vector norms

```
Manipulate[Sqrt[{1, 2, 3, 4}.DiagonalMatrix[{1, a, b, 3}].{1, 2, 3, 4}],
 {a, -3, 3, 1}, {b, -4, 4, 1}]
```

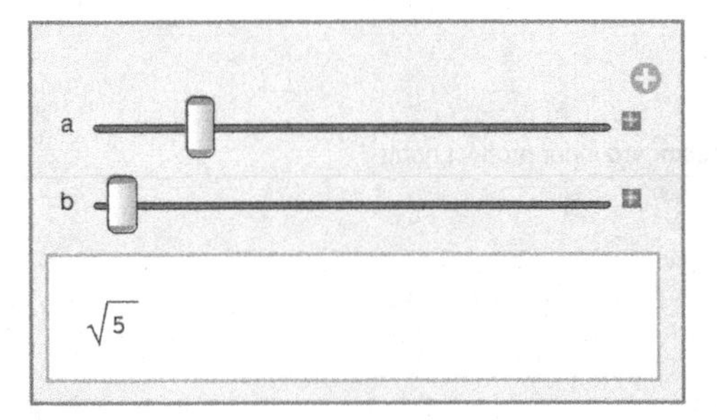

We use **Manipulate**, **Sqrt,** and **DiagonalMatrix** to explore the norms of vectors determined by different diagonal matrices. The display is the norm obtained by letting *a* = - 2 and *b* = -4.

Inner product space

An *inner product space* is a vector space equipped with an inner product.

Illustration

- The coordinate space $\mathbb{R}^2$ as an inner product space

```
Clear[x, y, z]
```

```
Dot[x, y] = < x, y >
```

```
x = {1, 2}; y = {-5, 3}; z = {-1, 12};
```

```
Dot[x, y]
```

1

```
Dot[x, y] == Dot[y, x]
```

```
True
```

```
Dot[-4 x, y] == -4 Dot[x, y]
```

```
True
```

```
Dot[x + y, z] == Dot[x, z] + Dot[y, z]
```

```
True
```

```
Dot[x, x] > 0
```

```
True
```

- The coordinate space $\mathbb{R}^3$ as an inner product space

```
A = RandomInteger[{0, 9}, {3, 3}];
```

```
MatrixForm[A]
```

$$\begin{pmatrix} 1 & 0 & 0 \\ 0 & 2 & 0 \\ 0 & 0 & 3 \end{pmatrix}$$

```
MatrixForm[M = Transpose[A].A]
```

$$\begin{pmatrix} 1 & 0 & 0 \\ 0 & 4 & 0 \\ 0 & 0 & 9 \end{pmatrix}$$

$$M = \begin{pmatrix} 110 & 90 & 14 \\ 90 & 106 & 14 \\ 14 & 14 & 2 \end{pmatrix};$$

```
SymmetricMatrixQ[M]
```

```
True
```

```
PositiveDefiniteMatrixQ[M]
```

```
True
```

A necessary and sufficient condition for a complex matrix to be positive-definite is that the Hermitian part

$$A_H = \frac{1}{2}\left(A + A^H\right) \tag{1}$$

is positive-definite. In the real case, this reduces to the requirement that the symmetric part

$$A_T = \frac{1}{2}\left(A + A^T\right) \tag{2}$$

is positive-definite.

```
PositiveSemidefiniteMatrixQ[1/2 (M + Transpose[M])]
```

```
True
```

The following sample calculations show that the function

```
⟨u_, v_⟩ := u.M.v
```

defines an inner product.

```
u = {1, 2, 3}; v = {4, -1, 3};
```

```
⟨u, v⟩
```

```
1128
```

```
⟨{1, 2, 3}, {4, -1, 3}⟩
```

```
1128
```

```
⟨u, v⟩ == ⟨v, u⟩
```

```
True
```

```
⟨{1, 2, 3}, {4, -1, 3}⟩ == ⟨{4, -1, 3}, {1, 2, 3}⟩
```

```
True
```

```
⟨4 u, v⟩ == 4 ⟨u, v⟩
```

```
True
```

```
⟨{4, 8, 12}, {4, -1, 3}⟩ == 4 ⟨{1, 2, 3}, {4, -1, 3}⟩
```

```
True
```

```
w = {3, 0, 2};
```

```
⟨u + w, v⟩ == ⟨u, v⟩ + ⟨w, v⟩
```

```
True
```

```
⟨{4, 2, 5}, {4, -1, 3}⟩ == ⟨{1, 2, 3}, {4, -1, 3}⟩ + ⟨{3, 0, 2}, {4, -1, 3}⟩
```

```
True
```

```
⟨u, u⟩ > 0
```

```
True
```

```
⟨{1, 2, 3}, {1, 2, 3}⟩ > 0
```

```
True
```

Interpolating polynomial

The polynomial p[t] of degree n whose graph passes through (n + 1) points in $\mathbb{R}^2$ is the *interpolating polynomial* in the variable t for the given points.

Illustration

- An interpolating polynomial of degree 3

```
points = {{-1, 5}, {0, -2}, {2, 7}, {3, 4}};
```

$p[t\_] := a_0 + a_1 t + a_2 t^2 + a_3 t^3;$

```
system = {p[-1] == 5, p[0] == -2, p[2] == 7, p[3] == 4}
```

$\{a_0 - a_1 + a_2 - a_3 == 5,\ a_0 == -2,\ a_0 + 2 a_1 + 4 a_2 + 8 a_3 == 7,\ a_0 + 3 a_1 + 9 a_2 + 27 a_3 == 4\}$

```
Solve[system]
```

$$\left\{\left\{a_0 \to -2,\ a_1 \to 0,\ a_2 \to \frac{65}{12},\ a_3 \to -\frac{19}{12}\right\}\right\}$$

$$p1 = -2 + \frac{65}{12} t^2 - \frac{19}{12} t^3;$$

```
plot1 = ListPlot[points];
```

```
plot2 = Plot[p1, {t, -2, 4}];
```

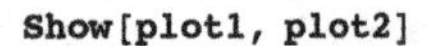

```
Show[plot1, plot2]
```

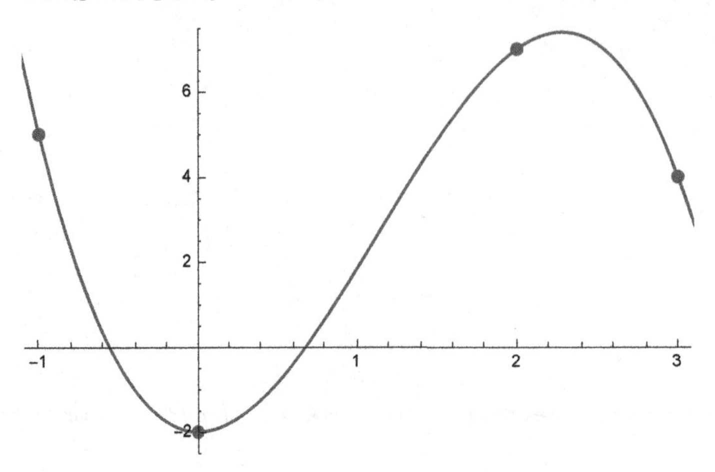

The *Mathematica* **Fit** function can also be used to calculate an interpolating polynomial with real coefficients.

```
p2 = Fit[points, {1, t, t^2, t^3}, t]
```

$-2. - 4.63083 \times 10^{-15}\, t + 5.41667\, t^2 - 1.58333\, t^3$

```
Chop[p1 == p2]
```

True

Intersection of subspaces

If U and V are two subspaces of a vector space W, then the *intersection* $(U \cap V)$ of U and V is the set of vectors of W belonging to both U and V. The intersection of two subspaces is always a subspace. If U and V have no nonzero vectors in common, then $(U \cap V)$ is the zero subspace of W. In that case, the subspaces U and V are said to be *disjoint*.

Illustration

- Two disjoint subspaces of $\mathbb{R}^4$

$U = \{a\{1, 3, 0, 1\} : a \in \mathbb{R}\}$

$V = \{a\{0, 3, 0, 2\} : a \in \mathbb{R}\}$

```
Reduce[{a, 3 a, 0, a} == {0, 3 a, 0, 2 a}, a]
```

a == 0

Hence

$U \cap V = \{\{0, 0, 0, 0\}\}$

- Two non-disjoint subspaces of $\mathbb{R}^4$

```
U = span[{{a, 3 a, 0, b} , {0, 0, 0, b}}];
```

```
V = span[{{a, 3 a, 0, a} }];
```

```
Reduce[{a, 3 a, 0, b} == {a, 3 a, 0, a}, {a, b}]
```

```
b == a
```

Hence

```
U ⋂ V = {{a, 3 a, 0, a} : a ∈ ℝ}
```

Invariant subspace

A subspace *W* of a vector space *V* is *invariant* under a linear operator $T : V \longrightarrow V$ (is *T-invariant*) if the vectors *T*[**v**] belong to *W* for all vectors **v** in *W*.

Illustration

- A subspace of $\mathbb{R}^3$ invariant under a projection operator

```
T[{x_, y_, z_}] := {x, y, 0}
```

```
Sy = {{0, y, 0} : y ∈ Reals}
```

```
T[{0, y, 0}]
```

```
{0, {-5, 3}, 0}
```

- A subspace of $\mathbb{R}^3$ not invariant under a linear operator

```
T[{x_, y_, z_}] := {0, x, y}
```

```
Sxy = {{x, y, 0} : x, y ∈ Reals}
```

```
T[{1, 1, 0}]
```

```
{0, 1, 1}
```

Since the vector {1, 1, 0} belongs to the subspace *Sxy*, but the vector *T*[{1, 1, 0)} = {0, 1, 1} does not, the subspace *Sxy* is not *T*-invariant.

A linear operator $T : \mathbb{R}^n \longrightarrow \mathbb{R}^n$ is representable by a diagonal matrix if and only if $\mathbb{R}^n$ is a direct sum of one-dimensional *T*-invariant subspaces of $\mathbb{R}^n$.

- Two one-dimensional *T*-invariant subspaces of $\mathbb{R}^3$

Suppose that *T* is linear operator represented in the standard basis of $\mathbb{R}^3$ by the matrix

$$A = \begin{pmatrix} 1 & 1 & 0 \\ 1 & 1 & 1 \\ 0 & 0 & 0 \end{pmatrix}$$

```
{{1, 1, 0}, {1, 1, 1}, {0, 0, 0}}
```

Then the following calculation

```
Eigensystem[A]
```

```
{{2, 0, 0}, {{1, 1, 0}, {-1, 1, 0}, {0, 0, 0}}}
```

shows that 2 is the only eigenvalue of *A* and that it has precisely two associated one-dimensional eigenspaces:

E_1 = {{a, a, 0} : a ∈ Reals}

E_2 = {{-a, a, 0} : a ∈ Reals}

Hence $\mathbb{R}^3$ is not a direct sum of one-dimensional *A*-invariant eigenspaces of the matrix *A*.

- Three one-dimensional T-invariant subspaces of $\mathbb{R}^3$

$$A = \begin{pmatrix} 4 & 3 & 9 \\ 0 & 1 & 3 \\ 0 & 0 & 2 \end{pmatrix};$$

```
Eigensystem[A]
```

```
{{4, 2, 1}, {{1, 0, 0}, {-9, 3, 1}, {-1, 1, 0}}}
```

The calculation

```
Solve[{x, y, z} == a {1, 0, 0} + b {-9, 3, 1} + c {-1, 1, 0}, {a, b, c}]
```

```
{}
```

shows that every vector {*x*, *y*, *z*} in $\mathbb{R}^3$ is a (unique) linear combination of the eigenvectors {1, 0, 0}, {-9, 3, 1}, {-1, 1, 0}.

Since eigenvectors belonging to distinct eigenvalues are linearly independent, the space $\mathbb{R}^3$ is a direct sum of the three associated *A*-invariant eigenspaces:

```
A.{a, 0, 0} == 4 {a, 0, 0}
```

```
True
```

```
A.{-9 b, 3 b, b} == 2 {-9 b, 3 b, b}
```

```
True
```

```
A.{-c, c, 0} == {-c, c, 0}
```

```
True
```

Inverse of a linear transformation

The inverse of a linear transformation $T : V \longrightarrow W$ between two real or complex vector spaces V and W is a linear transformation $S : W \longrightarrow V$ for which $S(T(\mathbf{v})) = \mathbf{v}$ for all vectors $\mathbf{v}$ in V and $T(S(\mathbf{w})) = \mathbf{w}$ for all vectors w in W.

If T from $\mathbb{R}^n$ to $\mathbb{R}^n$ is an invertible linear transformation represented by a matrix A, then the inverse matrix of A represents the *inverse* S of T.

Illustration

- Inverse of a linear transformation from $\mathbb{R}^2$ to $\mathbb{R}^2$

```
T[{x_, y_}] := {x + y, x - y}
```

```
S[{u_, v_}] := {1/2 (u + v), 1/2 (u - v)}
```

```
u = {3, 7};
```

```
{T[u], S[T[u]]}
```

```
{{10, -4}, {3, 7}}
```

In the standard basis of $\mathbb{R}^2$, the linear transformations T and S are represented by the following matrices:

```
MatrixForm[T = Transpose[{{1, 1}, {1, -1}}]]
```

$$\begin{pmatrix} 1 & 1 \\ 1 & -1 \end{pmatrix}$$

```
MatrixForm[S = Transpose[{{1/2, 1/2}, {1/2, -1/2}}]]
```

$$\begin{pmatrix} \frac{1}{2} & \frac{1}{2} \\ \frac{1}{2} & -\frac{1}{2} \end{pmatrix}$$

These matrices are inverses of each other:

```
{S.T == IdentityMatrix[2], T.S == IdentityMatrix[2]}
```

```
{True, True}
```

Inverse of a matrix

The *inverse* of a matrix A is a matrix B for which $A\,B = B\,A$ = **IdentityMatrix[*n*]**, where n is the dimension of A. Not every square matrix has an inverse. Matrix inverses have the following properties:

Properties of matrix inverses

```
Inverse[Inverse[A]] = A                                  (1)
Inverse[A.B] = Inverse[B].Inverse[A]                     (2)
Transpose[Inverse[A]] = Inverse[Transpose[A]]            (3)
Inverse[A].A = A.Inverse[A] = IdentityMatrix             (4)
Inverse[a A] = (1 / a) Inverse[A] if a ≠ 0               (5)
Det[Inverse[A]] = 1 / Det[A]                             (6)
```

Illustration

- The inverse of a 2-by-2 matrix

```
A = {{1, 2}, {3, 4}};
```

```
MatrixForm[Inverse[A]]
```

$$\begin{pmatrix} -2 & 1 \\ \frac{3}{2} & -\frac{1}{2} \end{pmatrix}$$

```
MatrixForm[id2 = IdentityMatrix[2]]
```

$$\begin{pmatrix} 1 & 0 \\ 0 & 1 \end{pmatrix}$$

```
{A.Inverse[A] == IdentityMatrix[2], A.Inverse[A] == IdentityMatrix[2]}
```

{True, True}

Not all matrices are invertible.

- A non-invertible 3-by-3 matrix

```
B = {{1, 2, 3}, {4, 5, 6}, {7, 8, 9}};
```

```
Inverse[B];
```

Inverse::sing : Matrix {{1, 2, 3}, {4, 5, 6}, {7, 8, 9}} is singular. ≫

Mathematica tells us that this matrix has no inverse. Matrices that have no inverse are called *singular.*

In addition to the Inverse function, *Mathematica* also supports the superscript notation of matrix inversion. For any given invertible matrix A, we can define the inverse of A by A^{-1}.

For any numerical *n*-by-*n* matrix *A* with real or complex scalars, the following statements are equivalent:

- The matrix *A* is invertible (nonsingular).
- The matrix *A* is a finite product of elementary matrices.
- The matrix *A* is equivalent to an identity matrix.
- The matrix equation *A***v** = **0** has only the trivial (zero) solution.
- The null space of *A* is a zero space: ***NullSpace***[A] = {*0*}.
- The matrix *A* has full rank: **MatrixRank**[*A*] = *n*.
- The columns of *A* are linearly independent.
- If *A* is a real matrix, the columns of *A* span $\mathbb{R}^n$.
- The transpose of *A* is invertible.
- The scalar *0* is not an eigenvalue of *A*.
- The matrix *A* has a left inverse *LA* and a right inverse *R A* and *L A* = *R A*.

The *inverse* of a matrix *A* is a matrix *B* for which *A B* = *B A* is an identity matrix. This definition forces the matrix *A* to be square. Not all square matrices have inverses. However, matrices that aren't square can have one-sided inverses: either *AB* or *BA* may be an identity matrix.

Illustration

- Inverse of a 2-by-2 matrix

```
A = {{1, 2}, {3, 4}};
```

```
B = Inverse[A]
```

$\left\{\{-2, 1\}, \left\{\frac{3}{2}, -\frac{1}{2}\right\}\right\}$

```
A.B == B.A == IdentityMatrix[2]
```

```
True
```

- One-sided inverse of a 2-by-3 matrix

```
A = {{1, 2, 3}, {4, 5, 6}};
```

```
B = PseudoInverse[A]
```

$\left\{\left\{-\frac{17}{18}, \frac{4}{9}\right\}, \left\{-\frac{1}{9}, \frac{1}{9}\right\}, \left\{\frac{13}{18}, -\frac{2}{9}\right\}\right\}$

```
A.B
```

```
{{1, 0}, {0, 1}}
```

```
B.A
```

$$\left\{\left\{\frac{5}{6}, \frac{1}{3}, -\frac{1}{6}\right\}, \left\{\frac{1}{3}, \frac{1}{3}, \frac{1}{3}\right\}, \left\{-\frac{1}{6}, \frac{1}{3}, \frac{5}{6}\right\}\right\}$$

These calculations show that the matrix *A* has a right inverse, but not a left inverse. The **PseudoInverse** function produces a one-sided inverse or an inverse, depending on whether or not the given matrix is invertible.

- Inverse of a 3-by-3 matrix

```
A = {{1, 2, 3}, {4, 5, 6}, {7, 8, 10}};
```

```
inverseA = Inverse[A]
```

$$\left\{\{-2, 1\}, \left\{\frac{3}{2}, -\frac{1}{2}\right\}\right\}$$

```
A.inverseA == inverseA.A == IdentityMatrix[3]
```

```
True
```

If *A* is an invertible matrix, both the **PseudoInverse** and the **Inverse** functions produce the inverse of *A*.

```
PseudoInverse[A] == Inverse[A]
```

```
True
```

Manipulation

- The inverse of a 3-by-3 matrix

```
MatrixForm[A = {{1, 2, 3}, {4, 5, 6}, {7, 8, 9 a}}]
```

$$\begin{pmatrix} 1 & 2 & 3 \\ 4 & 5 & 6 \\ 7 & 8 & 9\,a \end{pmatrix}$$

```
Manipulate[Evaluate[MatrixForm[PseudoInverse[A]]], {a, 2, 5, 1}]
```

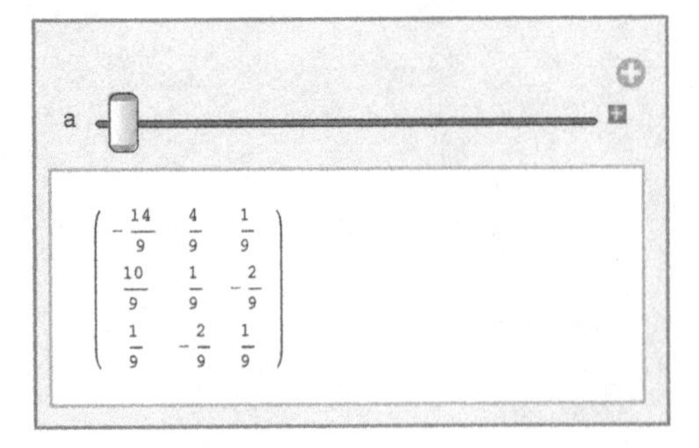

We can combine **Manipulate**, **Evaluate**, **MatrixForm,** and **PseudoInverse** to explore the pseudoinverses of the generated matrices. The displayed pseudoinverse is obtained by letting *a* = 2.

Invertible matrix

A matrix A is *invertible* if there exists a matrix *B*, called the inverse of A and usually denoted by A^{-1}, for which the products *(A B)* and *(B A)* are an identity matrix. An invertible matrix is always square.

There is an intimate connection between the Gaussian elimination steps for linear systems and the invertibility of matrix operations. Every invertible matrix is a (nonunique) product of elementary matrices and every elementary matrix is the result of a single application of a Gaussian elimination step to an identity matrix. This means that for small matrices, we have an excellent idea of the geometric meaning of invertibility.

Illustration

- An invertible 2-by-2 matrix

```
MatrixForm[A1 = {{1, 2}, {3, 4}}]
```

$$\begin{pmatrix} 1 & 2 \\ 3 & 4 \end{pmatrix}$$

```
MatrixForm[A2 = Inverse[A1]]
```

$$\begin{pmatrix} 1 & 2 \\ 3 & 4 \end{pmatrix}$$

$$\begin{pmatrix} -2 & 1 \\ \frac{3}{2} & -\frac{1}{2} \end{pmatrix}$$

```
A1.A2 == IdentityMatrix[2] == A2.A1
```

```
True
```

- Invertible 2-by-2 elementary matrices

$EA1 = \begin{pmatrix} 0 & 1 \\ 1 & 0 \end{pmatrix}$ (interchange of rows 1 and 2)

$EA2 = \begin{pmatrix} 1 & 0 \\ 0 & s \end{pmatrix}$ (multiplication of row 2 by a nonzero constant s)

$EA3 = \begin{pmatrix} s & 0 \\ 0 & 1 \end{pmatrix}$ (multiplcation of row 1 by a nonzero constant s)

$EA4 = \begin{pmatrix} 1 & s \\ 0 & 1 \end{pmatrix}$ (addition of s times row 2 to row 1)

$EA5 = \begin{pmatrix} 1 & 0 \\ s & 1 \end{pmatrix}$ (addition of s times row 1 to row 2)

- Invertible 3-by-3 matrices

$$A = \begin{pmatrix} 9 & 3 & 1 \\ -1 & 3 & 0 \\ 4 & 0 & 9 \end{pmatrix};$$

Adding 9 times the second row to the first row

$$E1 = \begin{pmatrix} 1 & 9 & 0 \\ 0 & 1 & 0 \\ 0 & 0 & 1 \end{pmatrix};$$

MatrixForm[A2 = E1. A]

$$\begin{pmatrix} 0 & 30 & 1 \\ -1 & 3 & 0 \\ 4 & 0 & 9 \end{pmatrix}$$

Adding 4 times the second row to the third row

$$E2 = \begin{pmatrix} 1 & 0 & 0 \\ 0 & 1 & 0 \\ 0 & 4 & 1 \end{pmatrix};$$

MatrixForm[A3 = E2. A2]

$$\begin{pmatrix} 0 & 30 & 1 \\ -1 & 3 & 0 \\ 0 & 12 & 9 \end{pmatrix}$$

Interchanging the first and second rows

$$E3 = \begin{pmatrix} 0 & 1 & 0 \\ 1 & 0 & 0 \\ 0 & 0 & 1 \end{pmatrix};$$

MatrixForm[A4 = E3. A3]

$$\begin{pmatrix} -1 & 3 & 0 \\ 0 & 30 & 1 \\ 0 & 12 & 9 \end{pmatrix}$$

Multiplying the first row by −1

$$E4 = \begin{pmatrix} -1 & 0 & 0 \\ 0 & 1 & 0 \\ 0 & 0 & 1 \end{pmatrix};$$

MatrixForm[A5 = E4. A4]

$$\begin{pmatrix} 1 & -3 & 0 \\ 0 & 30 & 1 \\ 0 & 12 & 9 \end{pmatrix}$$

Dividing the second row by 30

```
E5 = (1  0     0
      0  1/30  0
      0  0     1);

MatrixForm[A6 = E5. A5]
```

$$\begin{pmatrix} 1 & -3 & 0 \\ 0 & 1 & \frac{1}{30} \\ 0 & 12 & 9 \end{pmatrix}$$

Subtracting 12 times the second row from the third row

```
E6 = (1  0    0
      0  1    0
      0  -12  1);

MatrixForm[A7 = E6. A6]
```

$$\begin{pmatrix} 1 & -3 & 0 \\ 0 & 1 & \frac{1}{30} \\ 0 & 0 & \frac{43}{5} \end{pmatrix}$$

Adding 3 times the second row to the first row

```
E7 = (1  3  0
      0  1  0
      0  0  1);

MatrixForm[A8 = E7. A7]
```

$$\begin{pmatrix} 1 & 0 & \frac{1}{10} \\ 0 & 1 & \frac{1}{30} \\ 0 & 0 & \frac{43}{5} \end{pmatrix}$$

Multiplying the third row by 5/43

```
E8 = (1  0  0
      0  1  0
      0  0  5/43);

MatrixForm[A9 = E8. A8]
```

$$\begin{pmatrix} 1 & 0 & \frac{1}{10} \\ 0 & 1 & \frac{1}{30} \\ 0 & 0 & 1 \end{pmatrix}$$

Subtracting 1/30 times the third row from the second row

$$E9 = \begin{pmatrix} 1 & 0 & 0 \\ 0 & 1 & -1/30 \\ 0 & 0 & 1 \end{pmatrix};$$

```
MatrixForm[A10 = E9. A9]
```

$$\begin{pmatrix} 1 & 0 & \frac{1}{10} \\ 0 & 1 & 0 \\ 0 & 0 & 1 \end{pmatrix}$$

Subtracting 1/10 times the third row from the second row

$$E10 = \begin{pmatrix} 1 & 0 & -1/10 \\ 0 & 1 & 0 \\ 0 & 0 & 1 \end{pmatrix};$$

```
MatrixForm[A11 = E10. A10]
```

$$\begin{pmatrix} 1 & 0 & 0 \\ 0 & 1 & 0 \\ 0 & 0 & 1 \end{pmatrix}$$

The product of the ten elementary matrices used to convert the matrix *A* to the 3-by-3 identity matrix is the inverse *B* = matrix of the matrix *A*.

```
B = E10.E9.E8.E7.E6.E5.E4.E3.E2.E1 == Inverse[A]
```

```
True
```

Isometry

An *isometry* on a normed vector space is an invertible linear transformation that preserves the distances between the vectors of the space. Orthogonal transformations are isometries.

Illustration

- An orthogonal transformation preserving the Euclidean distances between vectors in $\mathbb{R}^2$

```
MatrixForm[A = {{Cos[x], Sin[x]}, {-Sin[x], Cos[x]}}]; x = π / 3;
```

```
u = {1, 3}; v = {-5, 4};
```

```
distance[u_, v_] := Norm[u - v]
```

```
distance[u, v]
```

$\sqrt{37}$

```
Simplify[distance[A.u, A.v]]
```

$\sqrt{37}$

The linear transformation represented by the orthogonal matrix A in the standard basis of $\mathbb{R}^2$ is an isometry.

Isomorphism of vector spaces

Many different vector spaces are *isomorphic*. This means that we can translate the vectors from one space one-by-one to vectors in another space and conversely in a way that preserves the vector space structure: that is, linear combinations are preserved. Here is the basic theorem that makes this statement precise.

Every n-dimensional real vector space is isomorphic to the real coordinate space $\mathbb{R}^n$ and every n-dimensional complex vector space is isomorphic to the complex coordinate space $\mathbb{C}^n$. These isomorphisms are the basis for being able to use matrix representations of linear transformations and matrix multiplication for the composition of linear transformations. Different choices of bases produce different isomorphisms.

In principle, this suggests that in finite-dimensional linear algebra, we can always work with column or row vectors and matrix operations. However, often the nature of the vectors and their additional structure is not captured by this basic isomorphism.

Illustration

- An isomorphism between the polynomial space $\mathbb{R}$[t,3] and the coordinate space $\mathbb{R}^4$

Consider the following two bases for $\mathbb{R}$[t,3] and $\mathbb{R}^4$:

```
Basis1 = {1, t, t^2, t^3};

Basis2 = {e1 = {1, 0, 0, 0}, e2 = {0, 1, 0, 0}, e3 = {0, 0, 1, 0}, e4 = {0, 0, 0, 1}};
```

The definition

```
{T[1] = e1, T[t] = e2, T[t^2] = e3, T[t^3] = e4}
{{1, 0, 0, 0}, {0, 1, 0, 0}, {0, 0, 1, 0}, {0, 0, 0, 1}}
```

determines a unique invertible linear transformation T from $\mathbb{R}$[t,3] to $\mathbb{R}^4$ whose inverse is defined by

```
{S[e1] = 1, S[e2] = t, S[e3] = t^2, S[e4] = t^3}
```

Hence the spaces $\mathbb{R}$[t,3] and $\mathbb{R}^4$ are isomorphic.

- An isomorphism between the matrix space $\mathbb{R}^{2\times 2}$ of 2-by-2 real matrices and the coordinate space $\mathbb{R}^4$

Consider the following basis for $\mathbb{R}^{2\times 2}$

```
B = {A11, A12, A21, A22};
```

where

```
A11 = {{1, 0}, {0, 0}};
A12 = {{0, 1}, {0, 0}};
A21 = {{0, 0}, {1, 0}};
A22 = {{0, 0}, {0, 1}};
```

Any real matrix $A = \{\{a, b\},\{c, d\}\}$ is a unique sum of the four basis matrices:

```
{{a, b}, {c, d}} == a A11 + b A12 + c A21 + d A22
```

```
True
```

We map the basis B to the standard basis for $\mathbb{R}^4$:

```
T[A11] = {1, 0, 0, 0};
T[A12] = {0, 1, 0, 0};
T[A21] = {0, 0, 1, 0};
T[A22] = {0, 0, 0, 1};
```

By definition, the function T defines an isomorphism from $\mathbb{R}^{2\times 2}$ to $\mathbb{R}^4$.

J

Jacobian determinant

If F and G are two differentiable functions in some region in $\mathbb{R}^4$ and u and v are differentiable functions in x and y, then determinants and Cramer's rule can be used to find the partial derivatives

$$u_x = -\frac{\text{Det}[\{\{F_x, F_v\}, \{G_x, G_v\}\}]}{\text{Det}[\{\{F_u, F_v\}, \{G_u, G_v\}\}]} \text{ and } v_x = -\frac{\text{Det}[\{\{F_u, F_x\}, \{G_u, G_x\}\}]}{\text{Det}[\{\{F_u, F_v\}, \{G_u, G_v\}\}]}$$

$$u_y = -\frac{\text{Det}[\{\{F_y, F_v\}, \{G_y, G_v\}\}]}{\text{Det}[\{\{F_u, F_v\}, \{G_u, G_v\}\}]} \text{ and } v_y = -\frac{\text{Det}[\{\{F_u, F_y\}, \{G_u, G_y\}\}]}{\text{Det}[\{\{F_u, F_v\}, \{G_u, G_v\}\}]}$$

at all points $\{x,y\}$ at which $\textbf{Det}[\{\{F_u, F_v\}, \{G_u, G_v\}\}] \neq 0$.

The four determinants involved are called the *Jacobians* of the functions F and G are denoted by

$$J\left[\frac{F,\ G}{x,\ v}\right],\ J\left[\frac{F,\ G}{y,\ v}\right],\ J\left[\frac{F,\ G}{u,\ x}\right],\ J\left[\frac{F,\ G}{u,\ v}\right]$$

Illustration

- The Jacobian of the polar coordinate transformation

```
F[r_, θ_] := r Cos[θ];
```

```
G[r_, θ_] := r Sin[θ];
```

```
Det[{{D[F[r, θ], r], D[F[r, θ], θ]}, {D[G[r, θ], r], D[G[r, θ], θ]}}]
```

$r\,\text{Cos}[\theta]^2 + r\,\text{Sin}[\theta]^2$

```
Simplify[r Cos[θ]^2 + r Sin[θ]^2]
```

r

Hence the Jacobian of the polar coordinate transformation is

$$J\left[\frac{F,\ G}{r,\ \theta}\right] = r \tag{1}$$

- Using Jacobians to calculate partial derivatives

```
F[x_, y_, u_, v_] := x^2 + y^3 x + u^2 + v^3;
```

```
G[x_, y_, u_, v_] := x^2 + 3 y x + u^4 - v^2;
```

```
J[F, G, u, v] =
  Det[
   {{D[F[x, y, u, v], u], D[F[x, y, u, v], v]}, {D[G[x, y, u, v], u], D[G[x, y, u, v], v]}}];

J[F, G, x, v] =
  Det[
   {{D[F[x, y, u, v], x], D[F[x, y, u, v], v]}, {D[G[x, y, u, v], x], D[G[x, y, u, v], v]}}];

J[F, G, u, x] =
  Det[
   {{D[F[x, y, u, v], u], D[F[x, y, u, v], x]}, {D[G[x, y, u, v], u], D[G[x, y, u, v], x]}}];

J[F, G, y, v] =
  Det[
   {{D[F[x, y, u, v], y], D[F[x, y, u, v], v]}, {D[G[x, y, u, v], y], D[G[x, y, u, v], v]}}];

J[F, G, u, y] =
  Det[
   {{D[F[x, y, u, v], u], D[F[x, y, u, v], y]}, {D[G[x, y, u, v], u], D[G[x, y, u, v], y]}}];
```

$$u_x = \text{Simplify}\left[\frac{-J[F, G, x, v]}{J[F, G, u, v]}\right]$$

$$-\frac{(4 + 6\,v)\,x + 9\,v\,y + 2\,y^3}{4\,\left(u + 3\,u^3\,v\right)}$$

$$v_x = \text{Simplify}\left[\frac{-J[F, G, u, x]}{J[F, G, u, v]}\right]$$

$$\frac{2\,x - 4\,u^2\,x + 3\,y - 2\,u^2\,y^3}{2\,v + 6\,u^2\,v^2}$$

$$u_y = \text{Simplify}\left[\frac{-J[F, G, y, v]}{J[F, G, u, v]}\right]$$

$$-\frac{3\,x\,\left(3\,v + 2\,y^2\right)}{4\,\left(u + 3\,u^3\,v\right)}$$

$$v_y = \text{Simplify}\left[\frac{-J[F, G, u, y]}{J[F, G, u, v]}\right]$$

$$\frac{3\,x - 6\,u^2\,x\,y^2}{2\,v + 6\,u^2\,v^2}$$

Jordan block

A *Jordan block* is a square matrix which has zero entries everywhere except on the diagonal, where the entries are a fixed scalar, and except on the superdiagonal, where the entries are either all 0s or all 1s.

Illustration

- A diagonal 2-by-2 Jordan block

```
J1 = MatrixForm[DiagonalMatrix[{3, 3}]]
```

$$\begin{pmatrix} 3 & 0 \\ 0 & 3 \end{pmatrix}$$

- A 2-by-2 Jordan block

```
J2 = MatrixForm[{{3, 1}, {0, 3}}]
```

$$\begin{pmatrix} 3 & 1 \\ 0 & 3 \end{pmatrix}$$

- A 3-by-3 Jordan block

```
J3 = {{3, 1, 0}, {0, 3, 1}, {0, 0, 3}}
```

$$\begin{pmatrix} 3 & 1 & 0 \\ 0 & 3 & 1 \\ 0 & 0 & 3 \end{pmatrix}$$

- A 4-by-4 Jordan block

```
J3 = {{5, 1, 0, 0}, {0, 5, 1, 0}, {0, 0, 5, 1}, {0, 0, 0, 5}}
```

$$\begin{pmatrix} 5 & 1 & 0 & 0 \\ 0 & 5 & 1 & 0 \\ 0 & 0 & 5 & 1 \\ 0 & 0 & 0 & 5 \end{pmatrix}$$

Jordan matrix

A matrix J is a *Jordan matrix* if it is a direct sum of *Jordan blocks*. The **JordanDecomposition** function yields the Jordan decomposition of a square matrix. The result is a list $\{s, j\}$ where s is a similarity matrix and j is a matrix in Jordan canonical form. Every square matrix is similar to a Jordan matrix, also called a matrix in *Jordan canonical form*. (*Mathematica* uses the lower-case letter s and j as names for the outputs of the **JordanDecomposition** function.)

Illustration

- A direct sum of Jordan blocks

The *direct sum* of the Jordan blocks J_1 and J_2 is the following matrix:

```
MatrixForm[J1⊕J2 = {{3, 1, 0, 0, 0, 0, 0}, {0, 3, 1, 0, 0, 0, 0}, {0, 0, 3, 0, 0, 0, 0},
   {0, 0, 0, 5, 1, 0, 0}, {0, 0, 0, 0, 5, 1, 0}, {0, 0, 0, 0, 0, 5, 1}, {0, 0, 0, 0, 0, 0, 5}}]
```

$$\begin{pmatrix} 3 & 1 & 0 & 0 & 0 & 0 & 0 \\ 0 & 3 & 1 & 0 & 0 & 0 & 0 \\ 0 & 0 & 3 & 0 & 0 & 0 & 0 \\ 0 & 0 & 0 & 5 & 1 & 0 & 0 \\ 0 & 0 & 0 & 0 & 5 & 1 & 0 \\ 0 & 0 & 0 & 0 & 0 & 5 & 1 \\ 0 & 0 & 0 & 0 & 0 & 0 & 5 \end{pmatrix}$$

(The direct sum symbol is created by typing Esc c+ Esc.)

- A Jordan decomposition

```
MatrixForm[A = {{27, 48, 81}, {-6, 0, 0}, {1, 0, 3}}]
```

$$\begin{pmatrix} 27 & 48 & 81 \\ -6 & 0 & 0 \\ 1 & 0 & 3 \end{pmatrix}$$

```
{S, J} = JordanDecomposition[A];
```

```
Map[MatrixForm, %]
```

$$\left\{ \begin{pmatrix} 3 & 18 & 2 \\ -3 & -9 & -\frac{1}{4} \\ 1 & 2 & 0 \end{pmatrix}, \begin{pmatrix} 6 & 0 & 0 \\ 0 & 12 & 1 \\ 0 & 0 & 12 \end{pmatrix} \right\}$$

```
MatrixForm[S.J.Inverse[S]]
```

$$\begin{pmatrix} 27 & 48 & 81 \\ -6 & 0 & 0 \\ 1 & 0 & 3 \end{pmatrix}$$

The matrix *S* is a similarity matrix and the *J* is the Jordan form of the matrix *A*.

Manipulation

- Exploring Jordan decompositions

```
MatrixForm[A = RandomInteger[{0, 9}, {3, 3}]]
```

$$A = \begin{pmatrix} 0 & 7 & 5 \\ 9 & 1 & 9 \\ 9 & 9 & 5 \end{pmatrix};$$

```
Manipulate[{S, J} = N[JordanDecomposition[{{a, 7, 5}, {9, 1, 9}, {9, 9, 5}}]], {a, -5, 5, 1}]
```

a

```
{{{-4.97792, -0.572864, 0.513742}, {3.34385, -0.592494, 0.841237}, {1., 1., 1.}},
 {{-9.70659, 0., 0.}, {0., -5.48822, 0.}, {0., 0., 17.1948}}}
```

We use **Manipulate**, **N,** and **JordanDecomposition** to explore the Jordan decompositions by varying the parameter *a.* If *a* = - *4*, for example, then the manipulation shows that the Jordan decomposition of the matrix

$$A = \begin{pmatrix} -4 & 7 & 5 \\ 9 & 1 & 9 \\ 9 & 9 & 5 \end{pmatrix}$$

is the product of *S*, *J*, and the inverse of *S*:

```
MatrixForm[Chop[S.J.Inverse[S]]]
```

$$\begin{pmatrix} -4. & 7. & 5. \\ 9. & 1. & 9. \\ 9. & 9. & 5. \end{pmatrix}$$

K

Kernel of a linear transformation

The *kernel* of a linear transformation $T: V \longrightarrow W$ from a vector space V to a vector space W is the set of all vectors **v** in V with the property that $T[\mathbf{v}] = \mathbf{0}$ in W. It is a subspace of V. The kernel of a matrix transformation T_A is the null space of the matrix A.

Illustration

- The kernel of a projection in $\mathbb{R}^3$

```
T[{x_, y_, z_}] := {x, y, 0}
```

```
T[{x, y, z}] == {0, 0, 0}
```

```
{x, y, 0} == {0, 0, 0}
```

Therefore the set of all vectors {*0, 0, z*} is the kernel of *T*.

- The kernel of a linear transformation from $\mathbb{R}^3$ to $\mathbb{R}^2$

```
T[{x_, y_, z_}] := {3 x, -4 z}
```

```
T[{x, y, z}] == {0, 0}
```

```
{3 x, -4 z} == {0, 0}
```

The set of all vectors {*0, y, 0*} is the kernel of *T*.

- The kernel of a matrix transformation from $\mathbb{R}^3$ to $\mathbb{R}^3$

$$A = \begin{pmatrix} 1 & 4 & 9 \\ 7 & 5 & 5 \\ 2 & 0 & 1 \end{pmatrix};$$

```
NullSpace[A]
```

```
{}
```

The kernel of the matrix transformation T_A is the zero subspace of $\mathbb{R}^3$. The transformation T_A is invertible since the matrix A is invertible.

```
MatrixForm[B = Inverse[A]]
```

$$\begin{pmatrix} -\frac{5}{73} & \frac{4}{73} & \frac{25}{73} \\ -\frac{3}{73} & \frac{17}{73} & -\frac{58}{73} \\ \frac{10}{73} & -\frac{8}{73} & \frac{23}{73} \end{pmatrix}$$

```
A.B == B.A == IdentityMatrix[3]
```

```
True
```

- A matrix transformation from $\mathbb{R}^3$ to $\mathbb{R}^3$ with a one-dimensional kernel

```
A = {{1, 2, 3}, {4, 5, 6}, {7, 8, 9}};
```

$$A = \begin{pmatrix} 1 & 2 & 3 \\ 4 & 5 & 6 \\ 7 & 8 & 9 \end{pmatrix};$$

```
NullSpace[A]
```

```
{{1, -2, 1}}
```

The kernel of the matrix transformation T_A is the span of the vector *{1, -2, 1}*.

```
A.(k * {1, -2, 1})
```

```
{0, 0, 0}
```

Kronecker delta

The Kronecker delta function $\delta_{\{n_1, n_2, \ldots\}}$ is a function on a list $\{n_1, n_2, \ldots\}$ of lists that equals 1 if all n_i are equal and 0 otherwise. The built-in **KroneckerDelta** function calculates Kronecker deltas.

Illustration

- The Kronecker delta function on a list of three unequal lists

```
data = {{1, 9}, {0, 1}, {5, 3}};
```

```
KroneckerDelta[{1, 9}, {0, 1}, {5, 3}]
```

```
0
```

- The Kronecker delta function on a list of two equal lists

```
data = {{1, 2}, {1, 2}};
```

```
KroneckerDelta[{1, 2}, {1, 2}]
```

1

- Using the Kronecker delta to construct an identity matrix

```
MatrixForm[A = Table[KroneckerDelta[i, j], {i, 1, 3}, {j, 1, 3}]]
```

$$\begin{pmatrix} 1 & 0 & 0 \\ 0 & 1 & 0 \\ 0 & 0 & 1 \end{pmatrix}$$

Manipulation

- The Kronecker delta of a list of three lists

```
Manipulate[KroneckerDelta[{1, b}, {a - 1, c}, {b, b}], {a, 1, 4, 1}, {b, 0, 3, 1}, {c, 1, 5, 1}]
```

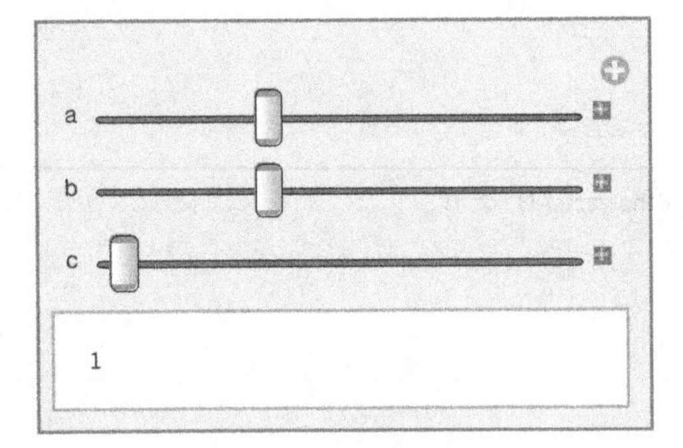

We use **Manipulate** and **KroneckerDelta** to calculate Kronecker deltas. The displayed value 1 is obtained by letting $a = 2$ and $b = c = 1$.

Kronecker product

The most intriguing and ingenious operation on matrices is their multiplication. One such operation is the Kronecker product. The built-in **KroneckerProduct** function calculates such products.

Illustration

- The Kronecker product of two 2-by-2 matrices

```
Clear[a, b, c, d, r, s, t, u]
```

```
MatrixForm[L = {{a, b}, {c, d}}]
```

$$\begin{pmatrix} a & b \\ c & d \end{pmatrix}$$

```
MatrixForm[R = {{r, s}, {t, u}}]
```

$$\begin{pmatrix} r & s \\ t & u \end{pmatrix}$$

```
MatrixForm[KroneckerProduct[L, R]]
```

$$\begin{pmatrix} ar & as & br & bs \\ at & au & bt & bu \\ cr & cs & dr & ds \\ ct & cu & dt & du \end{pmatrix}$$

Manipulation

```
A = {{2, 3}, {4, 6}}; B = {{a, 5}, {7, b}};
```

```
Manipulate[MatrixForm[KroneckerProduct[{{2, 3}, {4, 6}}, {{a, 5}, {7, b}}]],
 {a, -3, 3, 1}, {b, -4, 4, 1}]
```

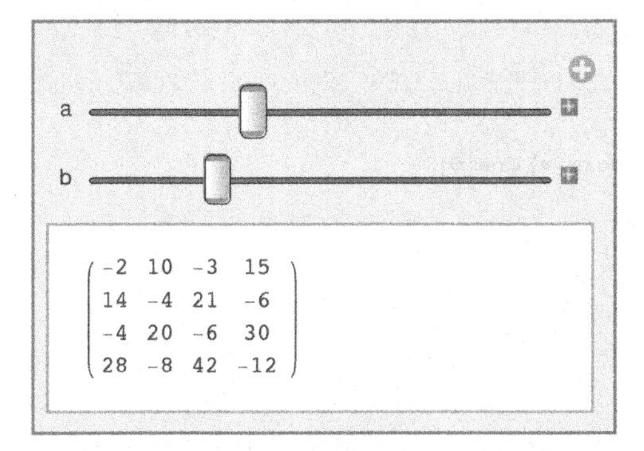

We can combine **Manipulate**, **MatrixForm,** and **KroneckerProduct** to explore the Kronecker products of different matrices. The displayed matrix is obtained by letting a = - 1 and b = - 2.

L

Law of cosines

If θ is an angle between two vectors **u** and **v** in $\mathbb{R}^2$ or $\mathbb{R}^3$, then the *law of cosines* says that

$$\|\mathbf{u} - \mathbf{v}\|^2 == \|\mathbf{u}\|^2 + \|\mathbf{v}\|^2 - 2\,\|\mathbf{u}\|\,\|\mathbf{v}\|\,\text{Cos}[\theta] \tag{1}$$

where $\|\mathbf{w}\|$ denotes the Euclidean norm of a vector **w**.

This law can be used to determine the angle between two vectors.

Illustration

- The law of cosines for two vectors in $\mathbb{R}^2$

```
u = {1, 2}; v = {-3, 4};
```

```
Norm[u - v]^2 == Norm[u]^2 + Norm[v]^2 - 2 Norm[u] Norm[v] Cos[θ]
```

$20 == 30 - 10\sqrt{5}\,\text{Cos}[\theta]$

```
Solve[%, Cos[θ]]
```

$\left\{\left\{\text{Cos}[\theta] \to \frac{1}{\sqrt{5}}\right\}\right\}$

```
θ = N[ArcCos[1/Sqrt[5]]]
```

1.10715

The angle θ is measured in radians.

- The law of cosines for two vectors in $\mathbb{R}^3$

```
Clear[θ]
```

```
u = {1, 2, 3}; v = {-3, 4, -5};
```

```
Norm[u - v]^2 == Norm[u]^2 + Norm[v]^2 - 2 Norm[u] Norm[v] Cos[θ]
```

$84 == 64 - 20\sqrt{7}\,\text{Cos}[\theta]$

```
Solve[%, Cos[θ]]
```

$$\left\{\left\{\text{Cos}[\theta] \to -\frac{1}{\sqrt{7}}\right\}\right\}$$

$$\theta = \text{N}\left[\text{ArcCos}\left[-\frac{1}{\sqrt{7}}\right]\right]$$

```
1.95839
```

Let us use the *Mathematica* **VectorAngle** function to verify the validity of these calculations.

```
VectorAngle[u, v]
```

$$\text{ArcCos}\left[-\frac{1}{\sqrt{7}}\right]$$

As expected, the two calculations produce the same result.

Least squares

For any given linear system $A\mathbf{v} = \mathbf{b}$, the method of least squares produces a solution of the system $A^T A\mathbf{v} = A^T \mathbf{b}$. If $A\mathbf{v} = \mathbf{b}$ has a solution, then the equation $A^T A\mathbf{v} = A^T \mathbf{b}$ has the same solution. If $A\mathbf{v} = \mathbf{b}$ has no solution, then the equation $A^T A\mathbf{v} = A^T \mathbf{b}$ produces a "best-possible" approximate solution for $A\mathbf{v} = \mathbf{b}$. The *Mathematica* **LeastSquares** function can be used to find least-squares solutions.

Illustration

- A least-squares solution of an inconsistent linear system

```
system = {3 x + 7 y - z == 12, x + 2 y + 5 z == 1, 2 x - y + 2 z == 1, x + y + z == 0};
```

```
Solve[system, {x, y, z}]
```

```
{}
```

```
MatrixForm[A = {{3, 7, -1}, {1, 2, 5}, {2, -1, 2}, {1, 1, 1}}]
```

$$\begin{pmatrix} 3 & 7 & -1 \\ 1 & 2 & 5 \\ 2 & -1 & 2 \\ 1 & 1 & 1 \end{pmatrix}$$

```
MatrixForm[b = {12, 1, 1, 0}]
```

$$\begin{pmatrix} 12 \\ 1 \\ 1 \\ 0 \end{pmatrix}$$

```
LinearSolve[Transpose[A].A, Transpose[A].b]
```

$$\left\{\frac{704}{527}, \frac{543}{527}, -\frac{9}{17}\right\}$$

```
LeastSquares[A, b]
```

$$\left\{\frac{704}{527}, \frac{543}{527}, -\frac{9}{17}\right\}$$

The solutions found by the **LeastSquares** function agree with the solutions found by the **LinearSolve** function if the linear system is consistent. In other words, if the associated matrix equation Av = **b** has a solution.

- A least-squares solution of a matrix with full rank

```
Clear[b]
```

```
A = {{6, 3, 1}, {2, 0, 2}, {8, 7, 3}}; b = {1, 2, 3};
```

```
MatrixRank[A]
```

3

```
LeastSquares[A, b]
```

{0, 0, 1}

```
LinearSolve[A, b]
```

{0, 0, 1}

Manipulation

- Exploring the least-squares solutions of a linear system

```
A = {{3, 7, -1}, {1, 2, 5}, {2, -1, 2}, {1, 1, 1}}; b = {12, 1, 1, a};
```

```
Manipulate[Evaluate[LeastSquares[A, b]], {a, -3, 3, 1}]
```

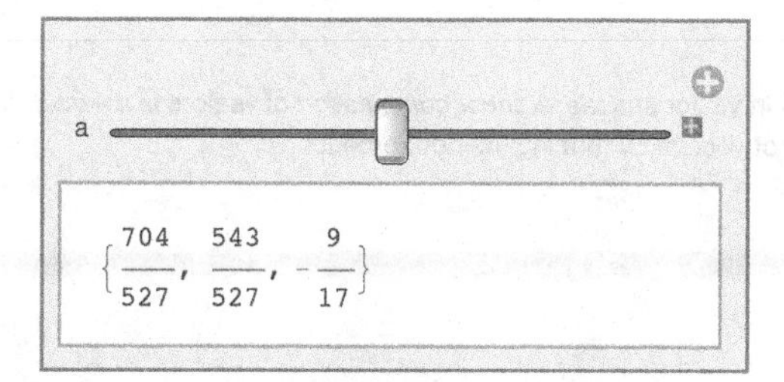

We combine **Manipulate**, **Evaluate,** and **LeastSquares** to explore the least-squares solutions of different linear systems. If let $a = 0$, for example, the displayed vector agrees with solution found earlier.

Left null space

The *left null space* of a matrix A consists of all vectors **v** for which $\mathbf{v}A = \mathbf{0}$. It is also the null space of the transpose of A.

Illustration

- A basis for the left null space of a 3-by-4 matrix

```
A = {{1, 2, 3, 4}, {5, 6, 7, 8}, {1, 1, 1, 1}};
```

```
leftnullspace = NullSpace[Transpose[A]]
```

```
{{1, -1, 4}}
```

```
k * {1, -1, 4}.A
```

```
{0, 0, 0, 0}
```

The left null space is a subspace of $\mathbb{R}^3$. Its dimension is 1 and its codimension is 3 - 1 = 2. In this example, all scalar multiples of the vector {1, -1, 4} are mapped to zero by left-multiplication.

Length of a vector

The *length* of a vector in a Euclidean vector space is based on the Pythagorean theorem. The measure is known as the Euclidean norm $\|\mathbf{v}\|$ of a vector **v**. Since the world "length" has a different meaning in *Mathematica*, the phrase "Euclidean norm" will be used in this guide.

See Euclidean norm

Linear combination

Linear combinations are the building blocks for vectors in vector spaces. A *linear combination* of vectors is a vector obtained from a list of scalars and a corresponding list of vectors by forming their dot product.

Illustration

- A linear combination of three vectors in $\mathbb{R}^2$

```
Clear[a, b, c, d, e, f]

scalars = {2, 3, 4};

vectors = {{a, b}, {c, d}, {e, f}};

lc = Dot[scalars, vectors]

{2 a + 3 c + 4 e, 2 b + 3 d + 4 f}

lc == 2 {a, b} + 3 {c, d} + 4 {e, f}

True
```

- A linear combination of vectors in $\mathbb{R}^3$

```
Clear[a, b, c, d, e, f, g, h, i]

u = {a, b, c}; v = {e, d, f}; w = {g, h, i};

lc1 = 3 u + 4 v + 5 w

{3 a + 4 e + 5 g, 3 b + 4 d + 5 h, 3 c + 4 f + 5 i}
```

- Different linear combinations representing the same vector

```
Clear[u, v, w, a, b, c]

u = {1, 2}; v = {3, 4}; w = {5, 6};

lc = a u + b v + c w;

Reduce[lc == 0, {a, b, c}]

b == -2 a && c == a

u - 2 v + w == 3 u - 6 v + 3 w

True
```

The idea of using linearly independent vectors as building blocks of linear combinations is to force the representations to be unique.

Linear dependence

A finite list of nonzero vectors is *linearly dependent* if the zero vector is a linear combination of the given vectors in which not all scalars are zero. This means that each vector in the list can be written as a linear combination of the others.

Illustration

- A linearly dependent list of vectors in $\mathbb{R}^3$

```
vectors = {{4, 6, 3}, {3, 4, 5}, {7, 10, 8}};
```

```
a {4, 6, 3} + b {3, 4, 5} + c {7, 10, 8} /. {a → -1, b → -1, c → 1}
```

```
{0, 0, 0}
```

```
{7, 10, 8} == {4, 6, 3} + {3, 4, 5}
```

```
True
```

Manipulation

- Linear dependence and independence of vectors in $\mathbb{R}^3$

```
Clear[a, b]
```

The following calculation shows that if c = 0, then the three vectors {a, 0, 0}, {0, b, 0}, and {-1, -1, 0} are linearly dependent if *a* = *b* = 1:

```
a = b = 1; c = 0;
```

```
Reduce[a {1, 0, 0} + b {0, 1, 0} + {-1, -1, c} == {0, 0, 0}]
```

```
True
```

On the other hand, the reduction

```
Reduce[a {1, 0, 0} + b {0, 1, 0} + {-a, -b, 1} == {0, 0, 0}]
```

```
False
```

shows that for all nonzero values of *a* and *b*, the three vectors {a, 0, 0}, {0, b, 0}, {-a, -b, 1} are linearly independent. For example,

```
1 {1, 0, 0} + 2 {0, 1, 0} + {-1, -1, 1} == {0, 0, 0}
```

```
False
```

is false.

We can combine these ideas into a manipulation:

```
Manipulate[Reduce[a {1, 0, 0} + b {0, 1, 0} + {-a, -b, c} == {0, 0, 0}],
 {a, -2, 2, 1}, {b, -2, 2, 1}, {c, -3, 3, 1}]
```

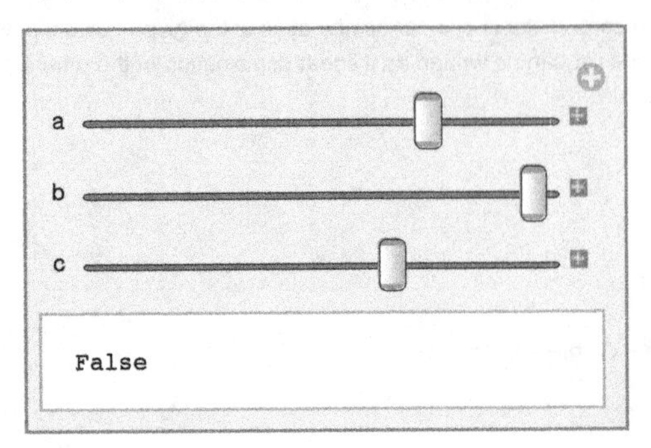

We combined **Manipulate** and **Reduce** to explore the linear dependence and independence of three generated vectors in $\mathbb{R}^3$. The manipulation shows that if $a = 1$, $b = 2$, and $c = 1$, for example, the three vectors {1, 0, 0}, {0, 2, 0}, and {- 1, - 2, 1} are linearly independent.

Linear dependence relation

A *linear dependence relation* among the columns $A_{[[\text{All},1,]]}, \ldots, A_{[[\text{All},n]]}$, of a matrix A is any nontrivial solution of the equation $x_1 A_{[[\text{All},1,]]} + \cdots + x_n A_{[[\text{All},n]]} = 0$.

Illustration

- A linear dependence relation among the columns of a 3-by-4 matrix

```
A = RandomInteger[{0, 9}, {2, 3}];
```

$$A = \begin{pmatrix} 9 & 1 & 6 \\ 5 & 7 & 7 \end{pmatrix};$$

```
dependenceRelations = Flatten[Solve[x1 A[[All,1]] + x2 A[[All,2]] + x3 A[[All,3]] == 0, {x1, x2, x3}]]
```

Solve::svars : Equations may not give solutions for all "solve" variables. ≫

$$\left\{\{a, b, c\}_2 \to \frac{33}{35}\{a, b, c\}_1,\ \{a, b, c\}_3 \to -\frac{58}{35}\{a, b, c\}_1\right\}$$

The columns of A have infinitely many dependence relations. One of them is *{-35, -33, 58}*.

```
x1 A[[All,1]] + x2 A[[All,2]] + x3 A[[All,3]] /. dependenceRelations
```

{0, 0}

Linear equation

A *linear equation* is an equation in the variables $x_1, \ldots, x_n$ and the scalars $a_0, a_1, \ldots, a_n, b$ of the form

$$a_0 + a_1 x_1 + \cdots + a_n x_n = b \tag{1}$$

Illustration

- A linear equation in three variables

```
3 x + 4 y - z == 5;
```

The *semicolon* indicates that the output of the equations is not required and should be suppressed. Otherwise both the input and output are displayed:

```
3 x + 4 y - z == 5
```

$3x+4y-z=5$

We can use dot products to create linear equations

- A linear equation created with a dot product

```
equation = Dot[{a0, a1, a2}, {1, x1, x2}] == b
```

$a_0 + a_1 x_1 + a_2 x_2 == b$

Manipulation

- Linear equations in three variables

```
Quit[]
```

```
Manipulate[Dot[{a, b, 5}, {x, y, z}] == 1, {a, 1, 5, 1}, {b, -5, 5, 1}]
```

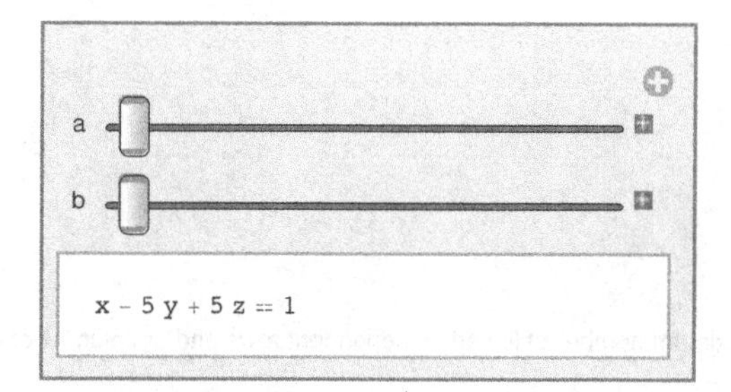

We combine **Manipulate** and **Dot** to build linear equations. If we let $a = 1$ and $b = -5$, the manipulation produces the linear equation $x - 5y + 5z = 1$.

Linear independence

A list of vectors in a vector space is *linearly independent* if the zero vector of the space can only be written as a trivial linear combination of the given vectors. A list of vectors is linearly independent if and only if it is not linearly dependent. A list of n vectors in $\mathbb{R}^n$ is linearly independent if and only if the n-by-n matrix whose rows are the given n vectors is invertible. The rank of an n-by-m matrix identifies the number of linearly independent row and/or columns of the matrix.

Illustration

- Two linearly independent vectors in $\mathbb{R}^2$

```
v = {1, 2}; w = {3, 4};
```

```
Solve[a v + b w == {0, 0}, {a, b}]
```

```
{{a → 0, b → 0}}
```

- Two linearly independent vectors in $\mathbb{R}^3$

```
v = {1, 2, 3}; w = {4, 5, 6};
```

```
Solve[a v + b w == {0, 0, 0}, {a, b}]
```

```
{{a → 0, b → 0}}
```

- Three linearly independent vectors in $\mathbb{R}^3$

```
u = {1, 2, 3}; v = {4, 5, 6}; w = {7, 8, 10};
```

```
Solve[a u + b v + c w == {0, 0, 0}, {a, b, c}]
```

```
{{a → 0, b → 0, c → 0}}
```

- Three linearly independent vectors in $\mathbb{R}^4$

```
A = {{8, 9, 0, 1}, {8, 2, 4, 9}, {7, 8, 1, 6}};
```

```
lc = a A[[1]] + b A[[2]] + c A[[3]];
```

```
Solve[lc == 0, {a, b, c}]
```

```
{{a → 0, b → 0, c → 0}}
```

- Using the **MatrixRank** function to calculate the maximum number of linearly independent rows and/or columns of a matrix

```
A = {{0, 0, 0, 0, 2, 0}, {1, 0, 0, 0, 0, 0}, {0, 0, 0, 0, 1, 0}, {1, 0, 0, 0, 1, 0}};
```

```
MatrixRank[A]
```

```
2
```

Manipulation

We can use the **Manipulate** function to explore the linear dependence and independence of vectors.

- Linear dependence and independence of two vectors in $\mathbb{R}^2$

```
Reduce[a {1, 2} + b {-1, 3} == {0, 0}, {a, b}]
```

```
a == 0 && b == 0
```

Therefore the vectors {1, 2} and {-1, 3} are linearly independent.

```
Manipulate[Evaluate[Reduce[a {2, 3} + b {4, c} == {0, 0}]],
 {a, -6, 6, 1}, {b, -6, 6, 1}, {c, -6, 6, 1}]
```

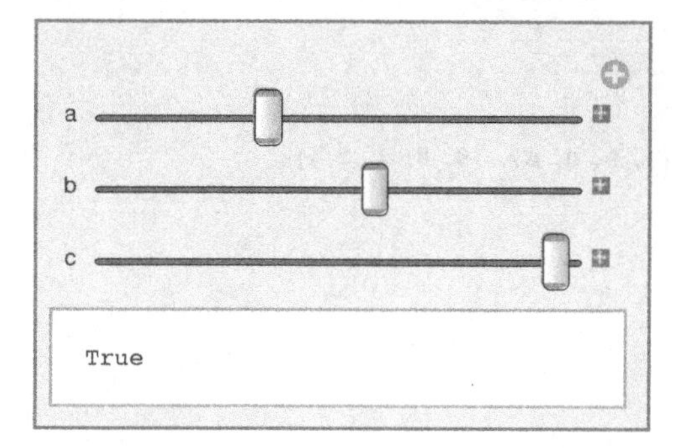

We combine **Manipulate**, **Evaluate,** and **Reduce** to generate both linearly dependent and linearly independent vectors. If we let $a = -2$, $b = 1$, and $c = 6$, the generated vectors are linearly dependent.

Linear operator

A *linear operator* T is a linear transformation from a vector space V to V. If V is finite-dimensional, then any matrix representing T is square.

Illustration

- A linear operator on $\mathbb{R}^2$

```
A = RandomInteger[{0, 9}, {2, 2}];
```

```
MatrixForm[A = {{7, 4}, {2, 5}}]
```

$$\begin{pmatrix} 7 & 4 \\ 2 & 5 \end{pmatrix}$$

We can check that A preserves linear combinations:

```
lin1 = Expand[A.(a {x, y} + b {u, v})]
```

$\{\{7\,a^2 + 23\,b + 4\,a\,e,\ 34\,b + 7\,a\,b + 4\,a\,d,\ 45\,b + 7\,a\,c + 4\,a\,f\},$
$\{2\,a^2 + 22\,b + 5\,a\,e,\ 29\,b + 2\,a\,b + 5\,a\,d,\ 36\,b + 2\,a\,c + 5\,a\,f\}\}$

```
lin2 = Expand[a A.{x, y} + b A.{u, v}]
```

$\{\{7\,a^2 + 23\,b + 4\,a\,e,\ 34\,b + 7\,a\,b + 4\,a\,d,\ 45\,b + 7\,a\,c + 4\,a\,f\},$
$\{2\,a^2 + 22\,b + 5\,a\,e,\ 29\,b + 2\,a\,b + 5\,a\,d,\ 36\,b + 2\,a\,c + 5\,a\,f\}\}$

```
lin1 == lin2
```

True

- A linear operator on ℝ[t,3]

```
A = RandomInteger[{0, 9}, {4, 4}];
```

```
MatrixForm[A = {{7, 3, 9, 4}, {1, 4, 7, 7}, {9, 9, 0, 6}, {5, 8, 2, 7}}]
```

$$\begin{pmatrix} 7 & 3 & 9 & 4 \\ 1 & 4 & 7 & 7 \\ 9 & 9 & 0 & 6 \\ 5 & 8 & 2 & 7 \end{pmatrix}$$

```
p = 1 + t - t^3;
```

```
q = 3 t + t^2;
```

```
linearcombination = Expand[4 p - 7 q]
```

$4 - 17\,t - 7\,t^2 - 4\,t^3$

```
T[p] = A.{1, 0, 0, 0}
```

{7, 1, 9, 5}

```
T[q] = A.{0, t, 0, 0}
```

{3 t, 4 t, 9 t, 8 t}

Linear system

A *linear system* is a list of linear equations $\{eq_1, \ldots, eq_n\}$. Its matrix form is a matrix equation $A\mathbf{v} = \mathbf{b}$ consisting of the coefficient matrix A, the equations $\{eq_1, \ldots, eq_n\}$, the vector **v** of variables of the system, and the vector **b** of the constant values of the equations.

Illustration

- Two equations in three unknowns

```
Clear[x, y, z, v, b]
```

```
{eq1, eq2} = {3 x + y + z == 5, x - y == 2};
```

```
A = {{3, 1, 1}, {1, -1, 0}};
```

```
v = {x, y, z};
```

```
b = {5, 2};
```

In **MatrixForm**, the matrix equation corresponding to the given linear system is

$$\begin{pmatrix} 3 & 1 & 1 \\ 1 & -1 & 0 \end{pmatrix} . \begin{pmatrix} x \\ y \\ z \end{pmatrix} = \begin{pmatrix} 5 \\ 2 \end{pmatrix}$$

In **StandardForm**, the system becomes an equation between two vectors.

```
A.v == b
```

```
{3 x + y + z, x - y} == {5, 2}
```

- A matrix converted to a linear system

A matrix that combines the coefficients of the variables and the constants of a linear system can be used to represent a linear system:

```
Clear[x, y]
```

$$A = \begin{pmatrix} 1 & 2 & 3 \\ 4 & 5 & 6 \end{pmatrix};$$

```
s = Column[{x + 2 y == 3, 4 x + 5 y == 6}, Alignment → Right]
```

```
  x + 2 y == 3
4 x + 5 y == 6
```

- Consistent linear system

A linear system $A\mathbf{v} = \mathbf{b}$ is consistent if there exists a vector v satisfying the equation.

```
A = {{3, 1, 1}, {1, -1, 0}};
```

```
v = {x, y, z};
```

```
b = {5, 2};
```

```
solution = LinearSolve[A, b]
```

$\left\{\frac{7}{4}, -\frac{1}{4}, 0\right\}$

```
A.solution == b
```

True

- A linear system with two equations with one solution

```
Clear[x, y]
```

```
system1 = {3 x + 4 y == 9, x - y == 7};
```

is a system of two linear equations in two variables. It has one solution.

```
solution1 = Flatten[Solve[system1]]
```

$\left\{x \to \frac{37}{7}, y \to -\frac{12}{7}\right\}$

```
system1 /. solution1
```

{True, True}

- A linear system with three equations and no solution

```
system2 = {3 x + 4 y == 9, x - y == 7, 4 x + y == 6};
```

is a linear system of three equations in two variables. It has no solution.

```
solution2 = Flatten[Solve[system2]]
```

{}

- A linear system in two equations and three variables with infinitely many solutions

```
system3 = {3 x + 4 y + z == 9, x - y + 2 z == 7};
```

is a linear system of two equations in three variables. It has infinitely many solutions.

```
solution3 = Flatten[Solve[system3]]
```

$\left\{y \to \frac{11}{9} - \frac{5x}{9}, z \to \frac{37}{9} - \frac{7x}{9}\right\}$

```
Simplify[system3 /. solution3]
```

{True, True}

For every value assigned to *z*, the variables *x* and *y* take on specific values provided by the output rules so that the list {*x*, *y*, z} is a solution of the system. Since there are infinitely many scalars for z, there are infinitely many lists that satisfy the given system.

- Using **Reduce** to calculate the exact solutions of linear systems

```
system = {x + 4 y == 12, x - y == 1};
```

```
Reduce[system, {x, y}]
```

$$x == \frac{16}{5} \ \&\& \ y == \frac{11}{5}$$

```
lines = ContourPlot[{x + 4 y == 12, x - y == 1}, {x, -5, 5}, {y, -5, 5}, Axes → True]
```

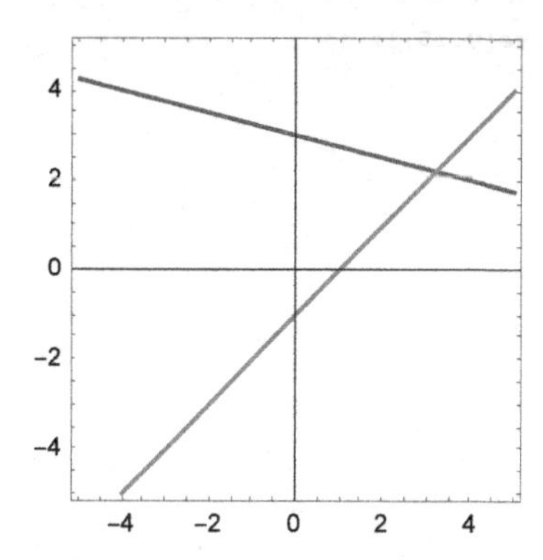

- Visualizing the solution of a linear system

The solution of a linear system with one solution is the point of intersection of the lines representing the equations in the system.

```
intersection = ListPlot[{{16 / 5, 11 / 5}}, PlotStyle → {Green, PointSize[0.05]}];
```

```
Show[lines, intersection]
```

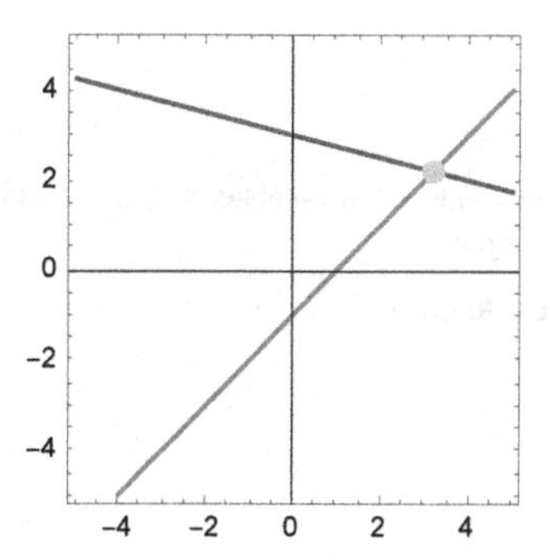

If the equations contain three variables, the situation is analogous, but a little harder to visualize.

- A linear system in three equations and three variables

```
Clear[x, y, z]
```

```
equations = {x + y + z == 5, 3 x + y - z == 9, 2 x - y + z == 1};
```

```
Solve[equations, {x, y, z}]
```

```
{{x → 2, y → 3, z → 0}}
```

The equations of the system determine three planes in $\mathbb{R}^3$ meeting at the point *{2, 3, 0}*. We can use the **ContourPlot3D** and the **ListPointPlot3D** functions to visualize the solution of the linear systems:

```
planes = ContourPlot3D[Evaluate[equations], {x, -5, 5}, {y, -5, 5}, {z, -5, 5}, Axes → True];
```

```
point = ListPointPlot3D[{{2, 3, 0}}, PlotStyle → PointSize[0.15]];
```

```
Show[planes, point]
```

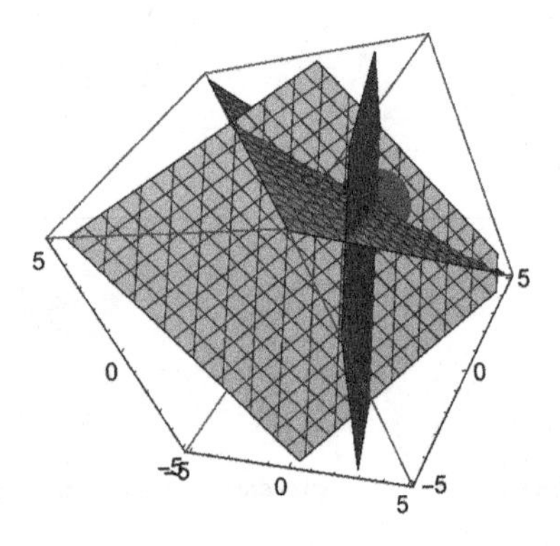

A matrix, together with a vector, can also represent a linear system. We require matrix multiplication for this purpose.

- A matrix and a vector combined to construct a linear system

```
coefficients = {{1, 2}, {4, 5}};
```

```
constants = {3, 6};
```

Using matrix multiplication, the given linear system can be built from the coefficients of the variables, written in matrix form, and the constants of the equations, written in vector form, using the **Thread** command:

```
Column[Thread[coefficients.{x, y} == constants], Alignment → Right]
```

```
  x + 2 y == 3
4 x + 5 y == 6
```

- Using matrices and **LinearSolve** to solve linear systems

```
equations = {x + 2 y == 5, 3 x + 4 y == 9};
```

```
StandardForm[coefficents = {{1, 2}, {3, 4}}];

constants = {5, 9};

LinearSolve[coefficents, constants]

{-1, 3}
```

If we replace *x* by -1 and *y* by 3 in the two equations of the given system, the first equation reduces to the identity 5 == 5 and the second to the identity 9 == 9. We therefore say that the pair {*-1, 3*} is a solution of the given system since it satisfies all of the equations in the system. In this notation it is understood that the first element of {*-1, 3*} corresponds to *x* and the second element to *y*.

The geometric interpretation of this example say that the two straight lines in $\mathbb{R}^2$ determined by the two equations of the system pass through the point {*-1. 3*}. We can use **Manipulate** to explore the impact of numerical changes in the coefficients on the solutions of linear systems.

Manipulation

The **Manipulate** function can be used to explore the solutions of systems obtained by introducing various parameters that change the numerical ranges of the variables involved.

- Exploring the solutions of linear systems

```
Clear[x, y, z, a, b, c]

equations = {x + 2 y + 3 z == 6, a x + b y + c z == 9};

coefficients = {{1, 2, 3}, {a, b, c}};

constants = {6, 9};

LinearSolve[coefficients, constants]
```

$$\left\{-\frac{6\,(-3+b)}{2\,a-b},\ \frac{3\,(-3+2\,a)}{2\,a-b},\ 0\right\}$$

```
Manipulate[Evaluate[Reduce[equations, {x, y, z}]], {a, -3, 3, 1}, {b, -3, 3, 1}, {c, -3, 3, 1}]
```

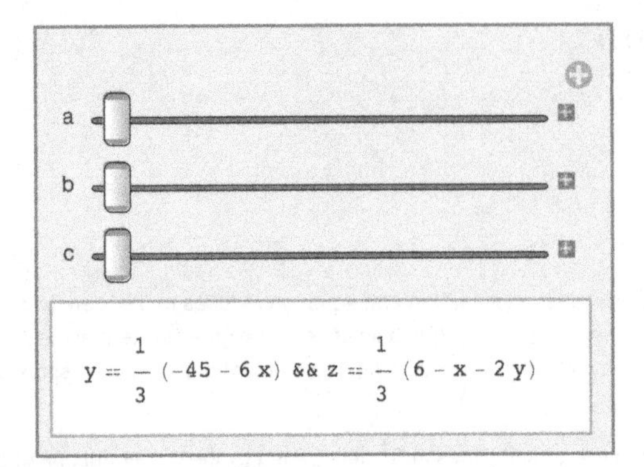

We combine the **Manipulate**, **Evaluate,** and **Reduce** commands to explore the solutions of different linear systems. The displayed solutions are obtained by letting $a = b = c = -3$.

Linear transformation

A function from one vector space to another is called *linear transformation* if it preserves linear combinations. All real m-by-n matrices determine linear transformations from $\mathbb{R}^n$ to $\mathbb{R}^m$, and all complex m-by-n matrices determine linear transformations from $\mathbb{C}^n$ to $\mathbb{C}^m$.

The matrix representations of the functions defining linear transformations are basis-dependent. The same linear transformations have different matrix representations if different bases are used for their definitions. In this guide we use standard bases most of the time to illustrate the ideas involved in the constructions. However, *orthonormal bases* also play a key role in the matrix representation of linear transformations.

The *image* of a transformation T from a vector space V to a vector space W is the set of all vectors T[**w**] in W and the *kernel* of T is the set all vectors **v** in V that are mapped to **0** in W by T.

Linear combinations are preserved by a transformation T if, for any vectors **u** and **v**, and scalars s and t:

```
T[s u + t v] = s T[u] + t T[v]
```
(1)

Illustration

- A linear transformation from $\mathbb{R}^{2\times3}$ to $\mathbb{R}^{2\times2}$

```
Clear[u, v, a, b, c, d, e, f]
```

```
MatrixForm[A = {{1, 2, 3}, {3, 4, 5}}]
```

$$\begin{pmatrix} 1 & 2 & 3 \\ 3 & 4 & 5 \end{pmatrix}$$

```
B = {{a, b}, {c, d}, {e, f}};
```

```
T[B_] := A.B
```

```
u = RandomInteger[{0, 9}, {3, 2}];
```

```
u = (4 3
     9 6
     0 3);

v = RandomInteger[{0, 9}, {3, 2}];

v = (1 2
     1 7
     2 7);

Simplify[T[3 u + 5 v] == 3 T[u] + 5 T[v]]

True
```

- A linear transformation from $\mathbb{R}^4$ to $\mathbb{R}[t,3]$

```
T[{a_, b_, c_, d_}] := a + b t + c t^2 + d t^3

T[{1, 2, 3, 4}]

1 + 2 t + 3 t^2 + 4 t^3

Plot[1 + 2 t + 3 t^2 + 4 t^3, {t, -4.5, 4.5}]
```

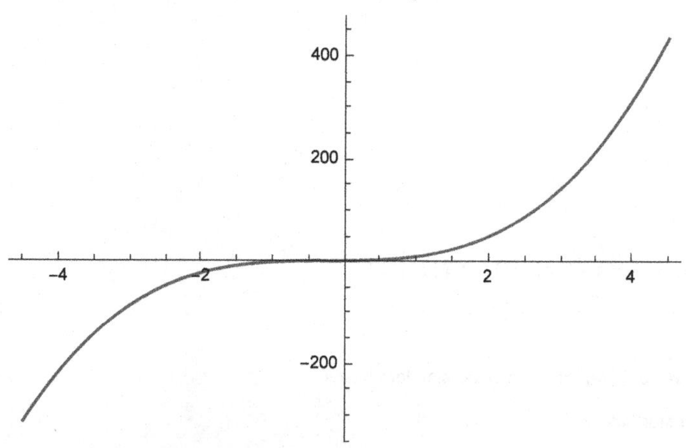

- A linear transformation from $\mathbb{R}^{3\times3}$ to $\mathbb{R}$

```
Clear[A, T, a, b, c, d, e, f, g, h, i]

T[A_] := Tr[A]

A = {{1, 2, 3}, {4, 5, 6}, {7, 8, 9}};

B = {{a, b, c}, {d, e, f}, {g, h, i}};
```

```
T[A]
```

```
15
```

```
T[B]
```

```
a + e + i
```

```
Clear[x, y]
```

```
Simplify[T[x A + y B] == x T[A] + y T[B]]
```

```
True
```

- A linear transformation from $\mathbb{R}^2$ to $\mathbb{R}^3$

```
Clear[x, y]
```

```
MatrixForm[A = {{1, 2}, {3, 4}, {5, 6}}]
```

$$\begin{pmatrix} 1 & 2 \\ 3 & 4 \\ 5 & 6 \end{pmatrix}$$

```
T[{x_, y_}] := A.{x, y}
```

```
T[{x, y}]
```

```
{x + 2 y, 3 x + 4 y, 5 x + 6 y}
```

```
Clear[a, b, w, x, y, z]
```

```
Simplify[T[a {x, y} + b {w, z}] == a T[{x, y}] + b T[{w, z}]]
```

```
True
```

- A linear transformation from $\mathbb{R}^3$ to $\mathbb{R}^2$ represented in the standard bases

Consider the following linear transformation:

```
T[{x_, y_, z_}] := {3 x + 2 y - 4 z, x - 5 y + 3 z}
```

```
T[{1, 2, 3}]
```

```
{-5, 0}
```

```
Clear[a, b, x, y, z]
```

```
Simplify[T[a {1, 2, 3} + b {x, y, z}] == a T[{1, 2, 3}] + b T[{x, y, z}]]
```

```
True
```

We represent *T* by a matrix *A* in the standard basis *sB* of $\mathbb{R}^3$ (basis for the domain of *T*) and the standard basis *sC* of $\mathbb{R}^2$ (basis for the codomain of *T*).

```
T[{x_, y_, z_}] := {3 x + 2 y - 4 z, x - 5 y + 3 z}
```

```
sB = {x1 = {1, 0, 0}, x2 = {0, 1, 0}, x3 = {0, 0, 1}};
```

```
sC = {y1 = {1, 0}, y2 = {0, 1}};
```

```
MatrixForm[A = Transpose[Join[{T[x1], T[x2], T[x3]}, 2]]]
```

$$\begin{pmatrix} 3 & 2 & -4 \\ 1 & -5 & 3 \end{pmatrix}$$

```
A.{a, b, c}
```

```
{3 a + 2 b - 4 c, a - 5 b + 3 c}
```

```
T[{a, b, c}] == A.{a, b, c}
```

```
True
```

- A linear transformation from $\mathbb{R}^3$ to $\mathbb{R}^2$ represented in two nonstandard bases

Consider the following linear transformation:

```
T[{x_, y_, z_}] := {3 x + 2 y - 4 z, x - 5 y + 3 z}
```

We represent *T* by a matrix *A* in a nonstandard basis *nB* of $\mathbb{R}^3$ (basis for the domain of *T*) and a nonstandard basis *nC* of $\mathbb{R}^2$ (basis for the codomain of *T*).

```
nB = {x1 = {1, 1, 1}, x2 = {1, 1, 0}, x3 = {1, 0, 0}};
```

```
nC = {y1 = {1, 3}, y2 = {2, 5}};
```

```
Solve[T[x1] == a y1 + b y2, {a, b}]
```

```
{{a → -7, b → 4}}
```

```
Solve[T[x2] == c y1 + d y2, {c, d}]
```

```
{{c → -33, d → 19}}
```

```
Solve[T[x3] == e y1 + f y2, {e, f}]
```

```
{{e → -13, f → 8}}
```

```
A = ( -7  -33  -13
       4   19    8 );
```

```
Solve[{1, 2, 3} == g x1 + h x2 + k x3, {g, h, k}]
```

```
{{g → 3, h → -1, k → -1}}
```

Hence the vector *{1, 2, 3}*, written in the basis *nB*, is *{3, -1, -1}*. We now multiply the matrix *A* (written in the two nonstandard bases) by the coordinate vectors of *{1, 2, 3}* to find its image in the basis *nC*:

```
Flatten[( -7  -33  -13     ( 3
           4   19    8 ) .  -1
                            -1 )]
```

```
{25, -15}
```

```
Clear[r, s]
```

```
Solve[25 {1, 3} - 15 {2, 5} == r {1, 0} + s {0, 1}, {r, s}]
```

```
{{r → -5, s → 0}}
```

As we can see, the image vector *{25, -15}* in the basis *nC* is the same as the image vector *{-5, 0}* in the standard basis.

Relative to suitable bases, the matrix *[T]* for a linear transformation *T* may have a particularly simple form. It may, for example, be diagonal, triangular, or symmetric.

- A linear transformation represented by a symmetric matrix

Let *T* be a linear transformation from $\mathbb{R}^3$ to $\mathbb{R}^3$ and let its values $T[\mathbf{e}_1]$, $T[\mathbf{e}_2]$, and $T[\mathbf{e}_3]$ have the following values on the standard basis of $\mathbb{R}^3$:

```
sB = {e1 = {1, 0, 0}, e2 = {0, 1, 0}, e3 = {0, 0, 1}};
```

```
T[e1] = 2 e1 + 3 e2 + 5 e3
```

```
{2, 3, 5}
```

```
T[e2] = 3 e1 - 7 e2 + 6 e3
```

```
{3, -7, 6}
```

```
T[e3] = 5 e1 + 6 e2 - 9 e3
```

```
{5, 6, -9}
```

Then the matrix representing *T* in the standard basis for both the domain and codomain of *T* is the symmetric matrix

```
A = ( 2   3   5
      3  -7   6
      5   6  -9 );
```

```
u = {1, 2, 3};
```

```
A.u
```

```
{23, 7, -10}
```

- A coordinate conversion matrix

If T is an identity transformation, its matrix in the standard basis *sB* and a nonstandard basis *nB* is called a *coordinate conversion* or *change-of-basis* matrix from *sB* to *nB*.

```
sB = {{1, 0, 0}, {0, 1, 0}, {0, 0, 1}};
```

```
nB = {{1, 2, 3}, {0, 1, 0}, {1, 0, 1}};
```

Calculating the coordinates of the vector $\mathbf{x}_{sB}$ in the standard basis *sB*:

```
xsB = {1, 1, 1};
```

```
Solve[xsB == a B[[1]] + b B[[2]] + c B[[3]], {a, b, c}]
```

```
{{a → 0, b → 1, c → 1}}
```

```
xnB = {0, 1, 1};
```

The coordinate conversion matrix *A* from *sB* to *nB* provides a formula for converting any vector $\mathbf{x}_{nB}$ in *sB* to its corresponding vector $\mathbf{x}_{sB}$ in the nonstandard basis *nB*.

```
Solve[{1, 0, 0} == a {1, 2, 3} + b {0, 1, 0} + c {1, 0, 1}, {a, b, c}]
```

$$\left\{\left\{a \to -\frac{1}{2},\ b \to 1,\ c \to \frac{3}{2}\right\}\right\}$$

```
Solve[{0, 1, 0} == a {1, 2, 3} + b {0, 1, 0} + c {1, 0, 1}, {a, b, c}]
```

```
{{a → 0, b → 1, c → 0}}
```

```
Solve[{0, 0, 1} == a {1, 2, 3} + b {0, 1, 0} + c {1, 0, 1}, {a, b, c}]
```

$$\left\{\left\{a \to \frac{1}{2},\ b \to -1,\ c \to -\frac{1}{2}\right\}\right\}$$

```
MatrixForm[A = Transpose[{{-1/2, 1, 3/2}, {0, 1, 0}, {1/2, -1, -1/2}}]]
```

$$\begin{pmatrix} -\frac{1}{2} & 0 & \frac{1}{2} \\ 1 & 1 & -1 \\ \frac{3}{2} & 0 & -\frac{1}{2} \end{pmatrix}$$

```
A.xsB == xnB
```

```
True
```

Manipulation

- Exploring the values of a linear transformation

```
T[{x_, y_, z_}] := {3 x + 2 y - 4 z, x - 5 y + 3 z};
```

```
Manipulate[Evaluate[ T[{x, y, z}]], {x, -2, 2, 1}, {y, -2, 2, 1}, {z, -2, 2, 1}]
```

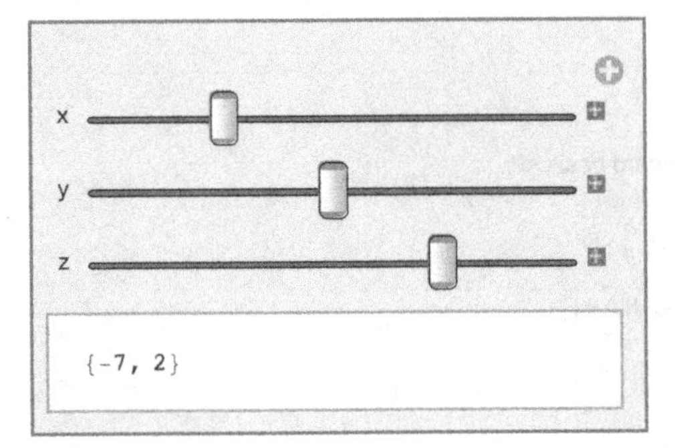

We use **Manipulate** and **Evaluate** to explore the values of a linear transformation. The displayed value of T is obtained by letting $x = -1$, $y = 0$, and $z = 1$.

Lower-triangular matrix

An n-by-n matrix $A = A_{[[i,j]]}$ is lower-triangular if $A_{[[i,j]]} = 0$ for all $i < j$. That is, if all entries above the main diagonal are 0.

Illustration

- A lower-triangular matrix

```
MatrixForm[A = {{1, 0, 0, 0}, {2, 5, 0, 0}, {3, 6, 8, 0}, {4, 7, 9, 10}}]
```

$$\begin{pmatrix} 1 & 0 & 0 & 0 \\ 2 & 5 & 0 & 0 \\ 3 & 6 & 8 & 0 \\ 4 & 7 & 9 & 10 \end{pmatrix}$$

- A lower-triangular matrix obtained by triangularization

```
MatrixForm[A = RandomInteger[{0, 9}, {5, 5}]]
```

$$\begin{pmatrix} 0 & 7 & 7 & 1 & 4 \\ 7 & 3 & 0 & 5 & 7 \\ 8 & 9 & 8 & 7 & 1 \\ 1 & 3 & 6 & 3 & 3 \\ 0 & 3 & 0 & 3 & 0 \end{pmatrix}$$

```
MatrixForm[B = LowerTriangularize[A]]
```

$$\begin{pmatrix} 0 & 0 & 0 & 0 & 0 \\ 7 & 3 & 0 & 0 & 0 \\ 8 & 9 & 8 & 0 & 0 \\ 1 & 3 & 6 & 3 & 0 \\ 0 & 3 & 0 & 3 & 0 \end{pmatrix}$$

LU decomposition

An *LU decomposition* of a matrix *A* is a product of a lower-triangular matrix *L* and an upper-triangular matrix *U*. If *A* is an *m*-by-*n* matrix that can be reduced to row echelon form without requiring a permutation of rows then there exist a lower-triangular matrix *L* with *1*s on the diagonal and an *m*-by-*n* row echelon matrix *U* such that *A* = *LU*. For every real *m*-by-*n* matrix *A* there exists a permutation matrix *P* for which the matrix product *PA* can be reduced to row echelon form without a permutation of rows.

Illustration

- An *LU* decomposition of a 3-by-3 matrix

```
A = {{1, 2, 3}, {2, 5, 7}, {2, 4, 1}};

L = {{1, 0, 0}, {2, 1, 0}, {2, 0, 1}};

U = {{1, 2, 3}, {0, 1, 1}, {0, 0, -5}};

A == L.U

True
```

- An *LU* decomposition of a 3-by-3 matrix using Mathematica

```
A = {{1, 2, 3}, {2, 5, 7}, {2, 4, 1}};

{lu, p, c} = LUDecomposition[A]

{{{1, 2, 3}, {2, 1, 1}, {2, 0, -5}}, {1, 2, 3}, 1}
```

The *lu* component combines the matrices *L* and *U*, the *p* component specifies the permutation of the rows of the matrix required (none in this example), and the 1 component is a condition number of the matrix. A condition number close to 1 indicates that the matrix is well conditioned so that its inverse can be computed with good accuracy.

The matrices *L* and *U* can be extracted from **lu** using sparse arrays.

```
L = lu SparseArray[{i_, j_} /; j < i → 1, {3, 3}] + IdentityMatrix[3]

{{1, 0, 0}, {2, 1, 0}, {2, 0, 1}}
```

```
U = UpperTriangularize[lu]
```

```
{{1, 2, 3}, {0, 1, 1}, {0, 0, -5}}
```

```
L.U
```

$$\begin{pmatrix} 1 & 2 & 3 \\ 2 & 5 & 7 \\ 2 & 4 & 1 \end{pmatrix}$$

- An *LU* decomposition requiring a permutation of rows

```
MatrixForm[A = {{1, 2, 3}, {2, 4, 1}, {2, 5, 7}}]
```

$$\begin{pmatrix} 1 & 2 & 3 \\ 2 & 4 & 1 \\ 2 & 5 & 7 \end{pmatrix}$$

```
{lu, p, c} = LUDecomposition[A]
```

```
{{{1, 2, 3}, {2, 1, 1}, {2, 0, -5}}, {1, 3, 2}, 0}
```

```
MatrixForm[L = lu SparseArray[{i_, j_} /; j < i → 1, {3, 3}] + IdentityMatrix[3]]
```

$$\begin{pmatrix} 1 & 0 & 0 \\ 2 & 1 & 0 \\ 2 & 0 & 1 \end{pmatrix}$$

```
MatrixForm[U = UpperTriangularize[lu]]
```

$$\begin{pmatrix} 1 & 2 & 3 \\ 0 & 1 & 1 \\ 0 & 0 & -5 \end{pmatrix}$$

```
MatrixForm[P = {{1, 0, 0}, {0, 0, 1}, {0, 1, 0}}]
```

$$\begin{pmatrix} 1 & 0 & 0 \\ 0 & 0 & 1 \\ 0 & 1 & 0 \end{pmatrix}$$

The matrix P is created by permuting rows *2* and *3* of the identity matrix.

```
A == P.L.U
```

```
True
```

$$A == \begin{pmatrix} 1 & 2 & 3 \\ 2 & 4 & 1 \\ 2 & 5 & 7 \end{pmatrix} == \begin{pmatrix} 1 & 0 & 0 \\ 0 & 0 & 1 \\ 0 & 1 & 0 \end{pmatrix} . \begin{pmatrix} 1 & 0 & 0 \\ 2 & 1 & 0 \\ 2 & 0 & 1 \end{pmatrix} . \begin{pmatrix} 1 & 2 & 3 \\ 0 & 1 & 1 \\ 0 & 0 & -5 \end{pmatrix} == P.L.U$$

```
True
```

M

Manhattan distance

The *Manhattan distance* between two vectors (city blocks) is equal to the one-norm of the distance between the vectors. The distance function (also called a "metric") involved is also called the "taxi cab" metric.

Illustration

- The Manhattan distance as the sum of absolute differences

```
ManhattanDistance[{a, b, c}, {x, y, z}]
```

```
Abs[a - x] + Abs[b - y] + Abs[c - z]
```

- The one-norm as Manhattan distance between two city blocks

```
block1 = {1, 2, 3, 4}; block2 = {5, 6, 7, 8};
```

```
Norm[block1 - block2, 1]
```

```
16
```

- The Manhattan length of two blocks

```
block1 = {5, 2, -3, 4}; block2 = {1, 6, -7, 8};
```

```
{Norm[block1, 1], Norm[block2, 1]}
```

```
{14, 22}
```

Markov matrix

See Stochastic matrix

Mathematica domain of a scalar

Mathematica classifies scalars into three domains: integers, reals, and complexes. They can be used for specifying assumptions about variables.

Illustration

$$\left\{3 \in \text{Integers}, \frac{1}{3} \in \text{Integers}, \frac{1}{3} \in \text{Rationals}, \pi \in \text{Integers}, \pi \in \text{Reals}, i \in \text{Reals}, i \in \text{Complexes}\right\}$$

```
{True, False, True, False, True, False, True}
```

Matrix

In its simplest form, a *matrix* is an array of scalars arranged in rows and columns. If the scalars are real numbers, we call the matrix a *real matrix* and if the scalars are complex numbers, we call it a *complex matrix*. The scalars in a matrix are called the elements of the matrix.

In *Mathematica*, the standard form of a matrix is represented internally in one-dimensional form as a list of rows. The **StandardForm** command is the default *Mathematica* command for representing matrices.

Mathematica can also be forced to output a matrix in the usual two-dimensional form by embedding the assignment statement in a **MatrixForm** or **TraditionalForm** command. Depending on the context, it might be useful to make the **TraditionalForm** option the default output option. This can be done in the *Mathematica* Preferences. As in the case of vectors, the **StandardForm** option produces rows of rows, and the **MatrixForm** and **TraditionalForm** options produce columns.

Illustration

- A matrix in standard form

$$\text{StandardForm}\left[A = \begin{pmatrix} 1 & 2 & 3 \\ 4 & 5 & 6 \end{pmatrix}\right]$$

```
{{1, 2, 3}, {4, 5, 6}}
```

- A matrix in traditional form

$$\text{MatrixForm}\left[A = \begin{pmatrix} 1 & 2 & 3 \\ 4 & 5 & 6 \end{pmatrix}\right]$$

$$\begin{pmatrix} 1 & 2 & 3 \\ 4 & 5 & 6 \end{pmatrix}$$

```
MatrixForm[A = {{7, 5, 2, 2, 4}, {8, 4, 5, 0, 9}, {6, 1, 5, 3, 9}}]
```

$$\begin{pmatrix} 7 & 5 & 2 & 2 & 4 \\ 8 & 4 & 5 & 0 & 9 \\ 6 & 1 & 5 & 3 & 9 \end{pmatrix}$$

Mathematica has several built-in symbols for generating matrices.

- A 2-by-3 matrix

```
A = {{1, 2, 3}, {4, 5, 6}}
```

$$\begin{pmatrix} 1 & 2 & 3 \\ 4 & 5 & 6 \end{pmatrix}$$

Since all built-in *Mathematica* symbols begin with capital letters, naming user-defined objects such as vectors, systems of equations, and matrices is best done by using either single or multiple lower-case letters.

- A 2-by-3 array of dimensions {2,3}

The array

$$A = \begin{pmatrix} 1 & 2 & 3 \\ 4 & 5 & 6 \end{pmatrix};$$

is a matrix consisting of the two rows

```
row1 = ( 1  2  3 ); row2 = ( 4  6  6 );
```

and the three columns

$$\text{column1} = \begin{pmatrix} 1 \\ 4 \end{pmatrix}; \text{column2} = \begin{pmatrix} 2 \\ 5 \end{pmatrix}; \text{column3} = \begin{pmatrix} 3 \\ 6 \end{pmatrix};$$

The number of rows and columns of *A* are called its *dimensions*.

```
Dimensions[{{1, 2, 3}, {4, 5, 6}}]
```

{2, 3}

The matrix has two rows and three columns.

- A 2-by-3 matrix verified abstractly to be a matrix

```
A = {{1, 2, 3}, {4, 5, 6}}
```

$$\begin{pmatrix} 1 & 2 & 3 \\ 4 & 5 & 6 \end{pmatrix}$$

```
MatrixQ[A]
```

True

Although this calculation tells us that, as expected, the given *list of lists* is a matrix, a more informative question might be whether the rows and columns are lists of the same lengths, as expected.

- A 2-by-3 matrix with three columns of height 2

```
MatrixForm[A = {{1, 2, 3}, {4, 5, 6}}]
```

$$\begin{pmatrix} 1 & 2 & 3 \\ 4 & 5 & 6 \end{pmatrix}$$

```
columns = {Length[A[[All, 1]]], Length[A[[All, 2]]], Length[A[[All, 2]]]}
```

```
{2, 2, 2}
```

More compactly, we can use the **Table** function.

```
columns = Table[Length[A[[All, i]]] == 2, {i, 1, 3}]
```

```
{True, True, True}
```

We could even use an if–then construction.

```
If[x == {1, 2, 3}, Length[A[[All, x]]] == 2, "Does not apply"]
```

```
True
```

We can also identify the length of each row separately:

```
rows = {Length[A[[1]]], Length[A[[2]]]}
```

```
{3, 3}
```

- A *3*-row array that is not a matrix

The **MatrixForm** option fails since *A* is not a matrix:

```
MatrixForm[A = {{1, 2, 3}, {4, 5}, {6, 7, 8}}]
```

$$\begin{pmatrix} \{1, 2, 3\} \\ \{4, 5\} \\ \{6, 7, 8\} \end{pmatrix}$$

```
Dimensions[A]
```

```
{3}
```

- A 3-by-3 array with pairs of integers as elements, generated by the **Table** command

```
MatrixForm[A = Table[{n, m}, {n, 1, 3}, {m, 1, 3}]]
```

$$\begin{pmatrix} \begin{pmatrix} 1 \\ 1 \end{pmatrix} & \begin{pmatrix} 1 \\ 2 \end{pmatrix} & \begin{pmatrix} 1 \\ 3 \end{pmatrix} \\ \begin{pmatrix} 2 \\ 1 \end{pmatrix} & \begin{pmatrix} 2 \\ 2 \end{pmatrix} & \begin{pmatrix} 2 \\ 3 \end{pmatrix} \\ \begin{pmatrix} 3 \\ 1 \end{pmatrix} & \begin{pmatrix} 3 \\ 2 \end{pmatrix} & \begin{pmatrix} 3 \\ 3 \end{pmatrix} \end{pmatrix}$$

- A 3-by-3 array with indexed labels as elements, generated by the **Table** command

```
MatrixForm[A = Table[a[n, m], {n, 1, 3}, {m, 1, 3}]]
```

$$\begin{pmatrix} a[1,1] & a[1,2] & a[1,3] \\ a[2,1] & a[2,2] & a[2,3] \\ a[3,1] & a[3,2] & a[3,3] \end{pmatrix}$$

- Another way of generating a table with indexed elements, generated by the **Array** command

```
MatrixForm[A = Array[a, {3, 3}]]
```

$$\begin{pmatrix} a[1,1] & a[1,2] & a[1,3] \\ a[2,1] & a[2,2] & a[2,3] \\ a[3,1] & a[3,2] & a[3,3] \end{pmatrix}$$

- A 3-by-4 array with indexed elements, generated by a pure (unnamed) function

```
MatrixForm[A = Array[a#1,#2 &, {3, 4}]]
```

$$\begin{pmatrix} a_{1,1} & a_{1,2} & a_{1,3} & a_{1,4} \\ a_{2,1} & a_{2,2} & a_{2,3} & a_{2,4} \\ a_{3,1} & a_{3,2} & a_{3,3} & a_{3,4} \end{pmatrix}$$

- A sparse array with elements suppressed and then revealed using the **MatrixForm** option

```
S = SparseArray[{{1, 1} → 1, {2, 2} → 2, {3, 3} → 3, {1, 3} → 4}]
```

SparseArray[<4>, {3, 3}]

```
SparseArray["<"4">", {3, 3}]
```

```
A = MatrixForm[S]
```

$$\begin{pmatrix} 1 & 0 & 4 \\ 0 & 2 & 0 \\ 0 & 0 & 3 \end{pmatrix}$$

- A 3-by-2 matrix generated by the **Partition** command

```
MatrixForm[A = Partition[{1, 2, 3, 4, 5, 6}, 2]]
```

$$\begin{pmatrix} 1 & 2 \\ 3 & 4 \\ 5 & 6 \end{pmatrix}$$

- A 3-by-3 array with comma-separated indexes, generated by the **Subscript** function

```
A = Array[Subscript[a, #1, #2] &, {3, 3}] // MatrixForm
```

$$\begin{pmatrix} a_{1,1} & a_{1,2} & a_{1,3} \\ a_{2,1} & a_{2,2} & a_{2,3} \\ a_{3,1} & a_{3,2} & a_{3,3} \end{pmatrix}$$

- The same array with comma-separated indexes, generated by the **Table** command

```
MatrixForm[A = Table[a_{n,m}, {n, 1, 3}, {m, 1, 3}]]
```

$$\begin{pmatrix} a_{1,1} & a_{1,2} & a_{1,3} \\ a_{2,1} & a_{2,2} & a_{2,3} \\ a_{3,1} & a_{3,2} & a_{3,3} \end{pmatrix}$$

- A matrix with variable elements

```
A = {{1, 2}, {4, 5}};
```

```
MatrixForm[A - t IdentityMatrix[2]]
```

$$\begin{pmatrix} 1 - t & 2 \\ 4 & 5 - t \end{pmatrix}$$

- A matrix with polynomial elements

```
A = {{1, 2}, {0, 4}}; id2 = IdentityMatrix[2];
```

```
MatrixForm[cA = A - id2 x]
```

$$\begin{pmatrix} 1 - x & 2 \\ 0 & 4 - x \end{pmatrix}$$

Matrix addition

See Addition of matrices

Matrix decomposition

Matrix decomposition involves expressing a matrix as a product of two or more matrices of a special type. Examples include the *LU* decomposition (expressing a matrix as a product of an upper- and lower-triangular matrix) and singular value decomposition.

Illustration

```
A = {{1, 2}, {3, 4}};
```

- Singular value decomposition

```
{u, w, v} = N[SingularValueDecomposition[A]]
```

```
{{{0.404554, 0.914514}, {0.914514, -0.404554}},
 {{5.46499, 0.}, {0., 0.365966}}, {{0.576048, -0.817416}, {0.817416, 0.576048}}}
```

```
A == u.w.Transpose[v]
```

```
True
```

- Diagonal decomposition

```
{evalsA, evecsA} = Eigensystem[A]
```

$$\left\{\left\{\frac{1}{2}\left(5+\sqrt{33}\right), \frac{1}{2}\left(5-\sqrt{33}\right)\right\}, \left\{\left\{\frac{1}{6}\left(-3+\sqrt{33}\right), 1\right\}, \left\{\frac{1}{6}\left(-3-\sqrt{33}\right), 1\right\}\right\}\right\}$$

```
A == Transpose[evecsA].DiagonalMatrix[evalsA].Inverse[Transpose[evecsA]]
```

```
True
```

Matrix equation

A *matrix equation* (also called a matrix-vector equation) is an equation of the form $A\mathbf{v} = \mathbf{b}$, where A is an m-by-n matrix, called the *coefficient matrix*, $\mathbf{v}$ is an n-by-1 column vector, and $\mathbf{b}$ is an m-by-1 column vector.

Illustration

- A matrix equation with a 2-by-3 coefficient matrix

```
A = RandomInteger[{0, 9}, {2, 3}];
```

$$A = \begin{pmatrix} 6 & 2 & 4 \\ 6 & 4 & 3 \end{pmatrix};$$

```
v = {{1}, {2}, {3}}; b = {{22}, {23}};
```

$$\begin{pmatrix} 6 & 2 & 4 \\ 6 & 4 & 3 \end{pmatrix}.v == b$$

```
True
```

- A matrix equation with a 3-by-2 coefficient matrix

```
A = RandomInteger[{0, 9}, {3, 2}];
```

$$A = \begin{pmatrix} 7 & 2 \\ 4 & 2 \\ 4 & 5 \end{pmatrix};$$

```
v = {{1}, {2}}; b = {{2}, {3}, {4}};
```

$$\begin{pmatrix} 7 & 2 \\ 4 & 2 \\ 4 & 5 \end{pmatrix}.v == b$$

```
False
```

Matrix norm

See Norm

Matrix space

A *matrix space* $\mathbb{R}^{m \times n}$ (or $\mathbb{C}^{m \times n}$ for complex numbers) is a vector space of dimension $m \times n$ that consists of the set of all m-by-n real (or complex) matrices.

Illustration

- Vector addition in $\mathbb{R}^{3 \times 4}$

```
A = RandomInteger[{0, 9}, {3, 4}];
```

$$\mathbf{A} = \begin{pmatrix} 2 & 8 & 7 & 1 \\ 0 & 3 & 7 & 9 \\ 3 & 0 & 0 & 4 \end{pmatrix};$$

```
B = RandomInteger[{0, 9}, {3, 4}];
```

$$\mathbf{B} = \begin{pmatrix} 0 & 0 & 7 & 1 \\ 4 & 3 & 2 & 4 \\ 4 & 3 & 1 & 8 \end{pmatrix};$$

```
MatrixForm[A + B]
```

$$\begin{pmatrix} 2 & 8 & 14 & 2 \\ 4 & 6 & 9 & 13 \\ 7 & 3 & 1 & 12 \end{pmatrix}$$

- Scalar multiplication in $\mathbb{R}^{3 \times 4}$

```
MatrixForm[5 A]
```

$$\begin{pmatrix} 10 & 40 & 35 & 5 \\ 0 & 15 & 35 & 45 \\ 15 & 0 & 0 & 20 \end{pmatrix}$$

Matrix-vector product

A *matrix-vector product* $A\mathbf{v}$ is a vector obtained by multiplying an m-by-n matrix A by an n-by-1 column vector $\mathbf{v}$.

Illustration

- Using a matrix equation to solve a linear system

```
system = {6 x + 2 y == 5, 3 x + 4 y == 9};
```

```
matrixequation = MatrixForm[{{6, 2}, {3, 4}}].MatrixForm[{{x}, {y}}] == MatrixForm[{{5}, {9}}]
```

$$\begin{pmatrix} 6 & 2 \\ 3 & 4 \end{pmatrix} . \begin{pmatrix} x \\ y \end{pmatrix} == \begin{pmatrix} 5 \\ 9 \end{pmatrix}$$

- Extracting a matrix-vector product from a linear system

```
system = {5 x - 7 y + 9 z == 4, x + y - 3 z == 5};
```

```
matrixvectorproduct = {{5, -7, 9}, {1, 1, -3}}.{x, y, z}
```

$\{5x-7y+9z, x+y-3z\}$

- Building and solving a linear system using a matrix-vector product

```
system = {5 x - 7 y + 9 z == 4, x + y - 3 z == 5, 2 x - y + 9 z == 8};
```

```
matrixvectorproduct = {{5, -7, 9}, {1, 1, -3}, {2, -1, 9}}.{x, y, z}
```

$\{5x-7y+9z, x+y-3z, 2x-y+9z\}$

```
newsystem = matrixvectorproduct == {4, 5, 8}
```

$\{5x-7y+9z, x+y-3z, 2x-y+9z\} = \{4, 5, 8\}$

```
Reduce[newsystem, {x, y, z}]
```

$$x = \frac{65}{18} \bigwedge y = \frac{89}{36} \bigwedge z = \frac{13}{36}$$

Manipulation

The **Manipulate** function can be used to explore the numerical properties of the solution of a linear system built from a matrix-vector equation.

```
Manipulate[Evaluate[Reduce[{{5, -7, 9 a}, {1, b, -3}, {2 c, -1, 9}}.{x, y, z} == {4, 5, 8}]],
  {a, 1, 5}, {b, 1, 2}, {c, 1, 9}]
```

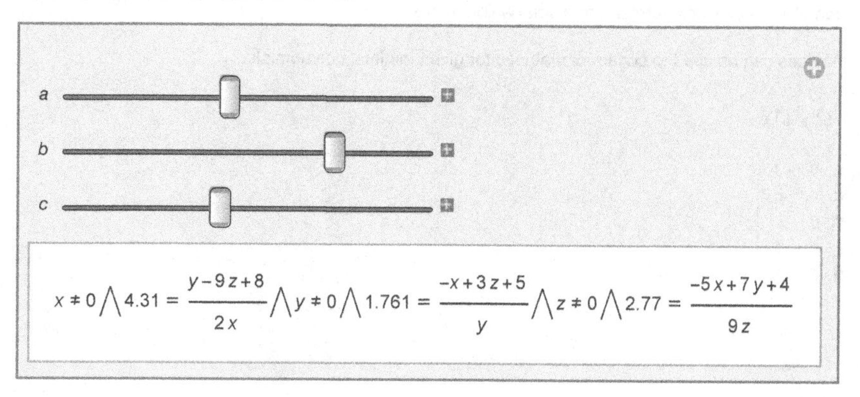

We use **Manipulate**, **Evaluate,** and **Reduce** to display the solutions of the matrix equation obtained by letting a = 2.77, b = 1.761, and c = 4.31.

Minimal polynomial

Every characteristic polynomial of a matrix has a *monic* polynomial (i.e., with leading coefficient equal to *1*) of least degree that also satisfies the matrix. The polynomial is called the *minimal polynomial* of the matrix.

In spite of its name, the **MinimalPolynomial** function in *Mathematica* does not calculate the minimal polynomial of a matrix. The *MinimalPolynomial* command serves an unrelated purpose. (The command **MinimalPolynomial**[*s*, *t*] produces the minimum polynomial in *t* for which the algebraic number *s* is a root.)

Illustration

- Minimal polynomial of a matrix

```
A = {{1, 1, 1, 1}, {1, 1, 1, 1}, {1, 1, 1, 1}, {1, 1, 1, 1}};
```

```
cpA = CharacteristicPolynomial[A, t]
```

$-4\,t^3 + t^4$

```
Factor[cpA]
```

$(-4 + t)\,t^3$

```
MatrixForm[A.(A - 4 IdentityMatrix[4])]
```

$$\begin{pmatrix} 0 & 0 & 0 & 0 \\ 0 & 0 & 0 & 0 \\ 0 & 0 & 0 & 0 \\ 0 & 0 & 0 & 0 \end{pmatrix}$$

Hence the minimum polynomial of A is

$(t - 4)\,t$

- Construction of a minimal polynomial with a companion matrix

Companion matrices can be used to construct matrices for given minimal polynomials.

$p = 3 - 4\,t + t^3 + t^5;$

$$A = \begin{pmatrix} 0 & 0 & 0 & 0 & -3 \\ 1 & 0 & 0 & 0 & 4 \\ 0 & 1 & 0 & 0 & 0 \\ 0 & 0 & 1 & 0 & -1 \\ 0 & 0 & 0 & 1 & -1 \end{pmatrix};$$

```
cpA = CharacteristicPolynomial[A, t]
```

$-3 + 4t - t^3 - t^4 - t^5$

```
Factor[cpA]
```

$-3 + 4t - t^3 - t^4 - t^5$

The polynomial *cpA* is the minimal polynomial of *A*.

- *Mathematica* code for constructing minimal polynomials

The following defined *Mathematica* function can be used to calculate the minimal polynomial of a matrix. We apply it to confirm that the minimum polynomial of the matrix

```
MatrixForm[A = {{1, 1, 1, 1}, {1, 1, 1, 1}, {1, 1, 1, 1}, {1, 1, 1, 1}}]
```

$$\begin{pmatrix} 1 & 1 & 1 & 1 \\ 1 & 1 & 1 & 1 \\ 1 & 1 & 1 & 1 \\ 1 & 1 & 1 & 1 \end{pmatrix}$$

is $t^2 - 4t$.

```
MatrixMinimalPolynomial[a_List?MatrixQ, x_] := Module[
  {i, n = 1, qu = {}, mnm = {Flatten[IdentityMatrix[Length[a]]]}},
  While[Length[qu] == 0, AppendTo[mnm, Flatten[MatrixPower[a, n]]];
   qu = NullSpace[Transpose[mnm]]; n++];
  First[qu].Table[x^i, {i, 0, n - 1}]]
```

This function is not built into *Mathematica* and needs to be activated by typing *Shift* + *Enter*. It can be found on the *MathWorld* website at http://mathworld.wolfram.com/MatrixMinimalPolynomial.html.

```
A = {{1, 1, 1, 1}, {1, 1, 1, 1}, {1, 1, 1, 1}, {1, 1, 1, 1}};
```

```
mpA = MatrixMinimalPolynomial[A, t]
```

$-4t + t^2$

The Cayley–Hamilton theorem says that the matrix *A* satisfies the minimal polynomial *mpA*.

```
MatrixFormas [MatrixPower[A, 2] - 4 A]
```

$$\begin{pmatrix} 0 & 0 & 0 & 0 \\ 0 & 0 & 0 & 0 \\ 0 & 0 & 0 & 0 \\ 0 & 0 & 0 & 0 \end{pmatrix}$$

Minor matrix

See Cofactor matrix

Multiplication of matrices

The result of combining two matrices using matrix multiplication is called a *matrix product*. In *Mathematica*, matrix multiplication is defined by placing a period between the two matrices to be multiplied.

The product of an *m*-by-*n* matrix *A* and an *n*-by-*p* matrix *B* is the *m*-by-*p* matrix whose elements are the dot products of the rows of *A* and the columns of *B*. The definition is designed to represent the composition of linear transformations represented by the matrices *A* and *B*.

In this guide, we write *AB or (A B)* in Text cells for the result of multiplying the matrix *A* by the matrix *B*. If needed, we denote the result of multiplying the three matrices *A*, *B*, and *C* by *(A B C)*. However, the letter C is never used as a name for a matrix in an Input cell since in *Mathematica* the letter C is a reserved symbol for various syntactic purposes.

Properties of matrix multiplication

```
A.(B.C) = (A.B).C                  (1)

A.(B + C) = A.B + A.C              (2)

(B + C).A = B.A + C.A              (3)

a (A.B) = (a A).B = A.(a B)        (4)

A.IdentityMatrix = A               (5)
```

One of the important features of matrix multiplication is the fact that it is not commutative. Multiplying matrix transformations is analogous to composing the associated linear transformations rather than multiplying them.

Illustration

- Noncommutativity of matrix multiplication

```
Clear[a, b, c, d]

A = {{1, 2}, {3, 4}};

B = {{a, b}, {c, d}};

MatrixForm[A.B]
```

$$\begin{pmatrix} a+2c & b+2d \\ 3a+4c & 3b+4d \end{pmatrix}$$

```
MatrixForm[B.A]
```

$$\begin{pmatrix} a+3b & 2a+4b \\ c+3d & 2c+4d \end{pmatrix}$$

```
A.B == B.A /. {a → 1, b → 1, c → 1, d → 1}

False
```

- Multiplication of two matrices

```
MatrixForm[matLeft = {row1, row2} = {{1, 2, 3}, {4, 5, 6}}]
```

$$\begin{pmatrix} 1 & 2 & 3 \\ 4 & 5 & 6 \end{pmatrix}$$

```
MatrixForm[matRight = {{a, b}, {c, d}, {e, f}}]
```

$$\begin{pmatrix} a & b \\ c & d \\ e & f \end{pmatrix}$$

```
MatrixForm[matLeft.matRight]
```

$$\begin{pmatrix} a + 2c + 3e & b + 2d + 3f \\ 4a + 5c + 6e & 4b + 5d + 6f \end{pmatrix}$$

```
MatrixForm[Transpose[matRight]]
```

$$\begin{pmatrix} a & c & e \\ b & d & f \end{pmatrix}$$

The rows of the transpose of the right matrix are the columns of the left matrix.

```
MatrixForm[matLeftmatRight =
  {{Dot[row1, {a, c, e}], Dot[row1, {b, d, f}]},
   {Dot[row2, {a, c, e}], Dot[row2, {b, d, f}]}}]
```

$$\begin{pmatrix} a + 2c + 3e & b + 2d + 3f \\ 4a + 5c + 6e & 4b + 5d + 6f \end{pmatrix}$$

```
matLeft.matRight == matLeftmatRight
```

```
True
```

We can use matrix multiplication to solve linear systems by representing the Gaussian elimination operations by matrix multiplication. We can also treat matrix multiplication as the *composition* of *linear transformations* (functions that preserve the linear combination of vectors).

- The product of a 3-by-4 and a 4-by-2 matrix

```
MatrixForm[A = {{4, 4, 7, 9}, {7, 8, 5, 1}, {8, 9, 8, 1}}]
```

$$\begin{pmatrix} 4 & 4 & 7 & 9 \\ 7 & 8 & 5 & 1 \\ 8 & 9 & 8 & 1 \end{pmatrix}$$

```
MatrixForm[B = {{3, 4}, {5, 9}, {9, 8}, {9, 3}}]
```

$$\begin{pmatrix} 3 & 4 \\ 5 & 9 \\ 9 & 8 \\ 9 & 3 \end{pmatrix}$$

```
MatrixForm[P = A.B]
```

$$\begin{pmatrix} 176 & 135 \\ 115 & 143 \\ 150 & 180 \end{pmatrix}$$

F is the product matrix of A and E.

- The product of a 2-by-2 and a 2-by-3 matrix as a matrix of dot products

```
MatrixForm[A = {{1, 2}, {3, 4}}]
```

$$\begin{pmatrix} 1 & 2 \\ 3 & 4 \end{pmatrix}$$

```
MatrixForm[B = {{2, 1, 3}, {4, 5, 6}}]
```

$$\begin{pmatrix} 2 & 1 & 3 \\ 4 & 5 & 6 \end{pmatrix}$$

```
MatrixForm[A.B]
```

$$\begin{pmatrix} 10 & 11 & 15 \\ 22 & 23 & 33 \end{pmatrix}$$

```
Dot[A[[1]], B[[All, 1]]]
```

```
10
```

```
Table[Dot[A[[i]], B[[All, j]]], {i, 1, 2}, {j, 1, 3}]
```

```
{{10, 11, 15}, {22, 23, 33}}
```

```
MatrixForm[A.B == Table[Dot[A[[i]], B[[All, j]]], {i, 1, 2}, {j, 1, 3}]]
```

```
True
```

Manipulation

- Multiplication of a 2-by-3 matrix A and a 3-by-2 matrix B

```
Clear[a]
```

```
A = {{3, 5 a, 6}, {1, 3, -2}}; B = {{4, a}, {5, 1}, {2, 3}};
```

```
Manipulate[Evaluate[A.B], {a, -6, 6, 1}]
```

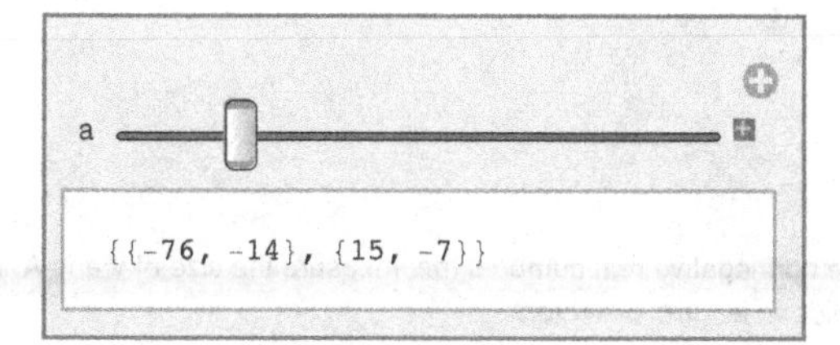

We use **Manipulate** and **Evaluate** to explore the product of two matrices. The displayed matrix is obtained by letting $a = -4$.

N

Norm

The *norms* $\|\mathbf{v}\|$ of a vector **v** and $\|A\|$ of a matrix A are nonnegative real numbers that measure the size of **v** and **A**. In the case of a vector, norms are generalizations of the Euclidean lengths of vectors.

Vector norm

The *norm* of a vector has various definitions, depending on the use to be made of the idea. The *standard norm* or *Euclidean norm* of a vector is based on the theorem of Pythagoras. A *vector norm* is a function $\mathbf{v} \longrightarrow \|\mathbf{v}\|$ on a vector space V which assigns a nonnegative real number $\|\mathbf{v}\|$ to every vector **v** in V with the following properties:

Properties of vector norms

$$\|\mathbf{v}\| > 0 \text{ if } \mathbf{v} \neq \mathbf{0} \quad (1)$$

$$\|\mathbf{v}\| = 0 \text{ if and only if } \mathbf{v} = \mathbf{0} \quad (2)$$

$$\|a\,\mathbf{v}\| = |a|\,\|\mathbf{v}\| \text{ for all scalars } a \quad (3)$$

$$\|\mathbf{v} + \mathbf{w}\| \leq \|\mathbf{v}\| + \|\mathbf{w}\| \quad (4)$$

The last inequality is known as the *triangle inequality*.

Illustration

- The Euclidean norm of a vector in $\mathbb{R}^2$

```
u = {1, 1};
```

```
Norm[u]
```

$\sqrt{2}$

As in the case of $\mathbb{R}^2$, the length of a vector is based on the theorem of Pythagoras. Here too, we use the word *norm* in order to avoid any conflict with the use of *length* in *Mathematica* for the number of elements of a list.

- The Euclidean norm of two vectors in $\mathbb{R}^3$

```
u = {1, 2, 3}; v = {a, b, c};
```

```
{Norm[u], Norm[v]}
```

$\left\{\sqrt{14}, \sqrt{\text{Abs}[a]^2 + \text{Abs}[b]^2 + \text{Abs}[c]^2}\right\}$

```
{Norm[u], Norm[v]} /. {a → 1, b → 1, c → 1}
```

$\{\sqrt{14}, \sqrt{3}\}$

The default norm is the *Euclidean norm*, also known as the two-norm. The parameter 2 in the command **Norm[v, 2]** can therefore be suppressed. Other norms exist and many of them are based on inner products.

- The Euclidean norm of a vector in $\mathbb{R}^4$ and $\mathbb{C}^4$

```
Clear[a, b, c, d]
```

```
w = {a, b, c, d};
```

```
Norm[w]
```

$\sqrt{(\text{Abs}[a]^2 + \text{Abs}[b]^2 + \text{Abs}[c]^2 + \text{Abs}[d]^2)}$

```
ComplexExpand[√(Abs[a]^2 + Abs[b]^2 + Abs[c]^2 + Abs[d]^2),
  {a, b, c, d}, TargetFunctions → Conjugate]
```

$\sqrt{(a\,\text{Conjugate}[a] + b\,\text{Conjugate}[b] + c\,\text{Conjugate}[c] + d\,\text{Conjugate}[d])}$

```
w1 = {1, 2, 3, 4}
```

{1, 2, 3, 4}

```
Norm[w1]
```

$\sqrt{30}$

```
w2 = {1, i, 3 + 4 i, 6}
```

{1, i, 3 + 4 i, 6}

```
Norm[w2]
```

$3\sqrt{7}$

In Euclidean spaces, a vector is a geometrical object that possesses both a magnitude and a direction defined in terms of the dot product. The associated norm is called the *two-norm*. The idea of a norm can be generalized.

- The two-norm of a vector in $\mathbb{R}^3$

```
vector = {1, 2, 3};
magnitude = Norm[vector, 2]
```

$\sqrt{14}$

```
Norm[vector] == Norm[vector, 2]
```

True

Other norms on finite-dimensional non-Euclidean spaces (except for $p = 2$) are the p-norms, for any real number p greater than or equal to 1. The Euclidean two-norm is one of them.

- Four p-norms of a vector in $\mathbb{R}^2$

```
u = {1, 2};
```

```
{Norm[u, 1], Norm[u, 2], Norm[u, 3], Norm[u, 4]}
```

$\left\{3, \sqrt{5}, 3^{2/3}, 17^{1/4}\right\}$

The **N** function can be used to calculate approximations of these norms:

```
N[{Norm[u, 1], Norm[u, 2], Norm[u, 3], Norm[u, 4]}]
```

{3., 2.23607, 2.08008, 2.03054}

For any given $p \geq 1$, the p-norm of a vector **v** in $\mathbb{R}^n$ is defined to be

$$\|u\|_p = \text{Norm}[x, p] = \left(\sum_{i=1}^{n} \text{Abs}[u_i]^p\right)^{1/p}$$

- The seven-norm of a vector in $\mathbb{R}^5$

```
u = {1, 2, -3, 4, -5};
```

$\|u\|_7 = \text{Total}\left[\text{Table}[\text{Abs}[u[[i]]], \{i, 1, 5\}]^7\right]^{1/7}$

$5^{2/7}\ 3873^{1/7}$

$\text{Norm}[u, 7] == \|u\|_7$

True

```
N[Norm[u, 7]]
```

5.15566

Manipulation

We can use the **Manipulate** function to compare the p-norms within specified ranges.

- p-norms of a vector in $\mathbb{R}^5$

```
Clear[p]
```

```
u = Range[5]
```

```
{1, 2, 3, 4, 5}
```

```
Table[N[Norm[u, p]], {p, 1, Length[u]}]
```

```
{15., 7.4162, 6.0822, 5.59365, 5.36022}
```

```
Manipulate[Table[N[Norm[u, p]], {p, a, b}], {a, 1, 10, 1}, {b, 1, 10, 1}]
```

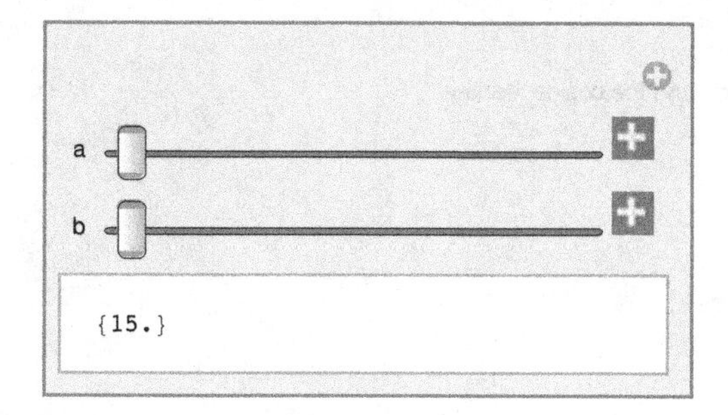

We use **Manipulate**, **Table**, and **Norm** to explore p-norms. If $a = b = 1$, for example, the manipulation shows that the norm of the vector **u** = **Range**[5] is approximately 15.0.

```
N[Norm[u, 1]]
```

```
15.
```

This shows that for $a = b = 7$, the seven-norm of the vector **u** calculated by the **Manipulate** function, is precisely the seven-norm obtained in the previous example using the definition of the seven-norm.

All vectors can be converted to vectors of length 1 by multiplying them by the reciprocals of their norms. The results are called *normal vectors*.

- Normalizing two vectors in $\mathbb{R}^3$

```
Clear[a, b, c]
```

```
u = {1, 2, 3}; v = {a, b, c};
```

```
{Norm[u], Norm[v]}
```

$$\left\{\sqrt{14}, \sqrt{\text{Abs}[a]^2 + \text{Abs}[b]^2 + \text{Abs}[c]^2}\right\}$$

```
normalu = u / Norm[u]
```

$$\left\{\frac{1}{\sqrt{14}}, \sqrt{\frac{2}{7}}, \frac{3}{\sqrt{14}}\right\}$$

```
Norm[normalu]
```

```
1
```

```
normalv = v / Norm[v];
```

All vectors obtained by assigning numerical values (not all zero) have norm 1:

```
Norm[normalv] /. {a → -3, b → 2, c → 1}
```

```
1
```

- Link between the dot product, the Euclidean norm, and the cosine identity

```
Dot[x, y] == Norm[x] Norm[y] cosine[x, y]     (1)
```

```
x = {1, 2}; y = {3, 4};
```

```
N[cos[x, y] = (Dot[x, y]) / (Norm[x] Norm[y])]
```

```
0.98387
```

```
Dot[x, y] == Norm[x] Norm[y] cos[x, y]
```

```
True
```

Matrix norm

A *matrix norm* is a measure of the size of the matrix. They include the Frobenius norm, the max norm, the one-norm, the two-norm, and the infinity-norm.

Properties of matrix norms

$$\|A\| \geq 0 \tag{1}$$

$$\|A\| = 0 \text{ if and only if } A = 0 \tag{2}$$

$$\|k\,A\| = |k|\,\|A\| \tag{3}$$

$$\|A + B\| \leq \|A\| + \|B\| \tag{4}$$

$$\|A.B\| \leq \|A\|\,\|B\| \tag{5}$$

The property $\|AB\| \leq \|A\|\,\|B\|$ assumes that the matrices A and B are square and that the product AB is defined.

Illustration

- The Frobenius norm of a 2-by-2 real matrix

```
A = {{9, -3}, {-2, 1}};
‖A‖F = Sqrt[Tr[Transpose[A].A]]
```

$\sqrt{95}$

- The max norm of an *n*-by-*n* orthogonal matrix

Since an *n*-by-*n* orthogonal matrix *A* preserves the Euclidean norm of a vector *v* in $\mathbb{R}^n$, $\|A.\mathbf{v}\|=\|\mathbf{v}\|$, for all vectors **v** in $\mathbb{R}^n$.

The max norm $\|A\|_{\text{Max}}$ of *A* induced by the Euclidean vector norm is 1 since it is defined to be the maximum of the ratios $\|A.\mathbf{v}\|/\|\mathbf{v}\| = \|\mathbf{v}\|/\|\mathbf{v}\| = 1$, taken over all nonzero vectors **v**.

- The Frobenius norm of the 4-by-4 Hilbert matrix

```
MatrixForm[H = HilbertMatrix[4]]
```

$$\begin{pmatrix} 1 & \frac{1}{2} & \frac{1}{3} & \frac{1}{4} \\ \frac{1}{2} & \frac{1}{3} & \frac{1}{4} & \frac{1}{5} \\ \frac{1}{3} & \frac{1}{4} & \frac{1}{5} & \frac{1}{6} \\ \frac{1}{4} & \frac{1}{5} & \frac{1}{6} & \frac{1}{7} \end{pmatrix}$$

```
‖H‖F = N[Sqrt[Tr[Transpose[H].H]]]
```

1.50973

The norm function of *Mathematica* has a "Frobenius" option:

```
N[Norm[H, "Frobenius"]]
```

1.50973

- The one-norm of the 4-by-4 Hilbert matrix

```
H = HilbertMatrix[4];
```

```
‖H‖1 = N[Max[Table[Total[H[[All, n]]], {n, 1, 4}]]]
```

2.08333

- The two-norm of the 4-by-4 Hilbert matrix

```
H = HilbertMatrix[4]; N[SingularValueList[H]];
```

```
‖H‖2 = Max[N[SingularValueList[H]]]
```

1.50021

- The infinity-norm of the 4-by-4 Hilbert matrix

```
H = HilbertMatrix[4];
```

```
||H||∞ = N[Max[Table[Total[H[[n]]], {n, 1, 4}]]]
```

```
2.08333
```

Every vector norm induces a matrix norm $\|A\|$ as the maximum of the vector norms $\|A\mathbf{v}\|$ taken over all vectors **v** for which $\|\mathbf{v}\| = 1$.

- The induced norm of a nonsingular *m*-by-*n* matrix

$$A = \begin{pmatrix} 5 & 4 \\ 6 & 2 \end{pmatrix};$$

```
A.{Cos[x], Sin[x]}
```

```
{5 Cos[x] + 4 Sin[x], 6 Cos[x] + 2 Sin[x]}
```

```
Norm[{5 Cos[x] + 4 Sin[x], 6 Cos[x] + 2 Sin[x]}] /. {Cos[x] → 1, Sin[x] → 0}
```

$$\sqrt{61}$$

Since orthogonal matrices preserve Euclidean norms, the induced matrix norm $\|A\|$ of an orthogonal matrix is 1.

- The induced norm of an orthogonal matrix

```
A = {{Cos[θ], Sin[θ]}, {-Sin[θ], Cos[θ]}};
```

```
θ = π / 3;
```

```
Simplify[Norm[A.{Cos[π / 4], Sin[π / 4]}]] == Norm[{Cos[π / 4], Sin[π / 4]}] == 1
```

```
True
```

As is pointed out in http://mathworld.wolfram.com/MatrixNorm.html, the task of computing an induced matrix norm is difficult in general since it involves "a nonlinear optimization problem with constraints." A geometric interpretation of the maximum value of $\|A\mathbf{v}\|/\|\mathbf{v}\|$ over all unit vectors **v** is the maximum stretching factor obtained when multiplying the vectors on the unit sphere by the matrix *A*.

- Stretching of a unit circle in $\mathbb{R}^2$

```
A = {{2, 0}, {0, 3}};
```

```
circle = Cos[x]^2 + Sin[x]^2 == 1;
```

```
A.{Cos[x], Sin[x]}
```

```
{2 Cos[x], 3 Sin[x]}
```

```
ParametricPlot[{2 Cos[x], 3 Sin[x]}, {x, 0, 2 π}]
```

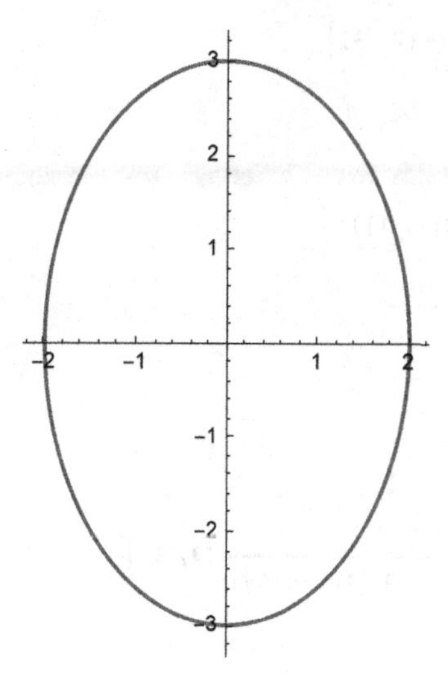


```
Norm[{2 Cos[π / 2], 3 Sin[π / 2]}]
```

3

This shows that the induced norm of the diagonal matrix A is 3.

Normal basis of a vector space

A *normal basis* of a normed vector space is a basis in which the vectors all have a norm of 1.

Illustration

- Normalization of a basis for $\mathbb{R}^2$ relative to the Euclidean norm on $\mathbb{R}^2$

```
basis = {{1, 2}, {3, 4}};
```

$$\text{normalizedbasis} = \left\{\frac{1}{\text{Norm}[\{1, 2\}]}\{1, 2\}, \frac{1}{\text{Norm}[\{3, 4\}]}\{3, 4\}\right\}$$

$$\left\{\left\{\frac{1}{\sqrt{5}}, \frac{2}{\sqrt{5}}\right\}, \left\{\frac{3}{5}, \frac{4}{5}\right\}\right\}$$

```
{Norm[normalizedbasis[[1]]], Norm[normalizedbasis[[2]]]}
```

{1, 1}

- Normalization of a basis for $\mathbb{R}^2$ relative to the one-norm on $\mathbb{R}^2$

```
basis = {{1, 2}, {3, 4}};
```

$$\text{normalizedbasis} = \left\{\frac{1}{\text{Norm}[\{1, 2\}, 1]}\{1, 2\}, \frac{1}{\text{Norm}[\{3, 4\}, 1]}\{3, 4\}\right\}$$

$$\left\{\left\{\frac{1}{3}, \frac{2}{3}\right\}, \left\{\frac{3}{7}, \frac{4}{7}\right\}\right\}$$

```
{Norm[normalizedbasis[[1]], 1], Norm[normalizedbasis[[2]], 1]}
```

{1, 1}

{1, 1}

- Normalization of a basis for $\mathbb{R}^2$ relative to the infinity-norm on $\mathbb{R}^2$

```
basis = {{1, 2}, {3, 4}};
```

$$\text{normalizedbasis} = \left\{\frac{1}{\text{Norm}[\{1, 2\}, \text{Infinity}]}\{1, 2\}, \frac{1}{\text{Norm}[\{3, 4\}, \text{Infinity}]}\{3, 4\}\right\}$$

$$\left\{\left\{\frac{1}{2}, 1\right\}, \left\{\frac{3}{4}, 1\right\}\right\}$$

```
{Norm[normalizedbasis[[1]], Infinity], Norm[normalizedbasis[[2]], Infinity]}
```

{1, 1}

{1, 1}

Normal equation

See Normalization of a matrix equation

Normal matrix

A complex square matrix A is *normal* if $AA^* = A^*A$, where A^* is the conjugate transpose of *A*.

Normality can be used to test for diagonalizability since a matrix is normal if and only if it is unitarily similar to a diagonal matrix and is therefore diagonalizable since all matrices of the form **ConjugateTranspose**[*A*].*A* and *A*.**ConjugateTranspose**[*A*] are diagonalizable.

Illustration

- A normal complex 2-by-2 matrix

```
MatrixForm[A = {{0, Cos[π/4] I}, {-Sin[π/4] I, 0}}]
```

$$\begin{pmatrix} 0 & \frac{i}{\sqrt{2}} \\ -\frac{i}{\sqrt{2}} & 0 \end{pmatrix}$$

```
NormalMatrixQ[A]
```

True

- A normal real 3-by-3 matrix

```
MatrixForm[A = {{1, 2, -1}, {-1, 1, 2}, {2, -1, 1}}]
```

$$\begin{pmatrix} 1 & 2 & -1 \\ -1 & 1 & 2 \\ 2 & -1 & 1 \end{pmatrix}$$

```
NormalMatrixQ[A]
```

True

Normal to a plane

In Euclidean geometry, a vector emanating from a point in a plane is a *normal* to the plane if it is perpendicular to every nonzero vector determined by two points in the plane. If $a x + b y + c z + d = 0$ is a plane, then the vector **n** = $\{a, b, c\}$ is normal to the given plane.

Illustration

- A normal to a plane

```
Clear[a, b, c, d]
```

```
plane = a x + b y + c z + d == 0;
```

```
point0 = {x0, y0, z0};
```

```
eq1 = a x0 + b y0 + c z0 + d == 0;
```

```
point1 = {x1, y1, z1};
```

```
eq2 = a x1 + b y1 + c z1 + d == 0;
```

```
eq3 = a (x0 - x1) + b (y0 - y1) + c (z0 - z1) == 0
```

$a (x_0 - x_1) + b (y_0 - y_1) + c (z_0 - z_1) == 0$

```
Dot[{a, b, c}, {(x0 - x1), (y0 - y1), (z0 - z1)}] == 0
```

```
a (x0 - x1) + b (y0 - y1) + c (z0 - z1) == 0
```

Therefore the vector **n** = *{a, b, c}* is normal to the plane $a\,x + b\,y + c\,z + d = 0$.

Manipulation

- Equations of planes determined by the point {1, 2, 3} and their normal {*a*, *b*, *c*}.

```
Manipulate[Expand[Dot[{a, b, c}, {a (x - 1), b (y - 2), c (z - 3)}] == 0],
 {a, -3, 3, 1}, {b, -3, 3, 1}, {c, 0, 4, 1}]
```

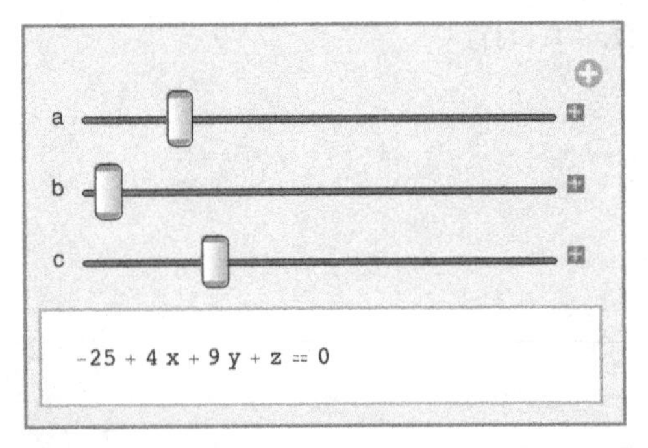

We use **Manipulate**, **Expand**, and **Dot** to explore equations of planes. The manipulation produces the linear equation also obtained with the following command:

```
Expand[Dot[{-2, -3, 1}, {-2 (x - 1), -3 (y - 2), 1 (z - 3)}] == 0]
```

```
-25 + 4 x + 9 y + z == 0
```

Normalization of a matrix equation

The *normalization* of a matrix equation $A\mathbf{v} = \mathbf{b}$ refers to the process of multiplying both sides of the equation by the transpose of the coefficient matrix A. Even if the original equation $A\mathbf{v} = \mathbf{b}$ has no solution, its normalized version $A^T A\mathbf{v} = A^T \mathbf{b}$ is always solvable. Normalization preserves the solutions of $A\mathbf{v} = \mathbf{b}$.

The *normal equation* of a matrix equation $A\mathbf{v} = \mathbf{b}$ is the equation obtained by normalizing the equation. It is called a normal equation because the vector $(\mathbf{b} - A\mathbf{v})$ is normal to the range of A.

Illustration

- Normalization of a vector equation

```
A = RandomInteger[{0, 9}, {2, 3}];
```

$$A = \begin{pmatrix} 4 & 0 & 8 \\ 6 & 2 & 5 \end{pmatrix};$$

```
MatrixForm[Transpose[A]]
```

$$\begin{pmatrix} 4 & 6 \\ 0 & 2 \\ 8 & 5 \end{pmatrix}$$

$$\mathbf{v} = \begin{pmatrix} x \\ y \\ z \end{pmatrix}; \mathbf{b} = \begin{pmatrix} 3 \\ 5 \end{pmatrix};$$

$$\text{equation} = \begin{pmatrix} 4 & 0 & 8 \\ 6 & 2 & 5 \end{pmatrix} . \begin{pmatrix} x \\ y \\ z \end{pmatrix} == \begin{pmatrix} 3 \\ 5 \end{pmatrix}$$

$$\text{normalizedequation} = \begin{pmatrix} 4 & 6 \\ 0 & 2 \\ 8 & 5 \end{pmatrix} . \begin{pmatrix} 4 & 0 & 8 \\ 6 & 2 & 5 \end{pmatrix} . \begin{pmatrix} x \\ y \\ z \end{pmatrix} == \begin{pmatrix} 4 & 6 \\ 0 & 2 \\ 8 & 5 \end{pmatrix} . \begin{pmatrix} 3 \\ 5 \end{pmatrix}$$

- Normalization preserves the solutions of consistent matrix equations.

```
A = {{4, 5, 6}, {1, 8, 7}}; v = {x, y, z}; b = {2, 3};

LinearSolve[A, b]
```

$$\left\{\frac{1}{27}, \frac{10}{27}, 0\right\}$$

```
LinearSolve[Transpose[A].A, Transpose[A].b]
```

$$\left\{\frac{1}{27}, \frac{10}{27}, 0\right\}$$

- Normalization produces “best possible” approximations of inconsistent matrix equations.

```
Clear[v, x, y, z, b, A]

A = {{4, 5, 6}, {8, 10, 12}}; v = {x, y, z}; b = {2, 3};

LinearSolve[A, b]
```

LinearSolve::nosol : Linear equation encountered that has no solution. ≫

```
LinearSolve[{{4, 5, 6}, {8, 10, 12}}, {2, 3}]

solution = LinearSolve[Transpose[A].A, Transpose[A].b]
```

$$\left\{\frac{2}{5}, 0, 0\right\}$$

```
Transpose[A].A.solution == Transpose[A].b

True
```

Manipulation

- Normalization of a matrix equation

```
Clear[u, v, A, a, b]
```

```
MatrixForm[A = {{1, 2}, {3, 4}, {4, 7}}]; b = {1, 2, 3};
```

```
Reduce[Transpose[A].A.{u, v} == Transpose[A].b]
```

$v == \frac{4}{15} \&\& u == \frac{3}{10}$

```
Clear[u, v]
```

```
Manipulate[Reduce[Transpose[A].A.{a u, v} == Transpose[A].b], {a, -3, 3, 1}]
```

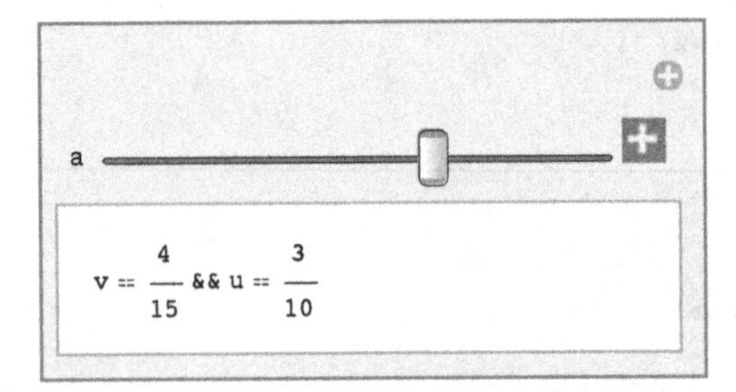

We use **Manipulate**, **Reduce**, and **Transpose** to visualize the normalization of matrix equations. The manipulation produces the result for $a = 1$.

Normalization of a vector

The *normalization* of a vector **v** in a normed vector space is the process of multiplying **v** by the reciprocal of its norm. This results in a vector in the same direction as **v**, but with a norm (or length) of 1.

Illustration

- Normalization of a vector in $\mathbb{R}^4$ with respect to the Euclidean norm on $\mathbb{R}^4$

```
v = {1, 2, 3, 4};
```

$nv = \frac{1}{\text{Norm}[v]} v$

$\left\{\frac{1}{\sqrt{30}}, \sqrt{\frac{2}{15}}, \sqrt{\frac{3}{10}}, 2\sqrt{\frac{2}{15}}\right\}$

```
Norm[%]
```

1

Normed vector space

A vector space equipped with a vector norm is a *normed vector space.*

Illustration

- The Euclidean norm for $\mathbb{R}^2$

The square root of **Dot[*u*, *u*]** is a vector norm called the *Euclidean* or *two-norm.*

```
u = {1, 2, 3};
```

```
Norm[u]
```

$\sqrt{14}$

```
Norm[u] == Norm[u, 2]
```

True

- The p-norms for $\mathbb{R}^3$

The Euclidean norm can be generalized to the family of so-called p-norms for all real numbers greater than or equal to 1.

```
u = {1, -2, 3}; p = 1;
```

```
Norm[u, 1]
```

6

```
Norm[u, p] == (Abs[1] + Abs[-2] + Abs[3])^(1/1)
```

True

```
u = {-1, 2, -3, 4}; p = 3;
```

```
Norm[u, p]
```

$10^{2/3}$

```
(Abs[-1]^3 + Abs[2]^3 + Abs[-3]^3 + Abs[4]^3)^(1/3)
```

$10^{2/3}$

- The one-norm for $\mathbb{R}^3$

The one-norm on the real vector space $\mathbb{R}^n$ is the sum of the absolute values of the coordinates of the vectors:

```
Quit[]

au = {a1 x1, a2 x2, a3 x3};

||au||1 = |a1 x1| + |a2 x2| + |a3 x3|

|a1 x1| + |a2 x2| + |a3 x3|

a = 3; u = {1, 2, 3};

a u

{3, 6, 9}

||a u||1 = Abs[3] + Abs[6] + Abs[9]

18
```

- The infinity-norm for $\mathbb{R}^3$

The infinity-norm on the real vector space $\mathbb{R}^n$ is the absolute value of the largest coordinate of the vectors:

```
ax = {a1 x1, a2 x2, a3 x3};

||ax||∞ = Max[|a1 x1|, |a2 x2|, |a3 x3|]

Max[|31 x1|, |32 x2|, |33 x3|]

a = 3; u = {1, 2, 3};

a u

{3, 6, 9}

||a u||∞ = Max[Abs[3], Abs[6], Abs[9]]

9
```

Null space

The *null space* of an *m*-by-*n* real matrix tells us which vectors **v** solve the homogeneous equation $A\mathbf{v} = \mathbf{0}$. *Mathematica* has a built-in command for computing the bases of null spaces. Eigenspaces are typical examples of null spaces.

Illustration

- A basis for the null space of a 2-by-2 matrix

$$A = \begin{pmatrix} 7 & 1 & 0 & 6 & 7 \\ 4 & 9 & 7 & 3 & 5 \\ 9 & 2 & 5 & 4 & 4 \end{pmatrix};$$

```
nspaceA = NullSpace[A]
```

```
{{-11, -14, 15, 0, 13}, {-199, -167, 217, 260, 0}}
```

```
A.nspaceA[[1]]
```

```
{0, 0, 0}
```

```
A.nspaceA[[2]]
```

```
{0, 0, 0}
```

The matrix *A* maps all linear combinations of the basis vectors nspaceA[[1]] and nspaceA[[2]] to zero:

```
Simplify[A.(a nspaceA[[1]] + b nspaceA[[2]])]
```

```
{0, 0, 0}
```

- A basis for the null space of a 3-by-4 matrix

```
MatrixForm[A = {{9, 9, 2, 8}, {6, 9, 4, 3}, {5, 7, 2, 8}}]
```

$$\begin{pmatrix} 9 & 9 & 2 & 8 \\ 6 & 9 & 4 & 3 \\ 5 & 7 & 2 & 8 \end{pmatrix}$$

```
nspaceA = NullSpace[A]
```

```
{{26, -52, 69, 12}}
```

```
u = a {26, -52, 69, 12}
```

```
{78, -156, 207, 36}
```

```
A.u
```

```
{0, 0, 0}
```

Thus the null space of the matrix *A* is the one-dimensional subspace of $\mathbb{R}^4$ consisting of all multiples of the vector {26, -52, 69, 12}.

- Eigenspaces considered as a null spaces

```
MatrixForm[A = {{9, 0, 0}, {6, 9, 0}, {5, 7, 2}}]
```

$$\begin{pmatrix} 9 & 0 & 0 \\ 6 & 9 & 0 \\ 5 & 7 & 2 \end{pmatrix}$$

The **Eigensystem** command produces a list of eigenvalues and eigenvectors of a matrix.

```
{evalues, evectors} = Eigensystem[A]
```

```
{{9, 9, 2}, {{0, 1, 1}, {0, 0, 0}, {0, 0, 1}}}
```

The eigenvalues of A are therefore 9 and 2, and the corresponding (nonzero) eigenvectors are {0, 1, 1} and {0, 0, 1}. *Mathematica* includes the zero vector of the eigenspace of 9 in its output, although the zero vector is not actually an eigenvector. However, it does of course belong to the eigenspace of an eigenvalue.

The command

```
NullSpace[A - 9 IdentityMatrix[3]]
```

```
{{0, 1, 1}}
```

produces a basis for the eigenspace of the eigenvalue 9 of *A*. The command,

```
NullSpace[A - 2 IdentityMatrix[3]]
```

```
{{0, 0, 1}}
```

on the other hand, produces a basis for the eigenspace of the eigenvalue 2 of A.

Manipulation

- Eigenspaces as null spaces

```
Clear[a, A]
```

```
MatrixForm[A = {{9 , 2}, {5 , a}}]
```

$$\begin{pmatrix} 9 & 2 \\ 5 & a \end{pmatrix}$$

```
evalues = Eigenvalues[A]
```

$$\left\{\frac{1}{2}\left(9 + a - \sqrt{121 - 18\,a + a^2}\right), \frac{1}{2}\left(9 + a + \sqrt{121 - 18\,a + a^2}\right)\right\}$$

```
NullSpace[A - evalues[[1]] IdentityMatrix[2]]
```

$$\left\{\left\{\frac{1}{10}\left(9 - a - \sqrt{121 - 18\,a + a^2}\right), 1\right\}\right\}$$

```
NullSpace[A - evalues[[2]] IdentityMatrix[2]]
```

$$\left\{\left\{\frac{1}{10}\left(9 - a + \sqrt{121 - 18 a + a^2}\right), 1\right\}\right\}$$

```
Manipulate[Evaluate[NullSpace[A - evalues[[1]] IdentityMatrix[2]]], {a, -1, 3, 1}]
```

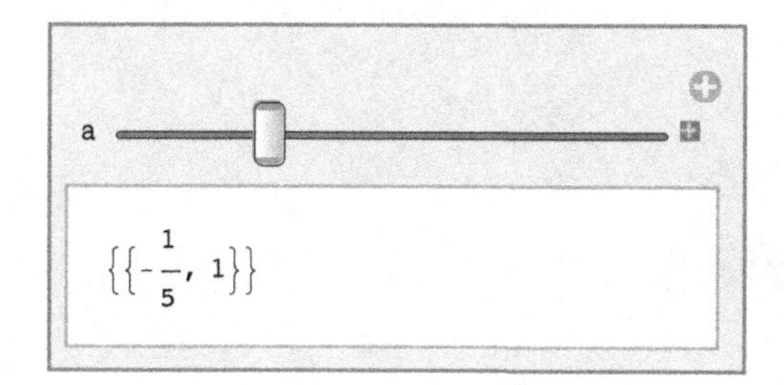

We use **Manipulate**, **Evaluate**, **NullSpace**, and **IdentityMatrix** to explore the eigenspace of first eigenvalue of the generated matrix as a null space.

```
Manipulate[Evaluate[NullSpace[A - evalues[[2]] IdentityMatrix[2]]], {a, -1, 3, 1}]
```

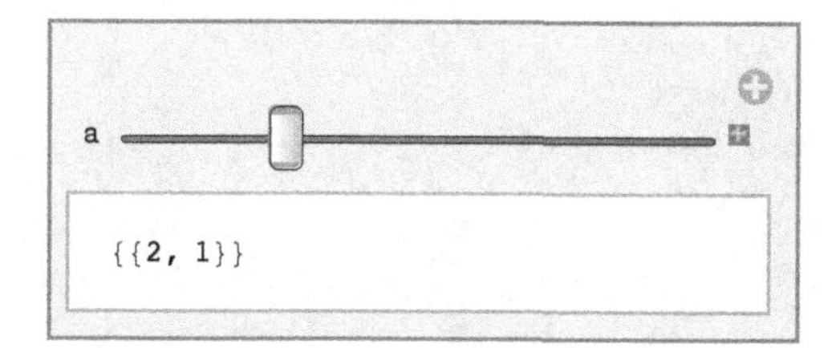

We use **Manipulate**, **Evaluate**, **NullSpace**, and **IdentityMatrix** to explore the eigenspace of second eigenvalue of the generated matrix as a null space.

If we let $a = 0$ in the matrix A, the two **Manipulate** illustrations display the bases of the two null spaces obtained with the Eigenvectors command, as expected:

```
Eigenvectors[A] /. {a → 0}
```

$$\left\{\left\{-\frac{1}{5}, 1\right\}, \{2, 1\}\right\}$$

Nullity of a matrix

The dimension of the null space of a matrix is called the *nullity* of the matrix.

Illustration

- The nullity of a random 3-by-4 matrix

```
A = RandomInteger[{0, 9}, {3, 4}];
```

$$A = \begin{pmatrix} 5 & 2 & 3 & 3 \\ 9 & 8 & 0 & 1 \\ 1 & 8 & 2 & 6 \end{pmatrix};$$

```
NullSpace[A]
```

```
{{76, -115, -286, 236}}
```

This shows that the null space of A has a basis consisting of one nonzero vector. Hence the nullity of the matrix A is 1.

- The nullity of a random 4-by-6 matrix

```
A = RandomInteger[{0, 9}, {4, 6}];
```

$$A = \begin{pmatrix} 9 & 8 & 2 & 9 & 5 & 8 \\ 3 & 3 & 8 & 9 & 4 & 4 \\ 5 & 4 & 1 & 7 & 6 & 8 \\ 6 & 5 & 6 & 4 & 2 & 2 \end{pmatrix};$$

```
NullSpace[A]
```

```
{{-351, 398, 62, -125, 0, 122}, {-713, 862, 46, -199, 244, 0}}
```

This shows that the null space of A has a basis consisting of two nonzero vectors. Hence the nullity of the matrix A is 2.

O

Orthogonal basis

A *basis* of an inner product space is *orthogonal* if all of its vectors are pairwise orthogonal.

Illustration

- An orthogonal basis of $\mathbb{R}^2$ in the Euclidean inner product
  - Linear independence

```
B = {{1, 2}, {-2, 1}};
```

```
Det[B]
```

5

  - Spanning

```
Solve[{x, y} == a {1, 2} + b {-2, 1}, {a, b}]
```

$\left\{\left\{a \to \frac{1}{5}(x + 2y), b \to \frac{1}{5}(-2x + y)\right\}\right\}$

  - Orthogonality

```
Dot[{1, 2}, {-2, 1}]
```

0

- An orthogonal basis of $\mathbb{R}^3$ in the Euclidean inner product
  - Linear independence

$B = \left\{\left\{21\sqrt{5}, 42\sqrt{5}, 0\right\}, \left\{-8\sqrt{105}, 4\sqrt{105}, 5\sqrt{105}\right\}, \left\{-10\sqrt{21}, 5\sqrt{21}, -20\sqrt{21}\right\}\right\};$

```
Det[B]
```

-1 157 625

  - Spanning

```
Solve[{x, y, z} == a B[[1]] + b B[[2]] + c B[[3]], {a, b, c}]
```

$$\left\{\left\{a \to \frac{1}{525}\left(\sqrt{5}\ x + 2\sqrt{5}\ y\right),\ b \to \frac{-8\sqrt{105}\ x + 4\sqrt{105}\ y + 5\sqrt{105}\ z}{11025},\ c \to \frac{-2\sqrt{21}\ x + \sqrt{21}\ y - 4\sqrt{21}\ z}{2205}\right\}\right\}$$

- Orthogonality

```
{Dot[B[[1]], B[[2]]], Dot[B[[1]], B[[3]]], Dot[B[[2]], B[[3]]]}

{0, 0, 0}
```

Orthogonal complement

The *orthogonal complement* S^c of a subset *S* of an inner product space *V* is the set of all vectors **v** in *V* with the property that <**v**, **w**> = 0 for all **w** in *S*.

Illustration

- Orthogonal complement of a subset of $\mathbb{R}^2$ in the Euclidean inner product

Let *S* be the set of all vectors in $\mathbb{R}^2$ of the form {*a*, *0*}. Then **y** belongs to S^c if **Dot[x, y]** = 0 for all real numbers *a*.

```
Solve[Dot[{a, 0}, {b, c}] == 0, b]

{{b → 0}}
```

Therefore the orthogonal complement S^c of *S* in $\mathbb{R}^2$ is the set of all vectors {*0*, *c*} in $\mathbb{R}^2$, for all real numbers *c*.

- A null space as the orthogonal complement of the row space of a matrix

```
A = RandomInteger[{0, 9}, {3, 5}];

MatrixForm[A = {{9, 0, 4, 1, 8}, {7, 3, 5, 3, 7}, {0, 5, 6, 9, 0}}]
```

$$\begin{pmatrix} 9 & 0 & 4 & 1 & 8 \\ 7 & 3 & 5 & 3 & 7 \\ 0 & 5 & 6 & 9 & 0 \end{pmatrix}$$

```
MatrixForm[S = NullSpace[A]]
```

$$\begin{pmatrix} -12 & -6 & 5 & 0 & 11 \\ 5 & 3 & -13 & 7 & 0 \end{pmatrix}$$

```
MatrixForm[T = RowReduce[A]]
```

$$\begin{pmatrix} 1 & 0 & 0 & -\frac{5}{7} & \frac{12}{11} \\ 0 & 1 & 0 & -\frac{3}{7} & \frac{6}{11} \\ 0 & 0 & 1 & \frac{13}{7} & -\frac{5}{11} \end{pmatrix}$$

```
Table[Dot[S[[i]], T[[j]]], {i, 1, 2}, {j, 1, 3}]
```

```
{{0, 0, 0}, {0, 0, 0}}
```

Since the set

```
S ⋃ T
```

$$\left\{\{-12, -6, 5, 0, 11\}, \left\{0, 0, 1, \frac{13}{7}, -\frac{5}{11}\right\}, \left\{0, 1, 0, -\frac{3}{7}, \frac{6}{11}\right\}, \left\{1, 0, 0, -\frac{5}{7}, \frac{12}{11}\right\}, \{5, 3, -13, 7, 0\}\right\}$$

is a basis for $\mathbb{R}^5$ and since each vector in *S* is orthogonal to each vector in *T*, the orthogonal complement S^c of *S* is *T* and the orthogonal complement, T^c of *T* is S.

We can also use the **DisjointQ** function to confirm that the spans of *S* and *T* are disjoint:

```
DisjointQ[S, T]
```

```
True
```

Orthogonal decomposition

Orthogonal projections and orthogonal complements can be used for an *orthogonal decomposition* of a vector into a sum of orthogonal vectors.

Properties of projections and orthogonal complements

$$\text{Proj}[\mathbf{u}, \mathbf{v}] = \frac{\langle \mathbf{u}, \mathbf{v} \rangle}{\langle \mathbf{v}, \mathbf{v} \rangle} \mathbf{v} \tag{1}$$

$$\text{Perp}[\mathbf{u}, \mathbf{v}] = \mathbf{u} - \frac{\langle \mathbf{u}, \mathbf{v} \rangle}{\langle \mathbf{v}, \mathbf{v} \rangle} \mathbf{v} \tag{2}$$

$$\text{Proj}[\mathbf{u}, \mathbf{v}] + \text{Perp}[\mathbf{u}, \mathbf{v}] = \mathbf{u} \tag{3}$$

$$\text{Proj}[\mathbf{u}, \mathbf{v}] = a\,\mathbf{v} \tag{4}$$

$$\langle \text{Perp}[\mathbf{u}, \mathbf{v}], \mathbf{v} \rangle = 0 \tag{5}$$

Illustration

- An orthogonal projection in $\mathbb{R}^2$

```
projection = Graphics[{Arrow[{{1, 2}, {10, 2}}], Arrow[{{1, 2}, {5, 8}}],
   Arrow[{{5, 8}, {5, 2}}], Arrow[{{1, 1.8}, {5, 1.8}}]}, Axes → True]
```

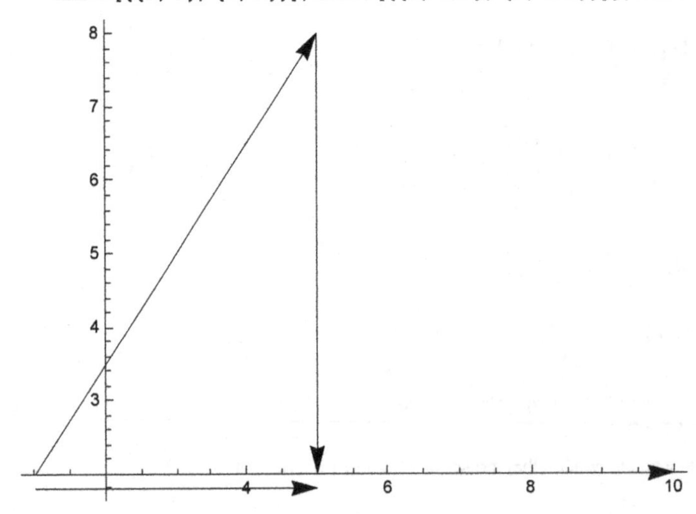

- The orthogonal projection Proj[u, v] of a vector u onto a vector v and the vector component Perp[u,v] of u orthogonal to v

```
Clear[a]
```

$$\text{Proj}[u\_, v\_] := \frac{\text{Dot}[u, v]}{\text{Dot}[v, v]} v$$

```
Perp[u_, v_] := u - Proj[u, v]
```

```
u = {1, 2, 3}; v = {5, 0, 3};
```

```
{Proj[u, v], Perp[u, v]}
```

$$\left\{\left\{\frac{35}{17}, 0, \frac{21}{17}\right\}, \left\{-\frac{18}{17}, 2, \frac{30}{17}\right\}\right\}$$

```
Proj[u, v] + Perp[u, v] == u
```

```
True
```

```
Solve[Proj[u, v] == a v, a]
```

$$\left\{\left\{a \to \frac{7}{17}\right\}\right\}$$

```
Dot[Perp[u, v], v] == 0
```

```
True
```

- Using the *Mathematica* **Projection** function to decompose a vector into a sum of two orthogonal vectors in $\mathbb{R}^3$

```
Clear[a]
```

```
u = {1, 2, 3}; v = {3, 5, 7};
```

```
perpComp[u, v] = u - Projection[u, v]
```

$\left\{-\frac{19}{83}, -\frac{4}{83}, \frac{11}{83}\right\}$

```
Projection[u, v]
```

$\left\{\frac{102}{83}, \frac{170}{83}, \frac{238}{83}\right\}$

```
Projection[u, v] + perpComp[u, v] == u
```

```
True
```

```
Solve[Projection[u, v] == a v, a]
```

$\left\{\left\{a \to \frac{34}{83}\right\}\right\}$

```
Dot[perpComp[u, v], v] == 0
```

```
True
```

The calculations show that the vectors **Projection[u, v]** and **perpComp[u, v]** combine into an orthogonal decomposition of the vector **u**.

Manipulation

- Orthogonal projections in $\mathbb{R}^2$

```
Manipulate[Projection[{a, b}, {1, 2}], {a, -3, 3, 1}, {b, -4, 4, 1}]
```

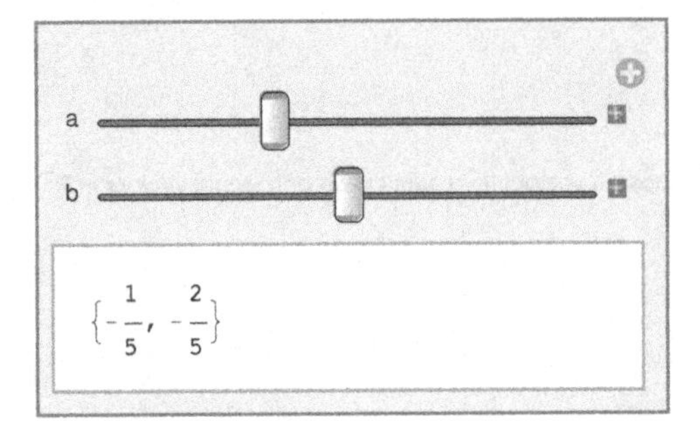

We use **Manipulate** and **Projection** to explore projections in $\mathbb{R}^2$. The manipulation produces the projection of the vector {-1, 0} onto the vector {1, 2}.

Orthogonal matrix

An *orthogonal matrix* is real square matrix whose inverse is its transpose. Much effort has gone into approximating invertible matrices with orthogonal ones because of the ease of computing transposes. The *singular value decomposition* is one technique for finding the best orthogonal approximation of a real invertible matrix.

Every invertible matrix or, more generally, every linearly independent set of real vectors, can be orthogonalized by the *Gram–Schmidt process.* The **Orthogonalize** function of *Mathematica* implements this process.

Orthogonal matrices are the matrices that preserve the standard inner product.

Illustration

- A 3-by-3 orthogonal matrix

$$A = \begin{pmatrix} \frac{6}{7} & -\frac{2}{7} & -\frac{3}{7} \\ -\frac{2}{7} & \frac{3}{7} & -\frac{6}{7} \\ -\frac{3}{7} & -\frac{6}{7} & -\frac{2}{7} \end{pmatrix}; \; u = \{1, 2, 3\}; \; v = \{4, 5, 6\};$$

```
A.Transpose[A] == Transpose[A].A == IdentityMatrix[3]
```

True

```
Dot[u, v] == Dot[A.u, A.v]
```

True

- Applying the Gram–Schmidt process to an invertible matrix

```
A = RandomInteger[{0, 9}, {3, 3}];
```

$$A = \begin{pmatrix} 6 & 5 & 6 \\ 4 & 3 & 9 \\ 7 & 3 & 1 \end{pmatrix};$$

```
Det[A]
```

97

As we can see, the matrix A is invertible and its columns and rows are therefore linearly independent.

```
oA = Orthogonalize[A]
```

$$\left\{\left\{\frac{6}{\sqrt{97}}, \frac{5}{\sqrt{97}}, \frac{6}{\sqrt{97}}\right\}, \left\{-\frac{170}{\sqrt{158401}}, -\frac{174}{\sqrt{158401}}, \frac{315}{\sqrt{158401}}\right\}, \left\{\frac{27}{\sqrt{1633}}, -\frac{30}{\sqrt{1633}}, -\frac{2}{\sqrt{1633}}\right\}\right\}$$

```
Transpose[oA] == Inverse[oA]
```

True

Orthogonal matrices preserve the Euclidean inner product.

- Orthogonal matrices and the Euclidean inner product

```
u = {1, 2, 3}; v = {4, 5, 6};
```

$$A = \begin{pmatrix} \frac{6}{\sqrt{97}} & \frac{5}{\sqrt{97}} & \frac{6}{\sqrt{97}} \\ -\frac{170}{\sqrt{158401}} & -\frac{174}{\sqrt{158401}} & \frac{315}{\sqrt{158401}} \\ \frac{27}{\sqrt{1633}} & -\frac{30}{\sqrt{1633}} & -\frac{2}{\sqrt{1633}} \end{pmatrix};$$

```
Transpose[A] == Inverse[A]
```

True

```
Dot[u, v] == Dot[A.u, A.v]
```

True

- An orthogonal matrix is a matrix whose inverse equals its transpose.

```
A = {{4, 7}, {1, 3}, {1, 5}};
```

```
tA = Transpose[A];
```

```
oA = Orthogonalize[tA.A]
```

$$\left\{\left\{\frac{1}{\sqrt{5}}, \frac{2}{\sqrt{5}}\right\}, \left\{-\frac{2}{\sqrt{5}}, \frac{1}{\sqrt{5}}\right\}\right\}$$

```
Inverse[oA] == Transpose[oA]
```

True

- Orthogonal matrices associated with a singular value decomposition

```
A = {{1, 0, 1}, {0, 3, 0}, {1, 0, 3}};
```

$$A = \begin{pmatrix} 1 & 0 & 1 \\ 0 & 3 & 0 \\ 1 & 0 & 3 \end{pmatrix};$$

```
{u, w, v} = SingularValueDecomposition[A];
```

```
MatrixForm[u]
```

$$\begin{pmatrix} \sqrt{\frac{2}{2+\left(2+\sqrt{2}\right)^2}} & 0 & -\sqrt{\frac{2}{2+\left(2-\sqrt{2}\right)^2}} \\ 0 & 1 & 0 \\ \frac{2+\sqrt{2}}{\sqrt{2+\left(2+\sqrt{2}\right)^2}} & 0 & \frac{2-\sqrt{2}}{\sqrt{2+\left(2-\sqrt{2}\right)^2}} \end{pmatrix}$$

```
MatrixForm[v]
```

$$\begin{pmatrix} \frac{-1+\sqrt{2}}{\sqrt{1+\left(-1+\sqrt{2}\right)^2}} & 0 & \frac{-1-\sqrt{2}}{\sqrt{1+\left(-1-\sqrt{2}\right)^2}} \\ 0 & 1 & 0 \\ \frac{1}{\sqrt{1+\left(-1+\sqrt{2}\right)^2}} & 0 & \frac{1}{\sqrt{1+\left(-1-\sqrt{2}\right)^2}} \end{pmatrix}$$

The matrices **u** and **v** are orthogonal:

```
Simplify[Transpose[u] == Inverse[u]]
```

```
True
```

```
Simplify[Transpose[v] == Inverse[v]]
```

```
True
```

The matrix **w** is diagonal:

```
MatrixForm[w]
```

$$\begin{pmatrix} \sqrt{2\left(3+2\sqrt{2}\right)} & 0 & 0 \\ 0 & 3 & 0 \\ 0 & 0 & \sqrt{2\left(3-2\sqrt{2}\right)} \end{pmatrix}$$

```
SingularValueList[A]
```

$$\left\{\sqrt{2\left(3+2\sqrt{2}\right)},\ 3,\ \sqrt{2\left(3-2\sqrt{2}\right)}\right\}$$

The matrix *A* is diagonalized by **u**, **w**, and **v**:

```
Simplify[A == u.w.Transpose[v]]
```

True

As expected, the singular values of *A* are the square roots of the eigenvalues of both the matrix $A\,A^T$ and $A^T A$.

```
Sqrt[Eigenvalues[A.Transpose[A]]]
```

$$\left\{\sqrt{2\left(3+2\sqrt{2}\right)},\ 3,\ \sqrt{2\left(3-2\sqrt{2}\right)}\right\}$$

```
Sqrt[Eigenvalues[Transpose[A].A]]
```

$$\left\{\sqrt{2\left(3+2\sqrt{2}\right)},\ 3,\ \sqrt{2\left(3-2\sqrt{2}\right)}\right\}$$

Orthogonal projection

The *orthogonal projection* of one vector onto another is the basis for the decomposition of a vector into a sum of orthogonal vectors. The *projection* of a vector **v** onto a second vector **w** is a scalar multiple of the vector **w**.

Illustration

- Orthogonal projections in $\mathbb{R}^2$

```
v = {1, 2}; w = {3, 4};
```

```
vpw = Projection[v, w]
```

$$\left\{\frac{33}{25},\ \frac{44}{25}\right\}$$

This is the projection of the vector v onto the vector w.

```
Solve[vpw == a w, a]
```

$$\left\{\left\{a \to \frac{11}{25}\right\}\right\}$$

This shows that when the vector **v** is projected onto **w**, the result is a vector that is in the same direction as **w**, but **w** is scaled by a factor of 11/25.

Similarly, we can project the vector **w** onto **v**.

```
wpv = Projection[w, v]
```

$\left\{\frac{11}{5}, \frac{22}{5}\right\}$

```
Solve[wpv == b v, b]
```

$\left\{\left\{b \to \frac{11}{5}\right\}\right\}$

- Orthogonal projections in $\mathbb{R}^3$

```
u = {1, 2}; v = {3, 4};
```

```
Projection[u, v]
```

$\left\{\frac{33}{25}, \frac{44}{25}\right\}$

The projection of the vector **v** onto the vector **w** can be used to decompose the vector **v** into a sum of orthogonal vectors.

- Orthogonal decomposition in $\mathbb{R}^3$

```
(v - vpw) + vpw == v
```

```
True
```

```
Dot[v - vpw, vpw]
```

$\frac{154}{25}$

The projection can be reversed by projecting the vector **w** onto the vector **v** and decomposing the vector **w** into a sum of orthogonal vectors.

```
(w - wpv) + wpv == w
```

```
True
```

```
Dot[w - wpv, wpv]
```

0

The **Projection** function can be applied to vectors of arbitrary length.

- Orthogonal projections in $\mathbb{R}^3$

```
u = {1, 2, 3}; v = {4, 5, 6};
```

```
upv = Projection[u, v]
```

$\left\{\frac{128}{77}, \frac{160}{77}, \frac{192}{77}\right\}$

```
vpu = Projection[v, u]
```

$\left\{\frac{16}{7}, \frac{32}{7}, \frac{48}{7}\right\}$

Orthogonal transformation

A linear transformation (linear operator) on a real inner product space V is an *orthogonal transformation* if it preserves the inner product <u, v> = <T[u], T[v]> for all vectors **u** and **v** in V.

If a matrix T_A represents a linear transformation $T: V \longrightarrow V$ in an orthonormal basis B, then the transpose Transpose[T_A] represents a linear transformation from V to V called the *adjoint* T^* of T.

Let T be an orthogonal transformation on a finite-dimensional inner product space. Then the following conditions are equivalent:

$< \mathbf{u}, \mathbf{v} > \; = \; < T[\mathbf{u}], T[\mathbf{v}] >$ for all **u** and **v** (1)

$\| \mathbf{u} \| = \| T[\mathbf{u}] \|$ for all **u** (2)

$T \circ T^* = T^* \circ T$, where T^* is the adjoint of T (3)

Illustration

- Two adjoint linear transformations on $\mathbb{R}^2$

```
Clear[a, b, c, d]

T = {{1, 2}, {3, 4}};

T* = Transpose[T];

u = {a, b}; v = {c, d};

Expand[Dot[T.u, v] == Dot[u, T*.v]]
True
```

- Two adjoint linear transformations on $\mathbb{R}^3$

```
T = {{1, 0, 2}, {4, 3, 0}, {0, 0, 3}};

T* = Transpose[T];

u = {1, 2, 3}; v = {4, 5, 6};
```

```
{Dot[T.u, v], Dot[u, T*.v]}
```

```
{132, 132}
```

Orthogonal vectors

Two vectors **u** and **v** in a vector space equipped with an inner product <**u**, **v**> are *orthogonal* if <**u**, **v**> = 0. Two numerical vectors are orthogonal in the Euclidean sense if their dot product is zero.

The orthogonality of vectors is a stronger relationship than their linear independence. Orthogonal vectors are always linearly independent and can be used to build unique linear combinations, but they are more. In $\mathbb{R}^3$, in particular, and nowhere else, *orthogonal vectors* can be constructed using an operation called the *cross product*. This operation has many powerful uses in scientific applications.

The *Mathematica* function for computing cross products is the **Cross** function.

Illustration

- Two orthogonal vectors in $\mathbb{R}^2$

```
u = {1, 2}; v = {-2, 1};
```

```
Dot[u, v]
```

```
0
```

Since orthogonal vectors are linearly independent, the calculation also shows that the two vectors are linearly independent.

```
Solve[a u + b v == 0 {a, b}]
```

```
{{a → 0, b → 0}}
```

As in the case of $\mathbb{R}^2$, orthogonality is a concept generalizing the idea of perpendicularity and two vectors may be orthogonal in one norm and not in another. The default is the Euclidean norm.

- Two orthogonal vectors in $\mathbb{R}^3$

```
u = {1, 2, 3}; v = {-3, 0, 1};
```

```
Dot[u, v]
```

```
0
```

```
Graphics3D[{Arrow[{{0, 0, 0}, {1, 2, 3}}], Arrow[{{0, 0, 0}, {-3, 0, 1}}]}, Axes -> True]
```

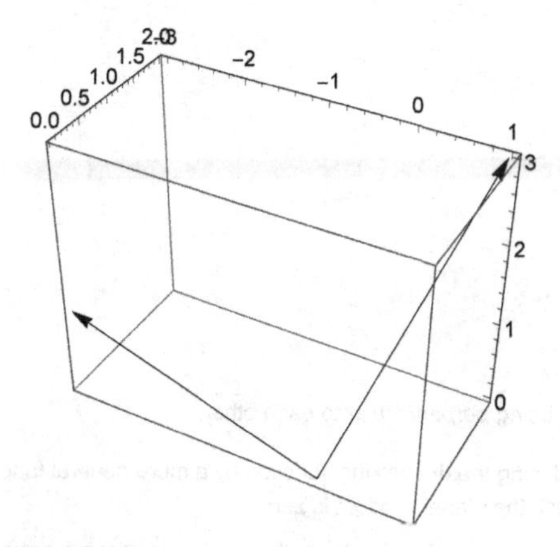

The vectors are also linearly independent:

```
Reduce[a {1, 2, 3} + b {-3, 0, 1} == 0, {a, b}]
```

a == 0 && b == 0

- Three orthogonal vectors in $\mathbb{R}^3$

```
u = {1, 2, 3}; v = {-3, 0, 1};
```

```
w = Cross[u, v]
```

{2, -10, 6}

The **Dot** function can be used to verify the pairwise orthogonality of the three vectors:

```
{Dot[u, v], Dot[u, w], Dot[v, w]}
```

{0, 0, 0}

- Graphing two orthogonal vectors in $\mathbb{R}^2$

```
u = {1, 2}; v = {-2, 1};
```

```
Dot[u, v]
```

0

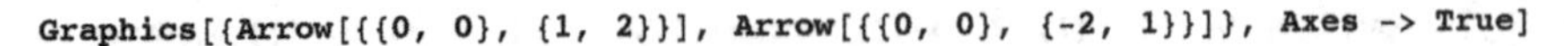

```
Graphics[{Arrow[{{0, 0}, {1, 2}}], Arrow[{{0, 0}, {-2, 1}}]}, Axes -> True]
```

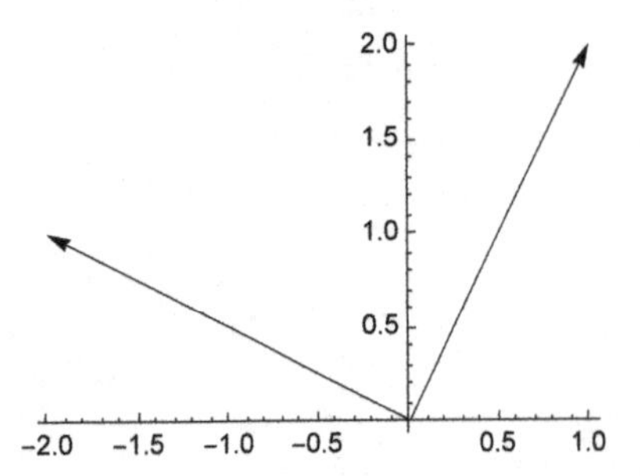

This image shows that orthogonal vectors can be thought of as being *perpendicular* to each other.

The idea of orthogonality of two vectors can be modified by replacing the dot product function by a more general function called an *inner product*. In that case, two vectors are orthogonal if their inner product is zero.

If **u.w.Transpose[v]** is a singular value decomposition of a matrix A, the columns of **u** and the columns of **v** are orthogonal.

- Singular vectors produced by a singular value decomposition are orthogonal

$$A = \begin{pmatrix} 9 & 6 & 8 \\ 9 & 2 & 6 \\ 7 & 8 & 2 \end{pmatrix};$$

```
{u, w, v} = SingularValueDecomposition[A];

N[u]

{{0.679681, 0.161157, 0.715585},
 {0.534613, 0.559111, -0.633706}, {0.502218, -0.813279, -0.29386}}

lsv = Table[N[u, 3][[All, i]], {i, 1, 3}]

{{0.680, 0.535, 0.502}, {0.161, 0.559, -0.813}, {0.716, -0.634, -0.294}}

Table[Dot[lsv[[i]], lsv[[j]]], {i, 1, 3}, {j, 1, 3}]
```

$\{\{1.00, 0.\times10^{-3}, 0.\times10^{-3}\}, \{0.\times10^{-3}, 1.00, 0.\times10^{-3}\}, \{0.\times10^{-3}, 0.\times10^{-3}, 1.00\}\}$

```
Floor[%]

{{1, 0, 0}, {0, 1, 0}, {0, 0, 1}}

rsv = Table[N[v, 3][[All, i]], {i, 1, 3}]

{{0.735, 0.467, 0.491}, {0.146, -0.817, 0.558}, {-0.662, 0.338, 0.669}}

Table[Dot[rsv[[i]], rsv[[j]]], {i, 1, 3}, {j, 1, 3}]
```

$\{\{1.0, 0.\times10^{-3}, 0.\times10^{-3}\}, \{0.\times10^{-3}, 1.00, 0.\times10^{-3}\}, \{0.\times10^{-3}, 0.\times10^{-3}, 1.00\}\}$

```
Floor[%]
```

```
{{0, 0, 0}, {0, 1, 0}, {0, 0, 1}}
```

As expected, the resulting vectors are orthogonal.

Manipulation

We can use manipulations to explore the orthogonality of vectors

- The orthogonality of two vectors in $\mathbb{R}^2$

```
Clear[a, b]

u = {a, 3}; v = {-2, b};

Reduce[Dot[u, v] == 0, {a, b}]

Manipulate[Evaluate[Dot[u, v] ], {a, -4, 7, 1}, {b, -3, 4, 1}]
```

$$b == \frac{2\,a}{3}$$

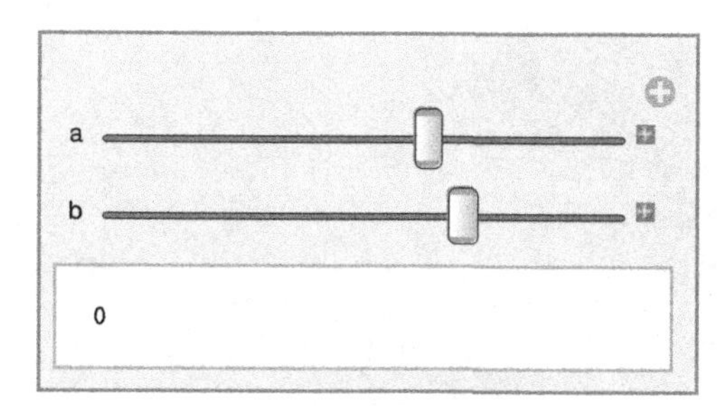

We use **Manipulate**, **Evaluate**, and **Dot** to explore the orthogonality of vectors. The manipulation shows that if $a = 3$ and $b = 2$, or if $a = 6$ and $b = 4$, for example, the generated vectors **u** and **v** are orthogonal. On the other hand, if $a = b = 2$, the corresponding vectors **u** = {2, 3} and **v** = {-2, 2} are not.

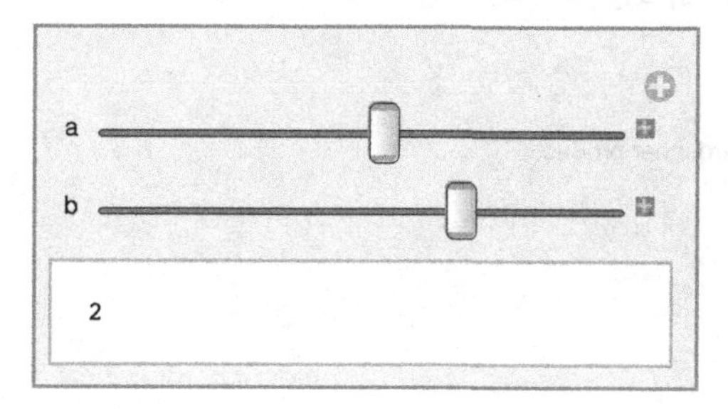

Orthogonality

See Orthogonal matrix, orthogonal projection, orthogonal vectors

Orthogonalization

See Gram–Schmidt process

Orthonormal basis

A basis is *orthonormal* if all of its vectors have a norm (or *length*) of 1 and are pairwise orthogonal.

One of the main applications of the Gram–Schmidt process is the conversion of bases of inner product spaces to *orthonormal bases.*

The **Orthogonalize** function of *Mathematica* converts any given basis of a Euclidean space $\mathbb{E}^n$ into an orthonormal basis. It can also be used to orthogonalize a set of vectors in a non-Euclidean space,

Illustration

- An orthonormal basis for $\mathbb{R}^3$

```
u = {1, 0, 1}; v = {1, 1, 1}; w = {0, 1, 1};
```

```
ou = Orthogonalize[{u, v, w}]
```

$$\left\{\left\{\frac{1}{\sqrt{2}}, 0, \frac{1}{\sqrt{2}}\right\}, \{0, 1, 0\}, \left\{-\frac{1}{\sqrt{2}}, 0, \frac{1}{\sqrt{2}}\right\}\right\}$$

```
Table[Norm[ou[[i]]], {i, 1, 3}]
```

```
{1, 1, 1}
```

```
Table[Dot[ou[[i]], ou[[j]]], {i, 1, 3}, {j, 1, 3}]
```

```
{{1, 0, 0}, {0, 1, 0}, {0, 0, 1}}
```

- An orthonormal basis for $\mathbb{R}^2$ relative to a nonstandard inner product.

```
Clear[x, y, u, v]
```

```
MatrixForm[A = DiagonalMatrix[{2, 3}]]
```

$$\begin{pmatrix} 2 & 0 \\ 0 & 3 \end{pmatrix}$$

The function

```
⟨u_, v_⟩ := u.A.v
```

defines a nonstandard inner product on $\mathbb{R}^2$.

Next we construct two vectors in $\mathbb{R}^2$ that are orthogonal in the inner product ip:

```
u = {1, 2}; v = {x, y};

Reduce[⟨u, v⟩ == 0, {x, y}]
```

$$y == -\frac{x}{3}$$

```
⟨{1, 2}, {3, -1}⟩
```

0

Now we take two vectors in $\mathbb{R}^2$ that are not orthogonal in the inner product ip and orthogonalize them:

```
w = {4, 6};

⟨u, w⟩
```

44

```
ip[u_, v_] := ⟨u, v⟩

{vector1, vector2} = Orthogonalize[{u, w}, ip]
```

$$\left\{\left\{\frac{1}{\sqrt{14}}, \sqrt{\frac{2}{7}}\right\}, \left\{\sqrt{\frac{3}{7}}, -\frac{1}{\sqrt{21}}\right\}\right\}$$

```
⟨vector1, vector2⟩
```

0

Manipulation

- Exploring orthonormal bases for $\mathbb{R}^2$

We can use manipulations to visualize the numerical properties of orthonormal bases generated from random matrices.

```
Manipulate[Orthogonalize[{{a, 9}, {3, 3}}], {a, -5, 5, 1}]
```

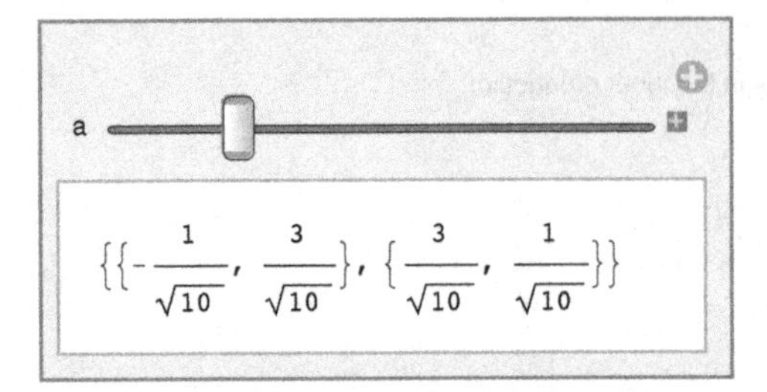

We use **Manipulate** and **Orthogonalize** to convert linearly independent vectors to orthogonal vectors. For each assignment of an integer between -5 and 5 to the parameter *a*, the **Orthogonalize** function produces an orthonormal basis for $\mathbb{R}^2$. If *a* = -3, for example, the **Manipulate** function displays the orthogonal matrix

$$A = \left\{\left\{-\frac{1}{\sqrt{10}}, \frac{3}{\sqrt{10}}\right\}, \left\{\frac{3}{\sqrt{10}}, \frac{1}{\sqrt{10}}\right\}\right\};$$

```
OrthogonalMatrixQ[A]
```

```
True
```

The rows of the generated matrix *A* form an orthonormal basis for $\mathbb{R}^2$ in the standard inner product.

Overdetermined linear system

A linear system is *overdetermined* if it has more equations than unknowns. When the system is expressed as matrix-vector equation, the matrix of coefficients will have more rows than columns.

Illustration

- An overdetermined inconsistent linear system with three equations in two variables

```
system = {x + y == 1, 2 x - y == 4, x + 5 y == 3};
```

```
Solve[system, {x, y}]
```

```
{}
```

- An overdetermined consistent linear system with three equations and two variables

```
system = {x + 2 y == 5, 5 x + 6 y == 6, 4.5 x + 6 y == 8.25};
```

```
A = {{1, 2}, {5, 6}, {4.5, 6}}; b = {5, 6, 8.25};
```

```
LinearSolve[A, b]
```

```
{-4.5, 4.75}
```

- An approximate solution for an overdetermined inconsistent linear system in matrix form

```
A = {{1., 2.}, {5., 6.}, {4.5, 6.}}; b = {5., 6., 5.};
```

```
LinearSolve[A, b];
```

LinearSolve::nosol : Linear equation encountered that has no solution. ≫

- Using **LeastSquares** to solve an inconsistent overdetermined linear system

We can sometimes find a best possible approximation of a solution of an inconsistent system by using the *method of least squares*:

```
LeastSquares[A, b]
```

```
{-3.35294, 3.60294}
```

This method is equivalent to applying the transpose of the coefficient matrix to both sides and then inverting the new coefficient matrix, if this is possible.

```
LinearSolve[Transpose[A].A, Transpose[A].b]
```

```
{-3.35294, 3.60294}
```

The resulting vector {-3.35294, 3.60294} is a best approximation in the sense of least squares.

P

Particular solution of a linear system

The solution of a linear system obtained by assigning numerical values to the free variables of a linear system is a *particular solution* of the linear system.

Illustration

- Particular solutions of a linear system

```
system = {3 x + 5 y - z - 1 == 0, x - 2 y + 4 z - 5 == 0};

A = Normal[CoefficientArrays[system, {x, y, z}]][[2]];

v = {{1}, {5}};

MatrixForm[augmentedmatrixA = Join[A, v, 2]]
```

$$\begin{pmatrix} 3 & 5 & -1 & 1 \\ 1 & -2 & 4 & 5 \end{pmatrix}$$

```
MatrixForm[pivotcolumns = RowReduce[augmentedmatrixA]]
```

$$\begin{pmatrix} 1 & 0 & \frac{18}{11} & \frac{27}{11} \\ 0 & 1 & -\frac{13}{11} & -\frac{14}{11} \end{pmatrix}$$

The associated linear system is

$$\text{reducedsystem} = \left\{x + \frac{18}{11} z == \frac{27}{11},\ y - \frac{13}{11} z == -\frac{14}{11}\right\}$$

$$\left\{x + \frac{18\,z}{11} == \frac{27}{11},\ y - \frac{13\,z}{11} == -\frac{14}{11}\right\}$$

This shows that the variables *x* and *y* can be considered to be the basic variables of the system and then *z* is free. By assigning numerical values to *z*, the corresponding numerical values of *x* and *y* produce a particular solution of the given system:

$$\text{system2} = \left\{x + \frac{18}{11} z == \frac{27}{11},\ y - \frac{13}{11} z == -\frac{14}{11}\right\} /.\ \{z \to 0\}$$

$$\left\{x == \frac{27}{11},\ y == -\frac{14}{11}\right\}$$

Manipulation

- Particular solutions of a linear system

```
Manipulate[Solve[{3 x + 5 y - z - 1 == 0, x - 2 y + 4 z - 5 == 0}, {x, y}], {z, -2, 2, 1}]
```

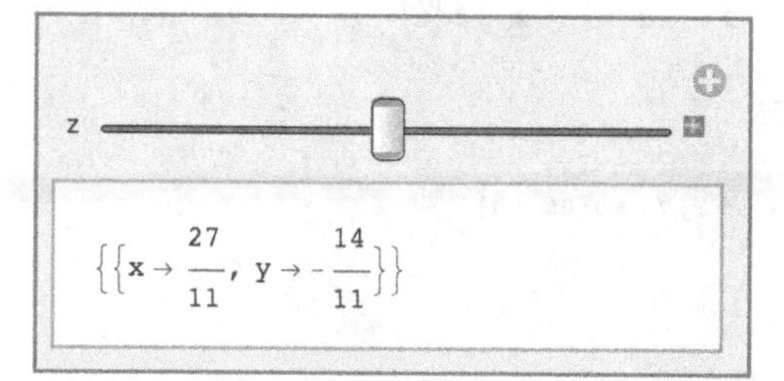

We use **Manipulate** and **Solve** to calculate particular solutions of linear systems. This manipulation displays the particular solution of the given linear system by assigning 0 to the free variable z.

Pauli spin matrix

The *Pauli spin matrices* are three complex matrices that arise in Pauli's treatment of spin in quantum mechanics. The Pauli matrices, together with the identity matrix, form a basis for the space of all 2-by-2 complex matrices.

Illustration

- The Pauli spin matrices

```
MatrixForm[P1 = {{0, 1}, {1, 0}}]
```

$$\begin{pmatrix} 0 & 1 \\ 1 & 0 \end{pmatrix}$$

```
MatrixForm[P2 = {{0, -i}, {i, 0}}]
```

$$\begin{pmatrix} 0 & -i \\ i & 0 \end{pmatrix}$$

```
MatrixForm[P3 = {{1, 0}, {0, -1}}]
```

$$\begin{pmatrix} 1 & 0 \\ 0 & -1 \end{pmatrix}$$

```
Clear[a, b, c, d, u, v, x, y]
```

```
A = {{u, v}, {x, y}};
```

```
Solve[A == a IdentityMatrix[2] + b P1 + c P2 + d P3, {a, b, c, d}]
```

$$\left\{\left\{a \to \frac{u+y}{2},\ b \to \frac{v+x}{2},\ c \to \frac{1}{2}\, i\, (v-x),\ d \to \frac{u-y}{2}\right\}\right\}$$

```
solution = Flatten[{{{a → (u + y)/2, b → (v + x)/2, c → (i v)/2 - (i x)/2, d → u/2 - y/2}}}]
```

$\{a \to \frac{u+y}{2}, b \to \frac{v+x}{2}, c \to \frac{i\,v}{2} - \frac{i\,x}{2}, d \to \frac{u}{2} - \frac{y}{2}\}$

```
A == Simplify[a IdentityMatrix[2] + b P1 + c P2 + d P3 /. solution]
```

True

Perfectly conditioned matrix

An invertible matrix A is *perfectly conditioned* relative to a matrix norm $\|A\|$ if its condition number $\|A\| \, \|A^{-1}\| = 1$.

Illustration

- A perfectly conditioned matrix in the infinity-norm

```
A = {{10^-6, 0}, {0, 10^-6}};
```

```
T = Map[Total, Abs[A]];
```

```
||A||∞ = Max[T]
```

$\frac{1}{1\,000\,000}$

```
B = Inverse[A];
```

```
iT = Map[Total, Abs[B]];
```

```
||B||∞ = Max[iT]
```

1 000 000

```
ConditionNumber = ||A||∞ ||B||∞
```

1

- A matrix that is not a perfectly conditioned matrix in the infinity-norm

```
A = {{2, -1, 0}, {2, -4, -1}, {-1, 0, 2}};
```

```
T = Map[Total, Abs[{{2, 1, 0}, {2, 4, 1}, {1, 0, 2}}]]
```

{3, 7, 3}

```
‖A‖∞ = Max[T]
```

7

```
B = Inverse[A]
```

$\left\{\left\{\frac{8}{13}, -\frac{2}{13}, -\frac{1}{13}\right\}, \left\{\frac{3}{13}, -\frac{4}{13}, -\frac{2}{13}\right\}, \left\{\frac{4}{13}, -\frac{1}{13}, \frac{6}{13}\right\}\right\}$

```
iT = Map[Total, Abs[B]]
```

$\left\{\frac{11}{13}, \frac{9}{13}, \frac{11}{13}\right\}$

```
‖B‖∞ = Max[iT]
```

$\frac{11}{13}$

```
ConditionNumber = N[‖A‖∞ ‖B‖∞]
```

5.92308

- A perfectly conditioned matrix in the Euclidean norm

```
MatrixForm[A = {{1, 0, 0}, {0, 1, 0}, {0, 0, 1.000001}}]
```

$\begin{pmatrix} 1 & 0 & 0 \\ 0 & 1 & 0 \\ 0 & 0 & 1. \end{pmatrix}$

```
ConditionNumber = Norm[A] Norm[Inverse[A]]
```

1.

Manipulation

- Changes in a condition number of a matrix

```
Clear[a, b]
```

```
MatrixForm[A = {{1 + a, 0}, {0, 1 + b}}]
```

$\begin{pmatrix} 1+a & 0 \\ 0 & 1+b \end{pmatrix}$

```
Manipulate[Evaluate[Norm[A] Norm[Inverse[A]]], {a, -0.001, 0.001}, {b, 0, 5}]
```

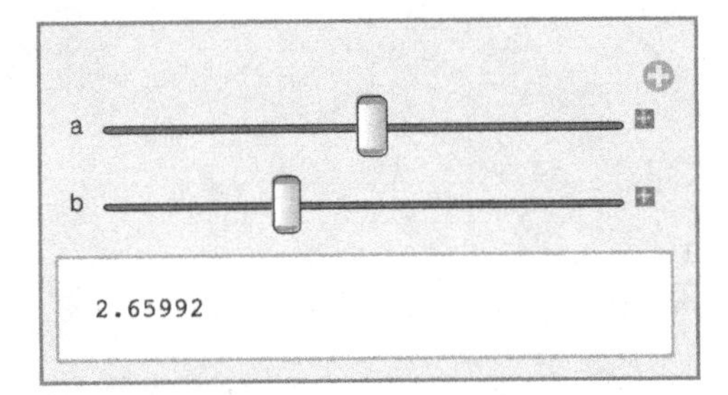

We use **Manipulate**, **Evaluate**, **Norm**, and **Inverse** to explore the effect of making small changes to the elements of a 2-by-2 matrix. The manipulation displays the condition number of the matrix obtained from the matrix A by adding $a = 0.00003$ and $b = 1.66$ to the diagonal elements of A.

Permutation matrix

A permutation matrix is a matrix obtained from an identity matrix by permuting the rows of the matrix.

Illustration

```
MatrixForm[A = IdentityMatrix[4]]
```

$$\begin{pmatrix} 1 & 0 & 0 & 0 \\ 0 & 1 & 0 & 0 \\ 0 & 0 & 1 & 0 \\ 0 & 0 & 0 & 1 \end{pmatrix}$$

- A permutation matrix involving one elementary row operation

```
MatrixForm[perm3214 = {A[[All,3]], A[[All,2]], A[[All,1]], A[[All,4]]}]
```

$$\begin{pmatrix} 0 & 0 & 1 & 0 \\ 0 & 1 & 0 & 0 \\ 1 & 0 & 0 & 0 \\ 0 & 0 & 0 & 1 \end{pmatrix}$$

The matrix *perm3214* is also an elementary matrix

- A permutation matrix involving more than elementary row operations

```
MatrixForm[perm3412 = {A[[All,3]], A[[All,4]], A[[All,1]], A[[All,2]]}]
```

$$\begin{pmatrix} 0 & 0 & 1 & 0 \\ 0 & 0 & 0 & 1 \\ 1 & 0 & 0 & 0 \\ 0 & 1 & 0 & 0 \end{pmatrix}$$

The matrix *perm3412* is not an elementary matrix since it involves more than one permutation of rows.

Pivot column of a matrix

The *pivot columns* of a matrix *A* are the columns that correspond to the pivots of a reduced row echelon matrix *B* obtained by row reducing *A*. The pivot columns of any reduced row echelon matrix are the columns of the matrix containing the pivot elements of the matrix. The pivot elements of a reduced row echelon matrix are the leading elements of the nonzero rows of the matrix. The leading element of a nonzero row of a matrix in reduced row echelon form is the leftmost nonzero element of the row.

Illustration

```
MatrixForm[A = {{4, 0, 8}, {-9, 0, 5}, {0, 0, 4}}]
```

$$\begin{pmatrix} 4 & 0 & 8 \\ -9 & 0 & 5 \\ 0 & 0 & 4 \end{pmatrix}$$

```
MatrixForm[B = RowReduce[A]]
```

$$\begin{pmatrix} 1 & 0 & 0 \\ 0 & 0 & 1 \\ 0 & 0 & 0 \end{pmatrix}$$

The 1s in the first and second rows of the reduced row echelon matrix *B* are the pivots of *B*. The first and third columns of *B* are the pivot columns of *B*. Hence the first and third columns of *A* are the pivot columns of *A* since *A* is row equivalent to *B*.

Manipulation

- Exploring the pivot columns of a matrix obtained by row reduction

```
Manipulate[RowReduce[{{1, 1, 1}, {2, 1, 1}, {a, b, c}}],
 {a, -5, 5, 1}, {b, -5, 5, 1}, {c, -5, 5, 1}]
```

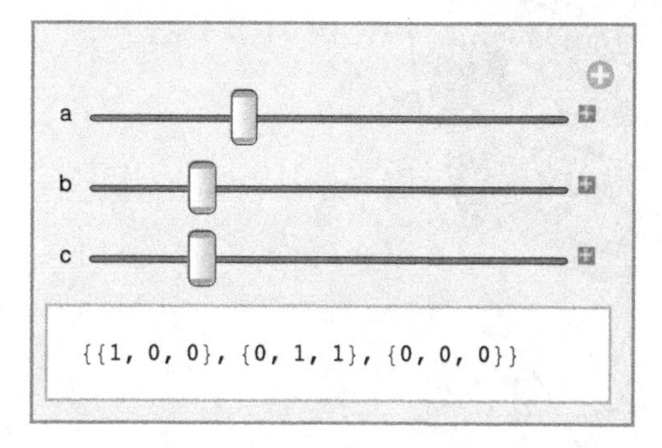

We use **Manipulate** and **RowReduce** to explore the pivot columns of 3-by-3 matrices. The manipulation displays the pivot columns of the matrix

```
MatrixForm[A = {{1, 1, 1}, {2, 1, 1}, {-2, -3, -3}}]
```

$$\begin{pmatrix} 1 & 1 & 1 \\ 2 & 1 & 1 \\ -2 & -3 & -3 \end{pmatrix}$$

obtained by letting $a = -2$, $b = c = -3$. The manipulation shows that the first and second columns of A are pivot columns.

Plane in Euclidean space

The set of solutions $\{x, y, z\}$ of a linear equation ax + by +cz + d = 0 forms a *plane* in the Euclidean space $\mathbb{E}^3$.

Illustration

- A plane in $\mathbb{E}^3$

```
plane = 3 x - 2 y + z - 5 == 0;
```

```
ContourPlot3D[Evaluate[plane], {x, -10, 10}, {y, -10, 10}, {z, -10, 10}]
```

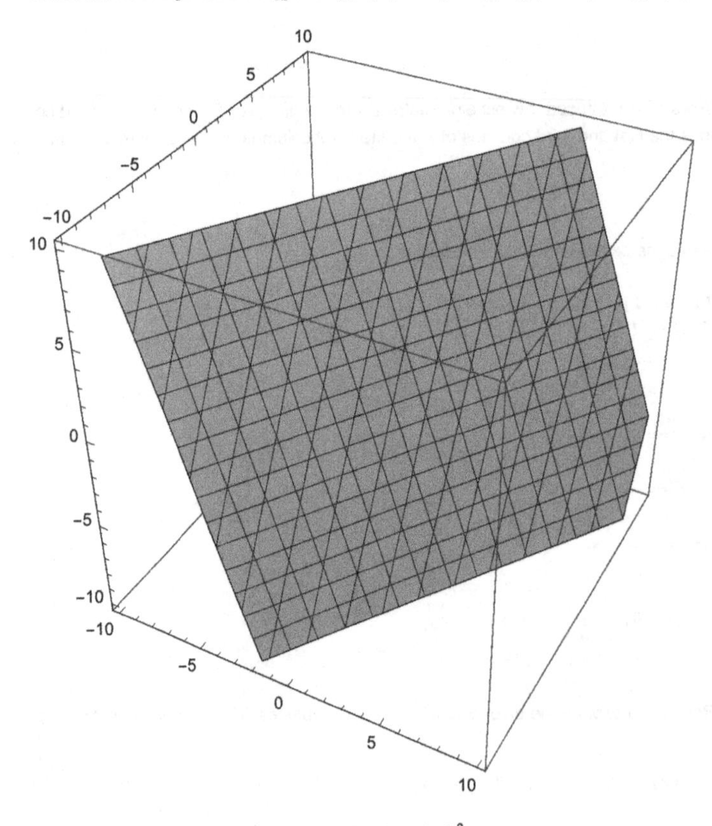

- Point-normal form of the equation of a plane in $\mathbb{E}^3$

The coefficients of the variables of x, y, and z form a vector called the *normal* to the given plane.

```
plane = 3 x - 2 y + z - 5 == 0;
```

```
normal = {3, -2, 1};
```

The normal is perpendicular to any vector $\{x - x_0, y - y_0, z - z_0\}$ formed with the arbitrary point $\{x, y, z\}$ and fixed point $\{x_0, y_0, z_0\}$ in the plane.

A fixed point in the plane is any solution of the linear equation of the plane.

```
Solve[plane, {x, y, z}]
```

Solve::svars : Equations may not give solutions for all "solve" variables. ≫

```
{{z → 5 - 3 x + 2 y}}
```

If $\{x_0, y_0, z_0\} = \{1, 1, z_0\}$ is a point in the plane, then z must be equal to 4.

```
plane /. {x → 1, y → 1, z → 4}
```

```
True
```

The normal vector {3, -2, 1} is perpendicular to the vector {x - 1, y - 1, z - 4} determined by the two points in the plane. Therefore

```
pnformplaneq = Expand[Dot[{3, -2, 1}, {x - 1, y - 1, z - 4}] == 0]
```

```
-5 + 3 x - 2 y + z == 0
```

As expected, the point-normal form of the equation is identical to the original equation.

```
TraditionalForm[plane - pnformplaneq == 0]
```

```
True
```

- Three-point form of the equation of a plane in $\mathbb{E}^3$

```
givenplane = 3 x - 2 y + z - 5 == 0;
```

```
Solve[givenplane, {x, y, z}]
```

Solve::svars : Equations may not give solutions for all "solve" variables. ≫

```
{{z → 5 - 3 x + 2 y}}
```

```
threepoints = {{0, 0, 5}, {1, 0, 2}, {1, 1, 4}};
```

```
equation = a x + b y + c z + d == 0;
```

```
eq1 = equation /. {x → 0, y → 0, z → 5}
```

```
5 c + d == 0
```

```
eq2 = equation /. {x → 1, y → 0, z → 2}
```

```
a + 2 c + d == 0
```

```
eq3 = equation /. {x → 1, y → 1, z → 4}
```

```
a + b + 4 c + d == 0
```

```
parameters = Flatten[Solve[{eq1, eq2, eq3}, {a, b, c, d}]]
```

Solve::svars : Equations may not give solutions for all "solve" variables. ≫

$$\left\{b \to -\frac{2a}{3}, c \to \frac{a}{3}, d \to -\frac{5a}{3}\right\}$$

```
parameters /. {a → 3}
```

```
{b → -2, c → 1, d → -5}
```

```
computedplane = equation /. {a → 3, b → -2, c → 1, d → -5}
```

```
-5 + 3 x - 2 y + z == 0
```

```
givenplane - computedplane == 0
```

```
True
```

Manipulation

The following manipulation shows that at least for the specified range of the parameters *c* and *d*, the point-normal form of the equation of a plane is independent of the choice of the fixed points in the plane.

- Point-invariance of the point-normal form

```
plane = 3 x - 2 y + z - 5 == 0;
```

```
Solve[plane, {x, y, z}]
```

Solve::svars : Equations may not give solutions for all "solve" variables. ≫

```
{{z → 5 - 3 x + 2 y}}
```

```
Clear[c, d]
```

```
normal = {3, -2, 1};
```

```
planarvector = {x - c, y - d, z - (5 - 3 c + 2 d)};

Manipulate[Evaluate[Expand[Dot[normal, planarvector] == 0]], {c, -5, 5, 1}, {d, -5, 5, 1}]
```

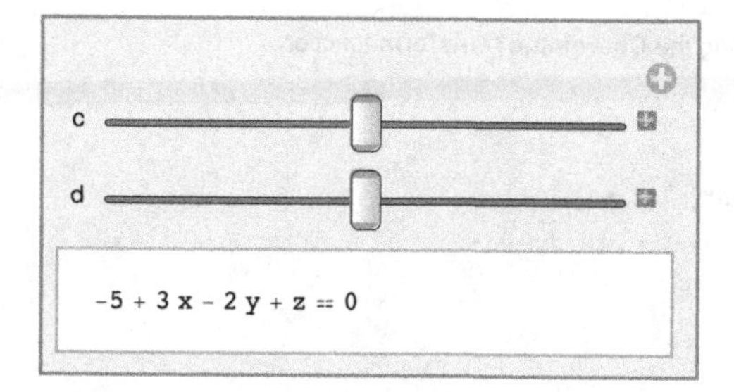

We use **Manipulate**, **Evaluate**, **Expand**, and **Dot** to explore the point-invariance of a plane in $\mathbb{R}^3$. By varying the values of *c* and *d* we can see that no matter which fixed points we choose, the point s = {-5, -1, 18} or {0, 0, 5}, for example, the equation of the plane remains unchanged.

Polar form of a complex number

The *polar form* of a nonzero complex number *z* = {*x*, *y*} is an expression of the form {r, θ}, where *x* = *r* **Cos**[θ] and y = *r* **Sin**[θ], determined by the modulus $r = \sqrt{x^2 + y^2}$ and the angle θ determined by *z* and the real coordinate axis.

Since there are infinitely many possible angles θ for which this definition makes sense, there are infinitely many representations of a complex number in polar form.

Illustration

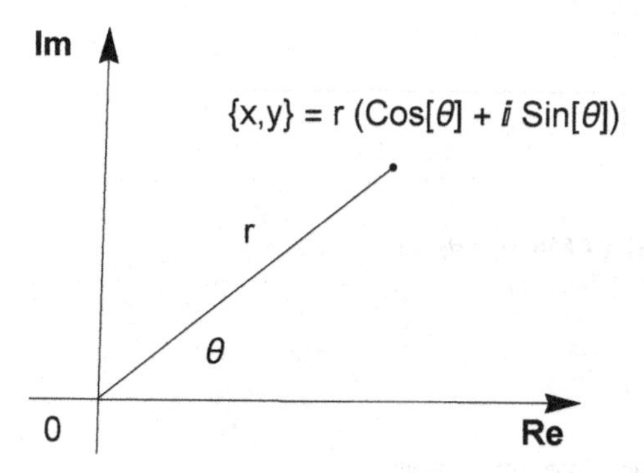

Mathematica has built-in transformation functions between the polar and Cartesian representations of complex numbers.

Illustration

- Transformation of a nonzero complex numbers using the **CoordinateTransform** function

```
Clear[r, θ]
```

```
CoordinateTransform[ "Polar" → "Cartesian", {r, θ}]
```

```
{r Cos[θ], r Sin[θ]}
```

```
Clear[x, y]
```

```
CoordinateTransform[{"Cartesian" -> "Polar"}, {x, y}]
```

$\left\{\sqrt{x^2+y^2}, \text{ArcTan}[x, y]\right\}$

- Transformation of a complex number in polar form to cartesian form

```
CoordinateTransform[ "Polar" → "Cartesian", {5, π/4}]
```

$\left\{\frac{5}{\sqrt{2}}, \frac{5}{\sqrt{2}}\right\}$

- Transformation of a complex number in cartesian form to polar form

```
CoordinateTransform[{"Cartesian" -> "Polar"}, {1, -1}]
```

$\left\{\sqrt{2}, -\frac{\pi}{4}\right\}$

- Multiplication of complex numbers in polar form

```
z1 = r1 (Cos[θ1] + I Sin[θ1]);
```

```
z2 = r2 (Cos[θ2] + I Sin[θ2]);
```

```
Simplify[z1 z2 == r1 r2 (Cos[θ1 + θ2] + I Sin[θ1 + θ2])]
True
```

Manipulation

- Using **Manipulate** to explore the product of complex numbers

```
Clear[a, b, ϕ, θ]
```

```
z1 = a (Cos[ϕ] + I Sin[ϕ]); z2 = b (Cos[θ] + I Sin[θ]);
```

```
Manipulate[a b (Cos[ϕ + θ] + I Sin[ϕ + θ]), {a, -2, 2}, {b, -2, 2}, {ϕ, 0, π}, {θ, 0, π}]
```

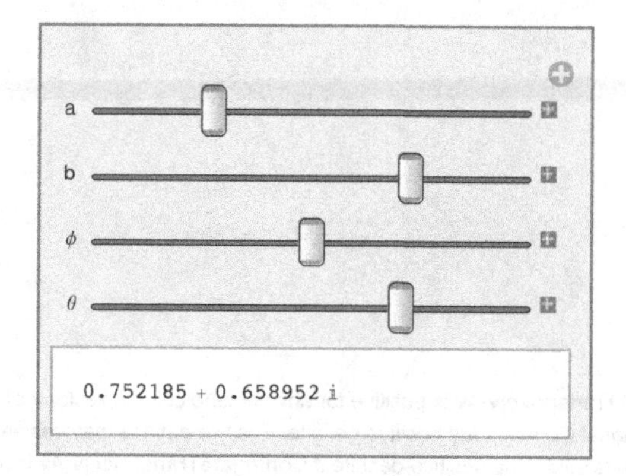


```
a = -1; b = 1; ϕ = 1.5708; θ = 2.29022;
```

```
z1 = a (Cos[ϕ] + I Sin[ϕ])
```

$3.67321 \times 10^{-6} - 1.\ i$

```
z2 = b (Cos[θ] + I Sin[θ])
```

-0.658951 + 0.752186 i

```
z1 z2
```

0.752183 + 0.658954 i

We use **Manipulate** to visualize the product complex numbers. The manipulation produces the product of the two complex numbers obtained by assigning numerical values to the parameters *a*, *b*, ϕ, and θ.

Polynomial space

The set of real polynomials ℝ[t,n] of degree less than or equal to *n* in the variable *t* forms a real vector space of dimension (*n* + 1) called a *polynomial space*.

Illustration

- Vector addition in ℝ[t,4]

ℝ[t,4] is the set of real polynomials of degree less than or equal to 4, in the variable *t*.

$p = 3 + t - 7 t^2 + t^3;$

$q = 5 t + t^4;$

```
p + q
```

$3 + 6 t - 7 t^2 + t^3 + t^4$

- Scalar multiplication in ℝ[t,6]

```
r = 3 + 6 x - 7 x^2 + x^3 + x^4 + 3 x^6;
Expand[-2 r]
```

$-6 - 12 x + 14 x^2 - 2 x^3 - 2 x^4 - 6 x^6$

Positive-definite matrix

A symmetric real *n*-by-*n* matrix *A* is positive-definite if **Transpose[v]***A***v** is positive for any nonzero column vector **v** of *n* real numbers. All real diagonal matrices with positive diagonal elements are positive-definite. Positive-definite *matrices* are used to define inner products. An *n*-by-*n* Hermitian matrix *A* is said to be positive-definite if **ConjugateTranspose[v]***A***v** is real and positive for all nonzero complex vectors **v**.

Illustration

- A positive-definite diagonal matrix

```
MatrixForm[A = DiagonalMatrix[{1, 2, 3}]]
```

$$\begin{pmatrix} 1 & 0 & 0 \\ 0 & 2 & 0 \\ 0 & 0 & 3 \end{pmatrix}$$

```
PositiveDefiniteMatrixQ[A]
```

True

- A 3-by-3 positive-definite matrix

All real symmetric matrices with positive diagonal elements are positive-definite.

$$A = \begin{pmatrix} 2 & -1 & 0 \\ -1 & 2 & 1 \\ 0 & 1 & 2 \end{pmatrix};$$

```
PositiveDefiniteMatrixQ[A]
```

True

Principal axis theorem

The principal axis theorem says that for every real quadratic form q[**x**] = $\mathbf{x}^T A\,\mathbf{x}$, there exist an orthogonal matrix Q and a diagonal matrix dM for which A = $Q\,dM\,Q^T$.

Illustration

- The principal axes of a quadratic form

```
q[x_, y_] := 3 x^2 - 2 x y + 3 y^2
```

```
Plot3D[q[x, y], {x, -5, 5}, {y, -5, 5}]
```

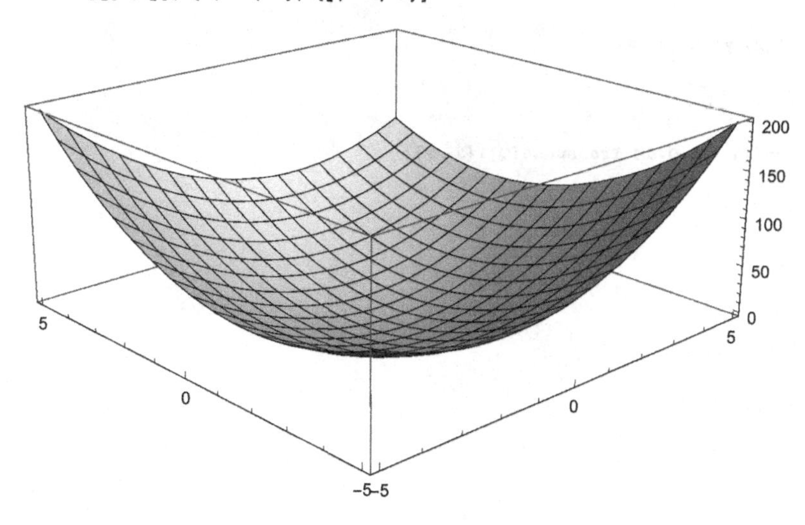


```
q[x, y] == Expand[{x, y}.(3  -1
                          -1  3).{x, y}]
```

```
True
```

```
q[x, y] == Expand[{x, y}.{{3, -1}, {-1, 3}}.{x, y}]
```

```
True
```

```
A = {{3, -1}, {-1, 3}};
```

```
evalsA = Eigenvalues[A]
```

```
{4, 2}
```

```
evecsA = Eigenvectors[A]
```

```
{{-1, 1}, {1, 1}}
```

Q = Orthogonalize[evecsA]

$$\left\{\left\{-\frac{1}{\sqrt{2}}, \frac{1}{\sqrt{2}}\right\}, \left\{\frac{1}{\sqrt{2}}, \frac{1}{\sqrt{2}}\right\}\right\}$$

dM = DiagonalMatrix[evalsA]

{{4, 0}, {0, 2}}

A == Q.dM.Transpose[Q]

True

q[x, y] == Expand[{x, y}.A.{x, y}]

$q[x, y] == 3 x^2 - 2 x y + 3 y^2$

q[x, y] == Simplify[{x, y}.Q.dM.Transpose[Q].{x, y}]

$q[x, y] == 3 x^2 - 2 x y + 3 y^2$

{u, v} = {x, y}.Q

$$\left\{-\frac{x}{\sqrt{2}} + \frac{y}{\sqrt{2}}, \frac{x}{\sqrt{2}} + \frac{y}{\sqrt{2}}\right\}$$

Transpose[Q].{x, y}

$$\left\{-\frac{x}{\sqrt{2}} + \frac{y}{\sqrt{2}}, \frac{x}{\sqrt{2}} + \frac{y}{\sqrt{2}}\right\}$$

q[u_, v_] := {u, v}.dM.{u, v}

q[u, v] == 4 u² + 2 v²

True

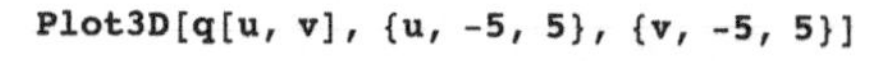

```
Plot3D[q[u, v], {u, -5, 5}, {v, -5, 5}]
```

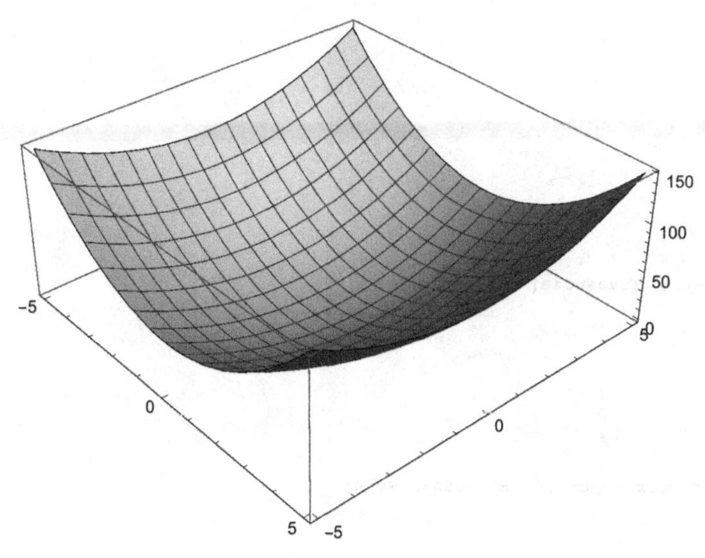

The calculation shows that q[*x*, *y*] and q[*u*, *v*] represent the same quadratic form.

Product of two vector spaces

See Cartesian product of two vector spaces

Pseudoinverse of a matrix

The pseudoinverse of a matrix (also called a Penrose matrix) is a generalization of an inverse matrix. An easy way to construct pseudoinverse matrices comes from the method of least squares. However, *Mathematica* also has a specific **PseudoInverse** function for this purpose. For any invertible matrix *A*,

```
PseudoInverse[A] = Inverse[A]
```

Illustration

- The pseudoinverse of a 3-by-3 matrix

```
RandomInteger[{0, 9}, {3, 3}]
```

$$\begin{pmatrix} 9 & 6 & 2 \\ 7 & 8 & 4 \\ 9 & 0 & 1 \end{pmatrix}$$

$$A = \begin{pmatrix} 9 & 6 & 2 \\ 7 & 8 & 4 \\ 9 & 0 & 1 \end{pmatrix}; b = \begin{pmatrix} 1 \\ 2 \\ 3 \end{pmatrix};$$

```
LinearSolve[A, b]
```

$$\begin{pmatrix} \frac{10}{51} \\ -\frac{55}{102} \\ \frac{21}{17} \end{pmatrix}$$

```
LinearSolve[Transpose[A].A, Transpose[A].b]
```

$$\begin{pmatrix} \frac{10}{51} \\ -\frac{55}{102} \\ \frac{21}{17} \end{pmatrix}$$

```
PseudoInverse[A] == Inverse[Transpose[A].A].Transpose[A]
```

True

- The pseudoinverse of a 3-by-4 matrix

$$A = \begin{pmatrix} 7 & 1 & 8 & 7 \\ 9 & 7 & 5 & 3 \\ 5 & 8 & 8 & 6 \end{pmatrix};$$

```
MatrixForm[B = PseudoInverse[A]]
```

$$\begin{pmatrix} \frac{6179}{127097} & \frac{18994}{127097} & -\frac{17488}{127097} \\ -\frac{30617}{254194} & \frac{3995}{127097} & \frac{28175}{254194} \\ \frac{5762}{127097} & -\frac{8349}{127097} & \frac{7861}{127097} \\ \frac{15159}{254194} & -\frac{10023}{127097} & \frac{12983}{254194} \end{pmatrix}$$

```
MatrixForm[A.B]
```

$$\begin{pmatrix} 1 & 0 & 0 \\ 0 & 1 & 0 \\ 0 & 0 & 1 \end{pmatrix}$$

- The pseudoinverse of an invertible matrix

```
A = RandomInteger[{0, 9}, {4, 4}];
```

$$A = \begin{pmatrix} 7 & 5 & 5 & 0 \\ 9 & 8 & 9 & 7 \\ 7 & 2 & 3 & 6 \\ 1 & 5 & 0 & 2 \end{pmatrix};$$

```
Det[A]
```

```
1099
```

```
PseudoInverse[A] == Inverse[A]
```

```
True
```

Pythagorean theorem

If **u** and **v** are two orthogonal vectors in $\mathbb{R}^n$ in the Euclidean inner product, then $\|\mathbf{u}+\mathbf{v}\|^2 = \|\mathbf{u}\|^2 + \|\mathbf{v}\|^2$, where $\|\mathbf{w}\|$ is the Euclidean norm of any vector **w**.

Illustration

- The Pythagorean theorem in $\mathbb{R}^2$

```
u = {1, -2}; v = {2, 1};
```

```
Dot[u, v]
```

```
0
```

```
Norm[u + v]^2 == Norm[u]^2 + Norm[v]^2
```

```
True
```

- The Pythagorean theorem in $\mathbb{R}^3$

```
u = {1, -2, 3}; v = {2, 1, 0};
```

```
Dot[u, v]
```

```
0
```

```
Norm[u + v]^2 == Norm[u]^2 + Norm[v]^2
```

```
True
```

Manipulation

- The Pythagorean theorem in $\mathbb{R}^3$

```
Clear[a, b]
```

```
u = {a, -2, 3}; v = {2, b, 0};
```

```
Manipulate[Norm[{a, -2, 3}]^2 + Norm[{2, b, 0}]^2 == Norm[{a, -2, 3} + {2, b, 0}]^2,
 {a, -10, 10, 1}, {b, -10, 10, 1}]
```

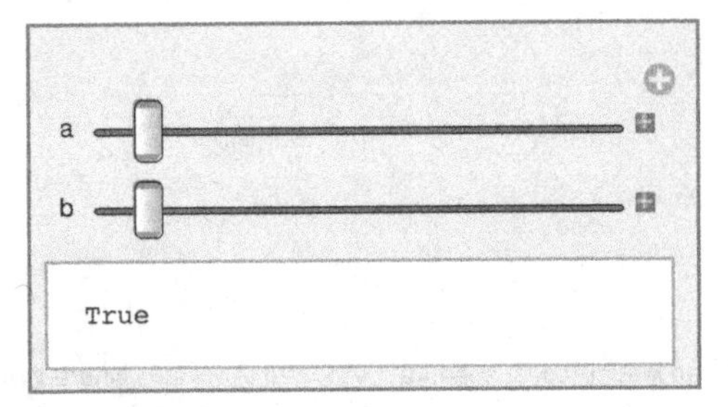

We use **Manipulate** and **Norm** to explore the Pythagorean theorem. This manipulation shows that if a = b, then the theorem holds since the vectors {a, -2, 3} and {2, b, 0} are orthogonal in the Euclidean space $\mathbb{E}^3$. For example, if a = b = - 9, then

```
Dot[{-9, -2, 3}, {2, -9, 0}]
```

0

Q

QR decomposition

The Gram–Schmidt process can be used to decompose a matrix A into a specific matrix product QR. If A is an m-by-n matrix whose n columns are linearly independent vectors in $\mathbb{R}^m$, then there exists an m-by-n matrix Q whose columns form an orthonormal set in $\mathbb{R}^m$ and an n-by-n upper-triangular matrix R such that $A = QR$.

The product matrix QR is the *QR decomposition* of the matrix A.

The built-in *Mathematica* function **QRDecomposition** makes it easy to decompose suitable matrices in to QR products.

Illustration

- A QR decomposition of a 3-by-3 matrix

```
MatrixForm[A = {{1, 3, 0}, {0, 5, 7}, {2, -8, 4}}]
```

$$\begin{pmatrix} 1 & 3 & 0 \\ 0 & 5 & 7 \\ 2 & -8 & 4 \end{pmatrix}$$

```
{q, r} = QRDecomposition[A]
```

$$\left\{\left\{\left\{\frac{1}{\sqrt{5}}, 0, \frac{2}{\sqrt{5}}\right\}, \left\{\frac{28}{\sqrt{1605}}, 5\sqrt{\frac{5}{321}}, -\frac{14}{\sqrt{1605}}\right\}, \left\{-\frac{10}{\sqrt{321}}, \frac{14}{\sqrt{321}}, \frac{5}{\sqrt{321}}\right\}\right\},\right.$$
$$\left.\left\{\left\{\sqrt{5}, -\frac{13}{\sqrt{5}}, \frac{8}{\sqrt{5}}\right\}, \left\{0, \sqrt{\frac{321}{5}}, 35\sqrt{\frac{5}{321}} - \frac{56}{\sqrt{1605}}\right\}, \left\{0, 0, \frac{118}{\sqrt{321}}\right\}\right\}\right\}$$

```
MatrixForm[Q = Transpose[q]]
```

$$\begin{pmatrix} \frac{1}{\sqrt{5}} & \frac{28}{\sqrt{1605}} & -\frac{10}{\sqrt{321}} \\ 0 & 5\sqrt{\frac{5}{321}} & \frac{14}{\sqrt{321}} \\ \frac{2}{\sqrt{5}} & -\frac{14}{\sqrt{1605}} & \frac{5}{\sqrt{321}} \end{pmatrix}$$

```
MatrixForm[R = r]
```

$$\begin{pmatrix} \sqrt{5} & -\frac{13}{\sqrt{5}} & \frac{8}{\sqrt{5}} \\ 0 & \sqrt{\frac{321}{5}} & 35\sqrt{\frac{5}{321}} - \frac{56}{\sqrt{1605}} \\ 0 & 0 & \frac{118}{\sqrt{321}} \end{pmatrix}$$

```
Q.R == A
```

True

- A *QR* decomposition of a 3-by-2 matrix

```
MatrixForm[A = {{1, 3}, {0, 5}, {2, -8}}]
```

$$\begin{pmatrix} 1 & 3 \\ 0 & 5 \\ 2 & -8 \end{pmatrix}$$

```
{q, r} = QRDecomposition[A];
```

```
Transpose[q].r == A
```

True

```
MatrixForm[q]
```

$$\begin{pmatrix} \frac{1}{\sqrt{5}} & 0 & \frac{2}{\sqrt{5}} \\ \frac{28}{\sqrt{1605}} & 5\sqrt{\frac{5}{321}} & -\frac{14}{\sqrt{1605}} \end{pmatrix}$$

```
MatrixForm[r]
```

$$\begin{pmatrix} \sqrt{5} & -\frac{13}{\sqrt{5}} \\ 0 & \sqrt{\frac{321}{5}} \end{pmatrix}$$

As we can see, the matrix *r* is upper-triangular. Since its diagonal elements are nonzero, we also know that it is invertible.

- A *QR* decomposition of a 2-by-3 matrix

```
MatrixForm[A = {{1, 3, 1}, {2, 5, 7}}]
```

$$\begin{pmatrix} 1 & 3 & 1 \\ 2 & 5 & 7 \end{pmatrix}$$

```
{q, r} = QRDecomposition[A];
```

```
q
```

$$\left\{\left\{\frac{1}{\sqrt{5}}, \frac{2}{\sqrt{5}}\right\}, \left\{\frac{2}{\sqrt{5}}, -\frac{1}{\sqrt{5}}\right\}\right\}$$

```
Transpose[q] == Inverse[q]
```

```
True
```

```
r
```

$$\left\{\left\{\sqrt{5}, \frac{3}{\sqrt{5}} + 2\sqrt{5}, 3\sqrt{5}\right\}, \left\{0, \frac{1}{\sqrt{5}}, -\sqrt{5}\right\}\right\}$$

The matrix **q** is orthogonal and the matrix **r** is upper-triangular, as expected.

Manipulation

- Exploring *QR* decompositions

```
MatrixForm[A = {{a, 3}, {2, b}}];
```

```
{q, r} = QRDecomposition[A];
```

```
Manipulate[Evaluate[q], {a, -4, 4, 1}, {b, -4, 4, 1}]
```

a

b

$$\left\{\left\{-\frac{2}{\sqrt{5}}, \frac{1}{\sqrt{5}}\right\}, \left\{-\frac{1}{\sqrt{5}}, -\frac{2}{\sqrt{5}}\right\}\right\}$$

```
Manipulate[Evaluate[r], {a, -4, 4, 1}, {b, -4, 4, 1}]
```

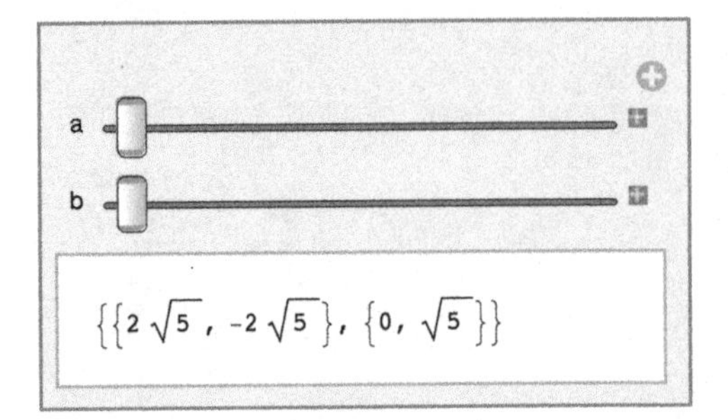

We use **Manipulate**, **QRDecomposition**, and **Evaluate** to explore the matrices q and r produced by a QR decomposition. If *a* = *b* = - 4, for example, the generated matrix **q** is orthogonal and the associated matrix **r** is upper-triangular:

$$q = \left\{\left\{-\frac{2}{\sqrt{5}}, \frac{1}{\sqrt{5}}\right\}, \left\{-\frac{1}{\sqrt{5}}, -\frac{2}{\sqrt{5}}\right\}\right\};$$

```
OrthogonalMatrixQ[q]
```

```
True
```

and

$$\text{MatrixForm}\left[r = \left\{\left\{2\sqrt{5}, -2\sqrt{5}\right\}, \left\{0, \sqrt{5}\right\}\right\}\right]$$

$$\begin{pmatrix} 2\sqrt{5} & -2\sqrt{5} \\ 0 & \sqrt{5} \end{pmatrix}$$

Quadratic form

A *quadratic form q*: *V* ⟶ ℝ on a real inner product space *V* is a function q[**v**] = f[**v**, **v**] for some symmetric bilinear form *f*: *V*× *V* ⟶ ℝ.

Illustration

- A quadratic form defined by a 2-by-2 symmetric matrix

```
Clear[x, y, v, q]
```

```
MatrixForm[A = {{1, 2}, {2, 3}}]
```

$$\begin{pmatrix} 1 & 2 \\ 2 & 3 \end{pmatrix}$$

```
v = {x, y};
```

```
q[v_] := Expand[v.A.v]
```

```
q[v]
```

$x^2 + 4 x y + 3 y^2$

- Converting a quadratic form to a matrix product

```
Clear[q, x, y, a, b, c]
```

```
q[{x_, y_}] := 7 x^2 - 9 y^2
```

```
Expand[{x, y}.{{a, b}, {b, c}}.{x, y}]
```

$a x^2 + 2 b x y + c y^2$

Thus $a = 7$, $2b = 0$, and $c = -9$, so that

```
q[{x, y}] == {x, y}.{{7, 0}, {0, -9}}.{x, y}
```

True

- A quadratic form on $\mathbb{R}^3$

```
Clear[x, y, z, a, b, c, d, e, f]
```

```
q[{x_, y_, z_}] := 3 x^2 - 2 x y + 3 x z + 6 y^2 + 5 y z - 8 z^2
```

```
MatrixForm[A = {{a, b, c}, {b, d, e}, {c, e, f}}]
```

$$\begin{pmatrix} a & b & c \\ b & d & e \\ c & e & f \end{pmatrix}$$

```
Expand[{x, y, z}.{{a, b, c}, {b, d, e}, {c, e, f}}.{x, y, z}]
```

$a x^2 + 2 b x y + d y^2 + 2 c x z + 2 e y z + f z^2$

```
MatrixForm[B = A /. {a → 3, b → -1, c → 3/2, d → 6, e → 5/2, f → -8}]
```

$$\begin{pmatrix} 3 & -1 & \frac{3}{2} \\ -1 & 6 & \frac{5}{2} \\ \frac{3}{2} & \frac{5}{2} & -8 \end{pmatrix}$$

```
q[{x, y, z}] == Expand[{x, y, z}.B.{x, y, z}]
```

True

Quintic polynomial

A *quintic polynomial* is a complex polynomial of degree 5.

The fundamental theorem of algebra guarantees that every polynomial equation of the form $p = 0$, involving a polynomial p of degree n, has n roots in the complex plane (counting multiplicity of the roots). According to the theorem of the unsolvability of the quintic, there is no algorithm for finding the solutions of arbitrary polynomials of degree 5 or higher consisting only of the four arithmetic operations together with the extraction of roots.

Numerical approximation techniques for solving polynomial equations involving polynomials of degree 5 or higher must usually be applied.

Illustration

```
MatrixForm[
 A = {{7, 1, 1, 9, 4}, {7, 7, 7, 2, 9}, {3, 2, 5, 6, 0}, {2, 1, 2, 2, 1}, {4, 0, 9, 1, 7}}]
```

$$\begin{pmatrix} 7 & 1 & 1 & 9 & 4 \\ 7 & 7 & 7 & 2 & 9 \\ 3 & 2 & 5 & 6 & 0 \\ 2 & 1 & 2 & 2 & 1 \\ 4 & 0 & 9 & 1 & 7 \end{pmatrix}$$

```
cpt = CharacteristicPolynomial[A, t]
```

$-1342 - 1508\,t + 968\,t^2 - 231\,t^3 + 28\,t^4 - t^5$

```
Roots[cpt == 0, t]
```

```
t == Root[1342 + 1508 #1 - 968 #1^2 + 231 #1^3 - 28 #1^4 + #1^5 &, 1] ||
 t == Root[1342 + 1508 #1 - 968 #1^2 + 231 #1^3 - 28 #1^4 + #1^5 &, 2] ||
 t == Root[1342 + 1508 #1 - 968 #1^2 + 231 #1^3 - 28 #1^4 + #1^5 &, 3] ||
 t == Root[1342 + 1508 #1 - 968 #1^2 + 231 #1^3 - 28 #1^4 + #1^5 &, 4] ||
 t == Root[1342 + 1508 #1 - 968 #1^2 + 231 #1^3 - 28 #1^4 + #1^5 &, 5]
```

The symbol # represents the values of the root. These values can be approximated numerically using the *Mathematica* **N[ ]** command:

```
N[Roots[cpt == 0, t]]
```

```
t == -0.61186 || t == 4.45378 || t == 17.7945 || t == 3.18179 - 4.18941 i || t == 3.18179 + 4.18941 i
```

R

Random matrix

Mathematica can generate a *random matrix* in several ways. The commands **RandomInteger** and **RandomReal** are the two basic commands.

In the commands **RandomInteger[**{a ,b}, {c, d}**]** and **RandomReal[**{a ,b}, {c, d}**]**, the interval {a, b} determines the range of integers or real numbers between a and b, and the pair {c, d} determines the dimensions of the matrix.

Illustration

- A random 3-by-4 matrix with integer elements

```
A = RandomInteger[{0, 9}, {3, 4}];
```

$$A = \begin{pmatrix} 0 & 8 & 4 & 2 \\ 4 & 6 & 8 & 1 \\ 8 & 6 & 5 & 1 \end{pmatrix};$$

```
Dimensions[%]
```

```
{3, 4}
```

- A random 4-by-3 matrix with real elements

```
A = RandomReal[{0, 9}, {4, 3}];
```

$$A = \begin{pmatrix} 7.016438851674735` & 5.594529085595454` & 1.7651835472965853` \\ 5.9705522429768525` & 1.1846711298028172` & 2.073325573818849` \\ 5.688608735820997` & 8.284427980487944` & 4.309072260274204` \\ 3.683700829298914` & 5.904113691672606` & 8.621603444105805` \end{pmatrix};$$

```
Dimensions[%]
```

```
{4, 3}
```

Manipulation

- A family of matrices with random integer elements

```
Manipulate[MatrixForm[RandomInteger[{0, 9}, {a, b}]], {a, 2, 5, 1}, {b, 2, 5, 1}]
```

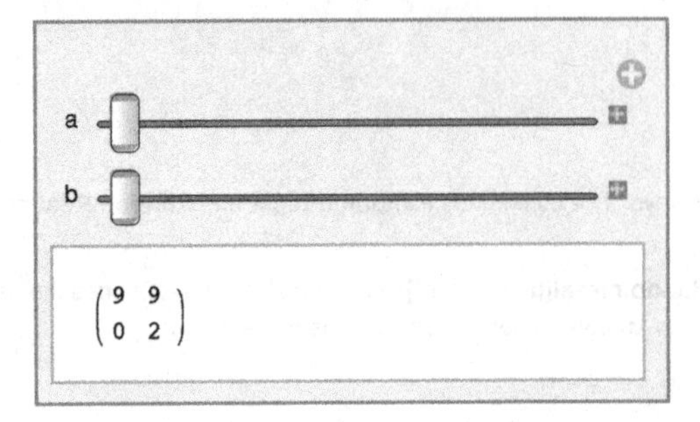

We use **Manipulate**, **MatrixForm**, and **RandomInteger** to generate random 2-by-2 matrices with integer elements displayed in two-dimensional form. Every evaluation of the Manipulate command generates a new matrix.

- A family of matrices with random real elements

```
Manipulate[MatrixForm[RandomReal[{0, 9}, {a, b}]], {a, 2, 5, 1}, {b, 2, 5, 1}]
```

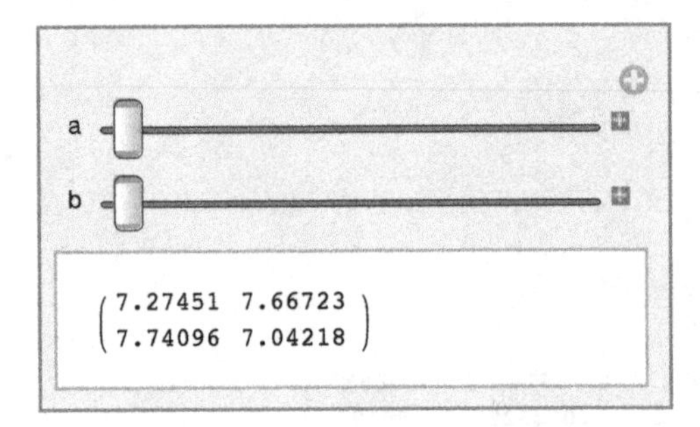

We use **Manipulate**, **MatrixForm**, and **RandomReal** to generate random 2-by-2 matrices with real elements displayed in two-dimensional form. Every evaluation of the **Manipulate** command generates a new matrix.

Range of a linear transformation

The *range* of a linear transformation $T : V \longrightarrow W$ from a vector space V to a vector space W is the set of all vectors $T[\mathbf{u}]$ in W. The range of T is a subspace of W. If T is a matrix transformation, then its range is the column space of T. The range of T is also called the *image* of the transformation.

Illustration

- The range of a 2-by-4 matrix transformation

```
A = {{7, 3, 8, 1}, {3, 8, 1, 0}};
```

```
v = {1, 2, 3, 4};
```

The vector **b** = {41, 22} is one of the vectors in the range of the transformation defined by the matrix *A* since it is a solution of the matrix equation *A***v** = **b**:

```
A.v == {41, 22}
```

```
True
```

Now consider a general vector in $\mathbb{R}^4$ and its image vectors A.**u**:

```
Clear[a, b, c, d]
```

```
u = {a, b, c, d};
```

```
A.u
```

```
{7 a + 3 b + 8 c + d, 3 a + 8 b + c}
```

The range of the transformation defined by *A* is the set of all vectors of the form

```
{7 a + 3 b + 8 c + d, 3 a + 8 b + c}
```

```
a = 1; b = 2; c = 3; d = 4;
```

```
{7 a + 3 b + 8 c + d, 3 a + 8 b + c} == {41, 22}
```

```
True
```

Notice that the vector {41, 22} satisfies this equation for the vector {1, 2, 3, 4}.

- The range of a 3-by-2 matrix transformation

```
MatrixForm[A = {{1, 6}, {7, 1}, {4, 2}}]
```

$$\begin{pmatrix} 1 & 6 \\ 7 & 1 \\ 4 & 2 \end{pmatrix}$$

```
A.{u, v}
```

```
{{7, 14, 21, 28}, {8, 16, 24, 32}, {6, 12, 18, 24}}
```

The range of the matrix transformation *A* consists of all vectors {*x*, *y*, *z*} in $\mathbb{R}^3$ satisfying the conditions

```
{x, y, z} = {u + 6 v, 7 u + v, 4 u + 2 v}
```

```
{{7, 14, 21, 28}, {8, 16, 24, 32}, {6, 12, 18, 24}}
```

for all real numbers *u* and *v*. Here are two such vectors:

```
{u + 6 v, 7 u + v, 4 u + 2 v} /. {u → 0, v → 0}
```

```
{{7, 14, 21, 28}, {8, 16, 24, 32}, {6, 12, 18, 24}}
```

```
{u + 6 v, 7 u + v, 4 u + 2 v} /. {u → 1, v → 1}
```

```
{{7, 14, 21, 28}, {8, 16, 24, 32}, {6, 12, 18, 24}}
```

- The column space of A

```
Clear[A, B, v, s, r, a, b, c, d]
```

$$A = \begin{pmatrix} 7 & 3 & 8 & 1 \\ 3 & 8 & 1 & 0 \end{pmatrix};$$

```
MatrixForm[B = RowReduce[Transpose[A]]]
```

$$\begin{pmatrix} 1 & 0 \\ 0 & 1 \\ 0 & 0 \\ 0 & 0 \end{pmatrix}$$

```
v = {1, 2, 3, 4};
```

```
solution = Flatten[Solve[A.v == r B[[1]] + s B[[2]], {s, r}]]
```

```
{s → 22, r → 41}
```

This calculation shows that the vector {22, 41} lies in the range of the linear transformation defined by the matrix A.

$$\text{Reduce}\left[\begin{pmatrix} 22 \\ 41 \end{pmatrix} == a \begin{pmatrix} 7 \\ 3 \end{pmatrix} + b \begin{pmatrix} 3 \\ 8 \end{pmatrix} + c \begin{pmatrix} 8 \\ 1 \end{pmatrix} + d \begin{pmatrix} 1 \\ 0 \end{pmatrix}, \{a, b, c, d\}\right]$$

```
c == 41 - 3 a - 8 b && d == -306 + 17 a + 61 b
```

$$\begin{pmatrix} 22 \\ 41 \end{pmatrix} == \begin{pmatrix} 7 \\ 3 \end{pmatrix} + \begin{pmatrix} 3 \\ 8 \end{pmatrix} + (41 - 3 - 8) \begin{pmatrix} 8 \\ 1 \end{pmatrix} - 228 \begin{pmatrix} 1 \\ 0 \end{pmatrix}$$

```
True
```

Rank-deficient matrix

An n-by-n matrix whose rank is not n is said to be *rank-deficient.*

Illustration

- A rank-deficient 5-by-5 matrix

```
A = Normal[SparseArray[{{1, 1} → 1, {2, 2} → 0, {3, 3} → 3, {5, 5} → 4}]]
```

```
{{1, 0, 0, 0, 0}, {0, 0, 0, 0, 0}, {0, 0, 3, 0, 0}, {0, 0, 0, 0, 0}, {0, 0, 0, 0, 4}}
```

```
{Dimensions[A], MatrixRank[A]}
```

```
{{5, 5}, 3}
```

The rank of *A* is 3, while its dimensions are 5-by-5. This means that the matrix *A* is rank-deficient.

Manipulation

```
Clear[A, a, b]
```

```
MatrixForm[A = {{8, a, 4, 6}, {4, 2, b, 6}, {a, 0, 0, 0}, {0, 0, 0, 0}}]
```

$$\begin{pmatrix} 8 & a & 4 & 6 \\ 4 & 2 & b & 6 \\ a & 0 & 0 & 0 \\ 0 & 0 & 0 & 0 \end{pmatrix}$$

```
Manipulate[MatrixRank[{{8, a, 4, 6}, {4, 2, b, 6}, {a, 0, 0, 0}, {0, 0, 0, c}}],
 {a, -2, 4, 1}, {b, -3, 6, 1}, {c, -2, 2, 1}]
```

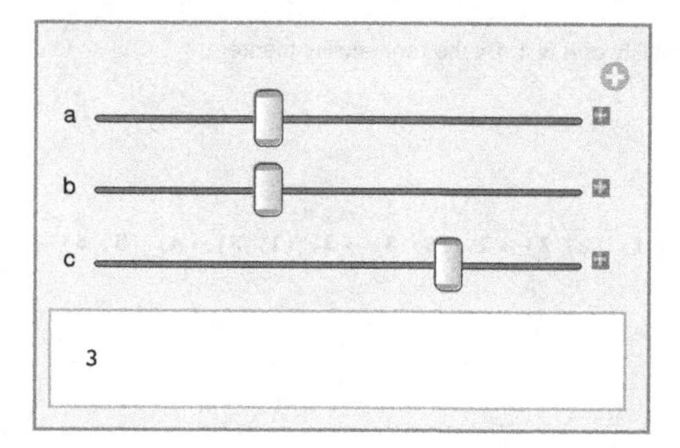

We use **Manipulate** and **MatrixRank** to explore rank-deficient matrices. The manipulation shows that if $a = b = c = 0$, the resulting matrix has rank 2. If $a = b = 0$, and $c = 1$, the resulting matrix has rank 3. If $a = 1$, $b = 0$, and $c = 1$, the resulting matrix has rank 4 and is of full rank.

Rank–nullity theorem

The *rank–nullity theorem* guarantees that for any linear transformation $T: V \longrightarrow W$ connecting two finite-dimensional vector spaces *V* and *W*, the dimension of *V* is the sum of the dimensions of the *kernel* and the *image* of *T*.

Illustration

- The rank and nullity of a 5-by-4 matrix

```
MatrixForm[A = {{0, 1, 1, 0, 0}, {1, 0, 1, 0, 0}, {0, 0, 1, 0, 0}, {1, 1, 1, 1, 1}}]
```

$$\begin{pmatrix} 0 & 1 & 1 & 0 & 0 \\ 1 & 0 & 1 & 0 & 0 \\ 0 & 0 & 1 & 0 & 0 \\ 1 & 1 & 1 & 1 & 1 \end{pmatrix}$$

The matrix A represents a linear transformation from $\mathbb{R}^5$ to $\mathbb{R}^4$

```
MatrixRank[A]
```

4

```
NullSpace[A]
```

{{0, 0, 0, -1, 1}}

```
Dimensions[%]
```

{1, 5}

Since the null space of *A* contains a single vector, the nullity of *A* is 1. By the rank–nullity theorem:

rank + nullity = 4 + 1 = 5 = dimension of $\mathbb{R}^5$

- The rank–nullity theorem for a 5-by-6 matrix

```
MatrixForm[A = Normal[SparseArray[{{1, 1} → 1, {2, 2} → 2, {3, 3} → 3, {1, 3} → 4, {5, 6} → -3}]]]
```

$$\begin{pmatrix} 1 & 0 & 4 & 0 & 0 & 0 \\ 0 & 2 & 0 & 0 & 0 & 0 \\ 0 & 0 & 3 & 0 & 0 & 0 \\ 0 & 0 & 0 & 0 & 0 & 0 \\ 0 & 0 & 0 & 0 & 0 & -3 \end{pmatrix}$$

The matrix *A* represents a linear transformation from $\mathbb{R}^6$ to $\mathbb{R}^5$

```
MatrixRank[A]
```

4

```
NullSpace[A]
```

{{0, 0, 0, 0, -1, 0}, {0, 0, 0, -1, 0, 0}}

```
Dimensions[%]
```

{2, 6}

Since the null space of *A* contains two vectors, the nullity of *A* is 2. By the rank–nullity theorem,

rank + nullity = 4 + 2 = 6 = dimension of $\mathbb{R}^6$

Rank of a matrix

The *rank* of a matrix is the number of linearly independent rows of the matrix. It is also called the *row rank* of the matrix. The row rank of a matrix equals its *column rank*, the number of linearly independent columns of the matrix.

The rank is also the number of pivots of a reduced row echelon matrix.

Illustration

- A 5-by-5 matrix of rank 5

```
A = {{6, 5, 7, 3, 6, 1, 0}, {5, 5, 2, 8, 8, 4, 5}, {6, 1, 8, 0, 9, 5, 4},
   {9, 7, 2, 6, 3, 1, 3}, {4, 2, 7, 5, 0, 2, 0}, {13, 9, 9, 11, 3, 3, 3}};
```

```
MatrixRank[A]
```

5

As expected, the reduced row echelon matrix *B* determined by the matrix *A* has five pivot columns:

```
MatrixForm[B = RowReduce[A]]
```

$$\begin{pmatrix} 1 & 0 & 0 & 0 & 0 & \frac{164}{313} & \frac{823}{939} \\ 0 & 1 & 0 & 0 & 0 & -\frac{2075}{1878} & -\frac{1085}{939} \\ 0 & 0 & 1 & 0 & 0 & -\frac{205}{3756} & -\frac{967}{1878} \\ 0 & 0 & 0 & 1 & 0 & \frac{625}{1252} & \frac{905}{1878} \\ 0 & 0 & 0 & 0 & 1 & \frac{709}{1878} & \frac{419}{939} \\ 0 & 0 & 0 & 0 & 0 & 0 & 0 \end{pmatrix}$$

- A 3-by-4 matrix of rank 2

```
A = {{0, 0, 0, 0}, {1, 0, 0, 0}, {0, 1, 0, 0}};
```

The reduced row echelon matrix *B* determined by the matrix *A* has two pivot columns:

```
MatrixForm[B = RowReduce[A]]
```

$$\begin{pmatrix} 1 & 0 & 0 & 0 \\ 0 & 1 & 0 & 0 \\ 0 & 0 & 0 & 0 \end{pmatrix}$$

Hence the rank of the matrix A must be 2:

```
MatrixRank[A]
2
```

- A 2-by-4 matrix of rank 0

```
A = {{0, 0, 0, 0}, {0, 0, 0, 0}};
```

The reduced row echelon matrix B determined by the matrix A has no pivot columns:

```
MatrixForm[B = RowReduce[A]]
```

$$\begin{pmatrix} 0 & 0 & 0 & 0 \\ 0 & 0 & 0 & 0 \end{pmatrix}$$

Hence the rank of the matrix A must be zero:

```
MatrixRank[A]
```

0

Manipulation

- Exploring the rank of 4-by-4 matrices

```
Manipulate[MatrixRank[{{8, a, 4, 6}, {4, 2, b, 6}, {a, 0, 0, 0}, {0, b, 0, 0}}],
 {a, -2, 4, 1}, {b, -3, 6, 1}]
```

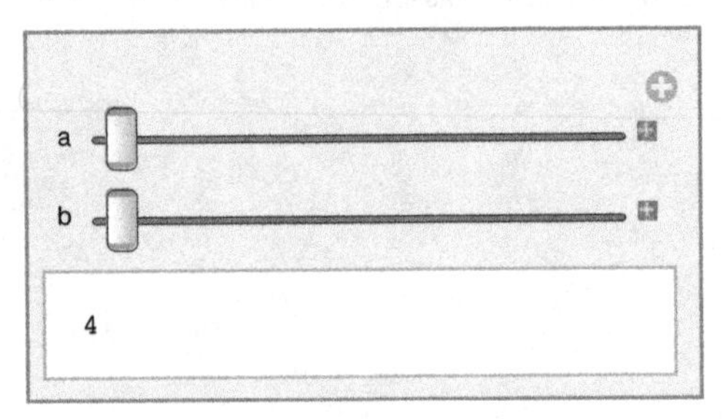

We use **Manipulate** and **MatrixRank** to generate 4-by-4 matrices with possibly different rank. The assignment of a = - 2 and b = - 3, for example, generates a matrix with rank 4. The assignment of $a = 0$ and $b = -3$ generates a matrix of rank 3.

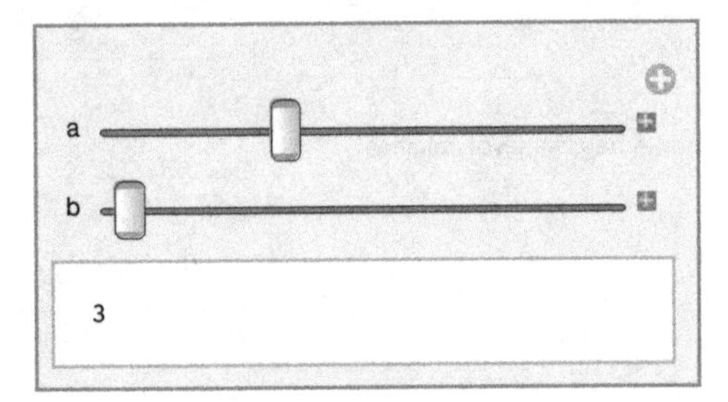

Rational canonical form

A matrix is in *rational canonical form* if it is a direct sum of companion matrices. Every square matrix is similar to a matrix in rational canonical form.

As discussed in the section on companion matrices, the following defined *Mathematica* function can be used to calculate companion matrices:

```
CompanionMatrix[p_, x_] := Module[{n, w = CoefficientList[p, x]}, w = -w / Last[w];
  n = Length[w] - 1;
  SparseArray[{{i_, n} :> w[[i]], {i_, j_} /; i == j + 1 → 1}, {n, n}]]
```

Illustration

- A matrix in rational canonical form

```
Clear[A, x]
```

```
p1 = 1 + 3 x^2 - 2 x^3 + x^4;
```

```
MatrixForm[array1 = Normal[CompanionMatrix[p1, x]]]
```

$$\begin{pmatrix} 0 & 0 & 0 & -1 \\ 1 & 0 & 0 & 0 \\ 0 & 1 & 0 & -3 \\ 0 & 0 & 1 & 2 \end{pmatrix}$$

```
p2 = 2 x + 5 x^2 + x^3;
```

```
MatrixForm[Normal[array2 = CompanionMatrix[p2, x]]]
```

$$\begin{pmatrix} 0 & 0 & 0 \\ 1 & 0 & -2 \\ 0 & 1 & -5 \end{pmatrix}$$

```
MatrixForm[ap1 = ArrayPad[array1, {0, 3}]]
```

$$\begin{pmatrix} 0 & 0 & 0 & -1 & 0 & 0 & 0 \\ 1 & 0 & 0 & 0 & 0 & 0 & 0 \\ 0 & 1 & 0 & -3 & 0 & 0 & 0 \\ 0 & 0 & 1 & 2 & 0 & 0 & 0 \\ 0 & 0 & 0 & 0 & 0 & 0 & 0 \\ 0 & 0 & 0 & 0 & 0 & 0 & 0 \\ 0 & 0 & 0 & 0 & 0 & 0 & 0 \end{pmatrix}$$

```
MatrixForm[ap2 = ArrayPad[array2, {4, 0}] ]
```

$$\begin{pmatrix} 0 & 0 & 0 & 0 & 0 & 0 & 0 \\ 0 & 0 & 0 & 0 & 0 & 0 & 0 \\ 0 & 0 & 0 & 0 & 0 & 0 & 0 \\ 0 & 0 & 0 & 0 & 0 & 0 & 0 \\ 0 & 0 & 0 & 0 & 0 & 0 & 0 \\ 0 & 0 & 0 & 0 & 1 & 0 & -2 \\ 0 & 0 & 0 & 0 & 0 & 1 & -5 \end{pmatrix}$$

```
MatrixForm[A = ap1 + ap2]
```

$$\begin{pmatrix} 0 & 0 & 0 & -1 & 0 & 0 & 0 \\ 1 & 0 & 0 & 0 & 0 & 0 & 0 \\ 0 & 1 & 0 & -3 & 0 & 0 & 0 \\ 0 & 0 & 1 & 2 & 0 & 0 & 0 \\ 0 & 0 & 0 & 0 & 0 & 0 & 0 \\ 0 & 0 & 0 & 0 & 1 & 0 & -2 \\ 0 & 0 & 0 & 0 & 0 & 1 & -5 \end{pmatrix}$$

The matrix *A* is in rational canonical form since it is a direct sum of companion matrices.

Rayleigh quotient

The *Rayleigh quotient* of a Hermitian matrix A and a nonzero vector v is the scalar

$$\frac{\texttt{Conjugate[v].A.v}}{\texttt{Conjugate[v].v}} \tag{1}$$

If *A* and **v** are real, then *A* is symmetric and the conjugate transpose of **v** is the transpose of **v** (or simply **v** in *Mathematica*). The Rayleigh coefficient occurs in problems in engineering, pattern recognition, and related fields.

Illustration

- The Rayleigh coefficient of a 2-by-2 Hermitian matrix and a vector v in $\mathbb{C}^2$

```
Clear[v, A, R]
```

```
MatrixForm[A = {{1, 3 + 4 I}, {3 - 4 I, 2}}]
```

$$\begin{pmatrix} 1 & 3+4i \\ 3-4i & 2 \end{pmatrix}$$

```
HermitianMatrixQ[A]
```

```
True
```

$$\texttt{R[A\_, v\_] :=} \frac{\texttt{Conjugate[v].A.v}}{\texttt{Conjugate[v].v}}$$

```
v = {3 - I, 5};
```

```
R[A, v]
```

$$\frac{22}{7}$$

Rectangular matrix

An *m*-by-*n* matrix is *rectangular* if $m \neq n$.

Illustration

- A random real rectangular 3-by-5 matrix

```
MatrixForm[A = RandomReal[{0, 9}, {3, 5}]]
```

$$\begin{pmatrix} 8.05463 & 4.62611 & 0.29784 & 6.95849 & 4.99793 \\ 5.69969 & 3.86907 & 4.76395 & 1.14268 & 4.36128 \\ 7.75929 & 2.2516 & 8.12218 & 3.29463 & 8.73757 \end{pmatrix}$$

$$A = \begin{pmatrix} 4.23662 & 0.922396 & 6.00001 & 0.17665 & 4.94943 \\ 3.32964 & 4.50523 & 6.1601 & 0.984526 & 5.48412 \\ 7.75887 & 6.37372 & 0.299192 & 6.55342 & 5.90837 \end{pmatrix}$$

```
Dimensions[%]
```

{3, 5}

- A random rectangular 6-by-2 matrix with integer elements

```
MatrixForm[A = RandomInteger[{0, 9}, {6, 2}]]
```

$$\begin{pmatrix} 0 & 6 \\ 6 & 8 \\ 3 & 0 \\ 5 & 6 \\ 8 & 7 \\ 7 & 0 \end{pmatrix}$$

```
Dimensions[%]
```

{6, 2}

Reduced row echelon matrix

The leading entry of a nonzero row of a matrix in row echelon form is called a *pivot* of the matrix. An *m*-by-*n* row echelon matrix is in reduced row echelon form if it has the following properties: Either the matrix is a zero matrix or all of its pivots are 1 and all entries above its pivots are 0.

The **RowReduce** function of *Mathematica* reduces a matrix to its reduced row echelon form.

Illustration

- A 4-by-5 reduced row echelon matrix

$$\begin{pmatrix} 0 & 0 & 1 & 8 & 1 \\ 0 & 0 & 0 & 5 & 4 \\ 0 & 0 & 0 & 0 & 6 \\ 0 & 0 & 0 & 0 & 0 \end{pmatrix}$$

The matrix is a matrix in row echelon form, but is not in reduced row echelon form.

- Reduction of a 4-by-5 matrix to reduced row echelon form

$$\mathbf{A} = \begin{pmatrix} 0 & 0 & 1 & 8 & 1 \\ 0 & 0 & 0 & 5 & 4 \\ 0 & 0 & 0 & 0 & 6 \\ 0 & 0 & 0 & 0 & 0 \end{pmatrix}$$

```
{{0, 0, 1, 8, 1}, {0, 0, 0, 5, 4}, {0, 0, 0, 0, 6}, {0, 0, 0, 0, 0}}
```

```
MatrixForm[B = RowReduce[A]]
```

$$\begin{pmatrix} 0 & 0 & 1 & 0 & 0 \\ 0 & 0 & 0 & 1 & 0 \\ 0 & 0 & 0 & 0 & 1 \\ 0 & 0 & 0 & 0 & 0 \end{pmatrix}$$

- Reduction of a random 3-by-3 real matrix to reduced row echelon form

```
A = RandomReal[{0, 9}, {3, 3}];
```

$$\mathbf{A} = \begin{pmatrix} 2.0716982088321423` & 2.497526720307203` & 7.236524211079711` \\ 5.554095004848561` & 1.9196532271480322` & 0.9904680439688676` \\ 5.341640232827222` & 1.875793841070399` & 8.202469374845403` \end{pmatrix};$$

```
MatrixForm[B = RowReduce[A]]
```

$$\begin{pmatrix} 1 & 0. & 0. \\ 0 & 1 & 0. \\ 0 & 0 & 1 \end{pmatrix}$$

Reflection

The left-multiplications $A\mathbf{v}$, $B\mathbf{v}$, and $R\mathbf{v}$ of a vector $\mathbf{v}$ in $\mathbb{R}^2$ by the matrices

```
MatrixForm[A = {{-1, 0}, {0, 1}}]
```

$$\begin{pmatrix} -1 & 0 \\ 0 & 1 \end{pmatrix}$$

```
MatrixForm[B = {{1, 0}, {0, -1}}]
```

$$\begin{pmatrix} 1 & 0 \\ 0 & -1 \end{pmatrix}$$

```
MatrixForm[R = {{0, 1}, {1, 0}}]
```

$$\begin{pmatrix} 0 & 1 \\ 1 & 0 \end{pmatrix}$$

respectively, represent *reflections* of **v** about the *x*-axis, the *y*-axis, and the line $y = x$.

These and more general reflections can be produced with the built-in **ReflectionMatrix** function.

```
MatrixForm[A == ReflectionMatrix[{1, 0}]]
```

```
True
```

```
MatrixForm[B == ReflectionMatrix[{0, 1}]]
```

```
True
```

```
MatrixForm[R == ReflectionMatrix[{1, -1}]]
```

```
True
```

Illustration

- Three reflections of a vector in $\mathbb{R}^2$

```
v = {3, 5};
```

```
{A.v, B.v, R.v}
```

```
{{-3, 5}, {3, -5}, {5, 3}}
```

As expected, the vector {3, 5} became {-3, 5} when reflected about the *x*-axis, {3, -5} when reflected about the *y*-axis, and {5, 3} when reflected about the line $y = x$.

- Three reflections of a general vector in $\mathbb{R}^2$

```
Clear[x, y]
```

```
v = {x, y};
```

```
{A.v, B.v, R.v}
```

```
{{-x, y}, {x, -y}, {y, x}}
```

- Reflections of vectors in $\mathbb{R}^3$ about a plane through the origin

```
vector1 = {1, 2, 3};
plane = Dot[{1, 1, 1}, {x, y, z}] == 0;
MatrixForm[R3D = ReflectionMatrix[{1, 1, 1}]]
```

$$\begin{pmatrix} \frac{1}{3} & -\frac{2}{3} & -\frac{2}{3} \\ -\frac{2}{3} & \frac{1}{3} & -\frac{2}{3} \\ -\frac{2}{3} & -\frac{2}{3} & \frac{1}{3} \end{pmatrix}$$

```
R3D.vector1
```

```
{-3, -2, -1}
```

```
vector2 = {1, 2, -3}
R3D.vector2
```

```
{1, 2, -3}
```

The last calculation illustrates the fact that reflections of points in the plane remain fixed.

Roots of unity

Solutions of equations of the form $x^n = 1$ are called *roots of unity.*

Illustration

- Second roots of unity

```
roots2 = x^2 == 1;
```

```
Solve[roots2]
```

```
{{x → -1}, {x → 1}}
```

- Third roots of unity

```
roots3 = x^3 == 1;
```

```
Solve[roots3]
```

$$\{\{x \to 1\}, \{x \to -(-1)^{1/3}\}, \{x \to (-1)^{2/3}\}\}$$

```
points1 = {Re[-(-1)^(1/3)], Im[-(-1)^(1/3)]}
```

$$\left\{-\frac{1}{2}, -\frac{\sqrt{3}}{2}\right\}$$

```
points2 = {Re[(-1)^(2/3)], Im[(-1)^(2/3)]}
```

$\left\{-\frac{1}{2}, \frac{\sqrt{3}}{2}\right\}$

- Fourth roots of unity

```
roots4 = x^4 == 1;
```

```
Solve[roots4]
```

```
{{x → -1}, {x → -i}, {x → i}, {x → 1}}
```

- Fifth roots of unity

```
roots5 = Table[{Cos[2 Pi * i / 5], Sin[2 Pi * i / 5]}, {i, 0, 4}];
```

```
pl1 := ListPlot[roots5, PlotMarkers → {Automatic, Medium},
   PlotLabel → Style["Roots of Unity 1^(1/5)", 16, Bold]];
```

```
pl2 := PolarPlot[1, {t, 0, 2 Pi}, PlotStyle → Red];
Show[{pl1, pl2}, PlotRange → {{-1.5, 1.5}, {-1.5, 1.5}}, AspectRatio → Automatic]
```

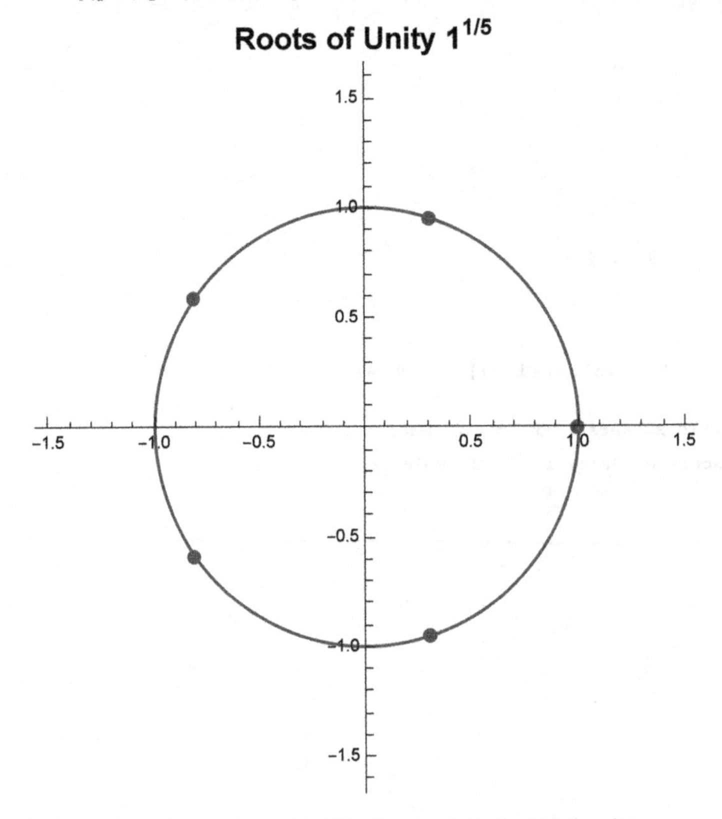

This plot shows the locations of the fifth roots of unity on the complex plane.

Manipulation

- Roots of unity

```
Manipulate[roots = Table[{Cos[2 Pi*n/a], Sin[2 Pi*n/a]}, {n, 0, (a + 1)}], {a, 1, 7, 1}]
```

a

$$\left\{\{1, 0\}, \left\{-\frac{1}{2}, \frac{\sqrt{3}}{2}\right\}, \left\{-\frac{1}{2}, -\frac{\sqrt{3}}{2}\right\}, \{1, 0\}, \left\{-\frac{1}{2}, \frac{\sqrt{3}}{2}\right\}\right\}$$

We use **Manipulate**, **Table**, **Cos,** and **Sin** to explore roots of unity. If $i = 3$ and $a = 5$, for example, then manipulation produces the conjugate of a root of the polynomial equation $x^5 == 1$.

```
Manipulate[{Cos[2 Pi*i/a], Sin[2 Pi*i/a]}, {i, 0, 5, 1}, {a, 1, 5, 1}]
```

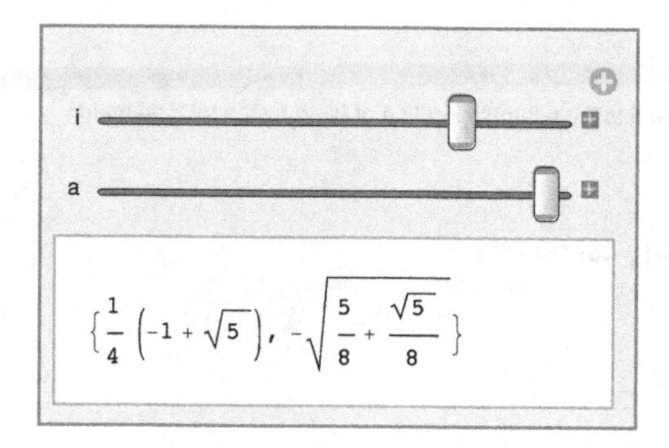

$$\left\{\frac{1}{4}\left(-1+\sqrt{5}\right), -\sqrt{\frac{5}{8}+\frac{\sqrt{5}}{8}}\right\}$$

```
roots5 = Solve[x^5 == 1]
```

$\{\{x\to 1\}, \{x\to -(-1)^{1/5}\}, \{x\to (-1)^{2/5}\}, \{x\to -(-1)^{3/5}\}, \{x\to (-1)^{4/5}\}\}$

```
{Re[x /. roots5[[3]]], -Im[x /. roots5[[3]]]}
```

$$\left\{\frac{1}{4}\left(-1+\sqrt{5}\right), -\sqrt{\frac{5}{8}+\frac{\sqrt{5}}{8}}\right\}$$

Rotation

Rotations in $\mathbb{R}^2$ are the result of the multiplication of points {*x*, *y*} in the plane by rotation matrices of the form:

```
Clear[θ]
```

```
MatrixForm[cw = {{Cos[θ], Sin[θ]}, {-Sin[θ], Cos[θ]}}]
```

$$\begin{pmatrix} \text{Cos}[\theta] & \text{Sin}[\theta] \\ -\text{Sin}[\theta] & \text{Cos}[\theta] \end{pmatrix}$$

for clockwise rotation, or

```
MatrixForm[ccw = {{Cos[θ], -Sin[θ]}, {Sin[θ], Cos[θ]}}]
```

$$\begin{pmatrix} \text{Cos}[\theta] & -\text{Sin}[\theta] \\ \text{Sin}[\theta] & \text{Cos}[\theta] \end{pmatrix}$$

for counterclockwise rotation.

The built-in **RotationMatrix** function can be used to calculate counterclockwise rotations:

```
ccw == RotationMatrix[θ]
```

```
True
```

Illustration

- A clockwise rotation of a point $p = \{x, y\}$ by an angle θ is the left-multiplication of $\{x, y\}$ by the rotation matrix

```
Clear[θ, x, y, cw, cwr]

MatrixForm[cw = {{Cos[θ], Sin[θ]}, {-Sin[θ], Cos[θ]}}];

cwr[{x_, y_}] := cw.{x, y}

θ = π / 2; {x, y} = {1, 0};

cwr[{x, y}]
```

```
{0, -1}
```

As expected, a clockwise rotation of the point {1, 0} by $\pi/2$ radians is the point {0, -1}.

- A counterclockwise rotation of a point $p = \{x, y\}$ by an angle θ is the left-multiplication of $\{x, y\}$ by the rotation matrix

```
Clear[θ, ccw, ccwr, x, y]

MatrixForm[ccw = {{Cos[θ], -Sin[θ]}, {Sin[θ], Cos[θ]}}];

ccwr[{x_, y_}] := ccw.{x, y}

θ = π / 2; {x, y} = {1, 0};

ccwr[{x, y}]
```

```
{0, 1}
```

As expected, a counterclockwise rotation of the point {1, 0} by $\pi/2$ radians produces the point {0, 1}.

The built-in *Mathematica* **RotationTransform** rotates vectors counterclockwise in the plane.

```
Clear[θ, x, y]

r = RotationTransform[θ]
```

$$\text{TransformationFunction}\left[\left(\begin{array}{cc|c} \text{Cos}[\theta] & -\text{Sin}[\theta] & 0 \\ \text{Sin}[\theta] & \text{Cos}[\theta] & 0 \\ \hline 0 & 0 & 1 \end{array}\right)\right]$$

```
θ = π / 2; {x, y} = {1, 0};
```

```
r[{x, y}]
```

```
{0, 1}
```

Manipulation

- Clockwise rotations of a vector in $\mathbb{R}^2$

```
Manipulate[{{Cos[θ], Sin[θ]}, {-Sin[θ], Cos[θ]}}.{1, 0}, {θ, 0, 4 π}]
```

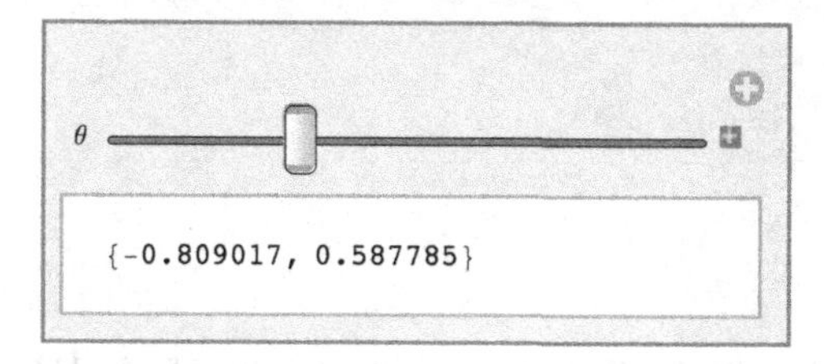

We use **Manipulate** to explore the clockwise rotations of the vector {1, 0} in $\mathbb{R}^2$. If $\theta = 3.76991$ radians, for example, then the vector {1, 0} is displaced to the vector {- 0.809017, 0.587785}.

```
Norm[{-0.809017, 0.587785}]
```

```
1.
```

Row echelon matrix

To define a row echelon matrix, we need the idea of a leading entry. The first nonzero entry from the left in a row of a matrix is called a leading entry of the matrix. A matrix is a *row echelon matrix* if it satisfies the following conditions:

Definition of a row echelon matrix

1. All zero rows of a row echelon matrix occur below all nonzero rows.
2. All entries below a leading entry of a row echelon are zero.
3. The leading entry of a nonzero row of a row echelon matrix occurs in a column to the right of the column containing the leading entry of the row above it.

Illustration

- A 4-by-5 row echelon matrix

$$\begin{pmatrix} 0 & 0 & 1 & 8 & 1 \\ 0 & 0 & 0 & 5 & 4 \\ 0 & 0 & 0 & 0 & 6 \\ 0 & 0 & 0 & 0 & 0 \end{pmatrix}$$

- Row reduction of a 4-by-5 matrix

$$A = \begin{pmatrix} 6 & 0 & 1 & 8 & 1 \\ 7 & 7 & 1 & 5 & 4 \\ 9 & 4 & 3 & 5 & 6 \\ 8 & 1 & 2 & 4 & 3 \end{pmatrix};$$

```
MatrixForm[B = RowReduce[A]]
```

$$\begin{pmatrix} 1 & 0 & 0 & 0 & -\frac{35}{274} \\ 0 & 1 & 0 & 0 & \frac{61}{137} \\ 0 & 0 & 1 & 0 & \frac{246}{137} \\ 0 & 0 & 0 & 1 & -\frac{1}{274} \end{pmatrix}$$

Manipulation

- Row reduction of 2-by-2 matrices

```
Manipulate[MatrixForm[RowReduce[{{a, -3, 5}, {b, 8, 2}}]], {a, -5, 5, 1}, {b, -5, 5, 1}]
```

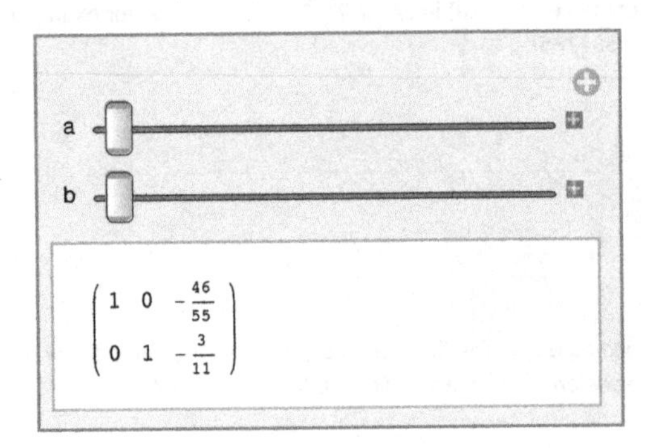

We use **Manipulate**, **MatrixForm**, and **RowReduce** to reduce 2-by-3 matrices to reduced row echelon form and display them two-dimensionally. If we let $a = b = -5$, for example, the manipulation produces the matrix

```
MatrixForm[RowReduce[{{-5, -3, 5}, {-5, 8, 2}}]]
```

$$\begin{pmatrix} 1 & 0 & -\frac{46}{55} \\ 0 & 1 & -\frac{3}{11} \end{pmatrix}$$

Row-equivalent matrices

Two matrices are *row equivalent* if they have the same reduced row echelon form. A matrix and its reduced row echelon form are row equivalent. The built-in **RowReduce** function of *Mathematica* reduces a matrix to its unique reduced row echelon form.

Row-equivalent matrices can be interpreted as augmented matrices of linear systems with the same solutions.

Illustration

- Two row-equivalent matrices

$$A = \begin{pmatrix} 6 & 0 & 1 & 8 & 1 \\ 7 & 7 & 1 & 5 & 4 \\ 9 & 4 & 3 & 5 & 6 \\ 8 & 1 & 2 & 4 & 3 \end{pmatrix}; B = \begin{pmatrix} 1 & 7 & 0 & -3 & 3 \\ 6 & 0 & 1 & 8 & 1 \\ 9 & 4 & 3 & 5 & 6 \\ 16 & 2 & 4 & 8 & 6 \end{pmatrix};$$

```
{MatrixForm[RowReduce[A]], MatrixForm[RowReduce[B]]}
```

$$\left\{\begin{pmatrix} 1 & 0 & 0 & 0 & -\frac{35}{274} \\ 0 & 1 & 0 & 0 & \frac{61}{137} \\ 0 & 0 & 1 & 0 & \frac{246}{137} \\ 0 & 0 & 0 & 1 & -\frac{1}{274} \end{pmatrix}, \begin{pmatrix} 1 & 0 & 0 & 0 & -\frac{35}{274} \\ 0 & 1 & 0 & 0 & \frac{61}{137} \\ 0 & 0 & 1 & 0 & \frac{246}{137} \\ 0 & 0 & 0 & 1 & -\frac{1}{274} \end{pmatrix}\right\}$$

```
RowReduce[A] == RowReduce[B]
```

```
True
```

- Two linear system with the same solutions

```
Clear[x, y, z, w]

system1 = {6 x + z + 8 w == 1, 7 x + 7 y + z + 5 w == 4, 9 x + 4 y + 3 z + 5 w == 6, 8 x + y + 2 z + 4 w == 3};

system2 = {x == -35 / 274, y == 61 / 137, z == 246 / 137, w == -1 / 274};

Solve[system1, {x, y, z, w}]
```

$$\left\{\left\{x \to -\frac{35}{274}, y \to \frac{61}{137}, z \to \frac{246}{137}, w \to -\frac{1}{274}\right\}\right\}$$

```
Solve[system2, {x, y, z, w}]
```

$$\left\{\left\{x \to -\frac{35}{274}, y \to \frac{61}{137}, z \to \frac{246}{137}, w \to -\frac{1}{274}\right\}\right\}$$

Row space

The *row space* of a matrix is the set of all linear combinations of the rows of the matrix.

Illustration

- A basis for the row space of a 3-by-4 matrix

```
A = {{1, 2, 3, 4}, {5, 6, 7, 8}, {0, 1, 3, 0}};
```

```
rowspace = RowReduce[A]
```

```
{{1, 0, 0, -5}, {0, 1, 0, 9}, {0, 0, 1, -3}}
```

The command **RowReduce** produces a basis for this space. The row space is a subspace of $\mathbb{R}^4$. Its dimension is 3 and its codimension is 4 -3 = 1.

- A basis for the row space of a 4-by-3 matrix

```
TraditionalForm[A = {{5, 1, 5}, {0, 5, 9}, {2, 8, 4}, {9, 9, 6}}]
```

$$\begin{pmatrix} 5 & 1 & 5 \\ 0 & 5 & 9 \\ 2 & 8 & 4 \\ 9 & 9 & 6 \end{pmatrix}$$

```
TraditionalForm[rA = RowReduce[A]]
```

$$\begin{pmatrix} 1 & 0 & 0 \\ 0 & 1 & 0 \\ 0 & 0 & 1 \\ 0 & 0 & 0 \end{pmatrix}$$

rowspace = a {1, 0, 0} + b {0, 1, 0} + c {0, 0, 1} for all scalars a, b, c

Row vector

A *row vector* is a list of scalars.

Illustration

- A row vector of length 3

```
rowvector = {1, 2, 3}
```

```
{1, 2, 3}
```

By default, *Mathematica* considers vectors to be row vectors. The **Flatten** command can be used to convert a column vector into a row vector.

- Conversion of a column vector to a row vector

```
columnvector = {{1}, {2}, {3}}
```

```
{{1}, {2}, {3}}
```

```
Flatten[columnvector]
```

```
{1, 2, 3}
```

Manipulation

- A linear combination of row vectors

```
u = Range[5]; v = Reverse[2 Range[5]]; w = {2, 2, 2, 2, 2};
```

```
u + v + w
```

```
{13, 12, 11, 10, 9}
```

```
Manipulate[Evaluate[a u + b v + c w], {a, -2, 2, 1}, {b, -2, 2, 1}, {c, -2, 2, 1}]
```

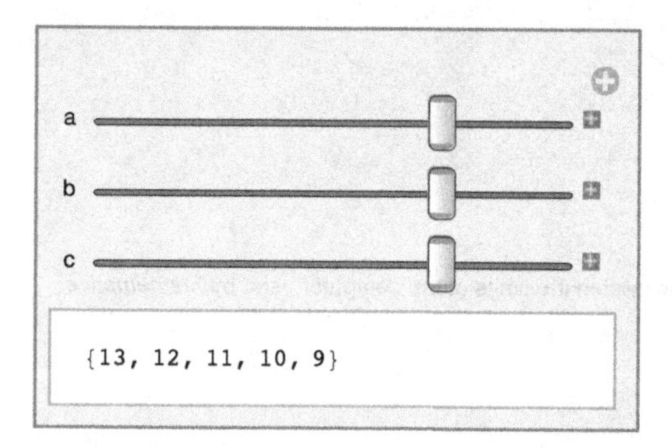

We use **Manipulate** and **Evaluate** to explore the linear combinations of three vectors in $\mathbb{R}^5$. If $a = b = c = 1$, for example, the manipulation displays the sum of the three given vectors.

S

Scalar

The constants used in most linear algebra contexts are either real numbers or complex numbers. They are jointly referred to as *scalars*. Complex numbers are often involved when roots of polynomials (eigenvalues, for example) are calculated. They are required extensively in engineering.

Mathematica divides the scalars into domains: integers, rational numbers, real numbers, and complex numbers. Integers are written in the usual notation: 0, 1, -1, 2, -2, and so on, and integers followed by a decimal point, such as 0., 1., -1., 2., -2., and so on, are considered to be decimal approximations of integers. Rational numbers can be written in two ways: both 2/3 and $\frac{2}{3}$ represent the same fraction. The **N** function converts exact and symbolic numbers into decimals and displays five places to the right of the decimal point. The underlying accuracy of the displayed numbers is unaffected by the display.

Real scalars

```
{N[5], N[Exp[1]], N[π]}
```

```
{5., 2.71828, 3.14159}
```

Real numbers can be displayed to any level of accuracy within the limits of the computer used by *Mathematica*.

```
N[π, 30]
```

```
3.14159265358979323846264338328
```

- Real addition

```
3.0 + 5.918
```

```
8.918
```

- Real subtraction

```
12.583 - 25.9999
```

```
-13.4169
```

- Real multiplication

In *Mathematica*, a single blank space invokes multiplication. *Mathematica* automatically inserts the *times* symbol:

```
4.689 × 6
```

```
28.134
```

Alternately, a star can be used to designate multiplication

```
4.689*6
```

```
28.134
```

The times symbol × can be inserted explicitly by typing Esc * Esc.

- Real division

```
9.174555/126.4
```

```
0.0725835
```

- The real number 1

The number 1 is a scalar. *Mathematica* thinks of it as an integer (an exact real number). If we want to think of it as a real number as such, we use the *N* function to convert it.

```
N[1]
```

```
1.
```

Appending a decimal point to an integer signals to *Mathematica* that this number is an approximate number.

```
N[1, 29]
```

```
1.0000000000000000000000000000
```

Although the number 1 can be represented as an infinitely repeating decimal expression, most finite approximations are considered to be good enough to stand for the number 1.

```
1 == .9999999999999999999999999999999999999999
```

```
True
```

Internally, the precision of the approximations are recorded and can be modified for specific calculations, if required.

- Precision of the real number 1.

```
Precision[1.]
```

```
MachinePrecision
```

```
Precision[.9999999999999999999999999999999999999999]
```

```
40.
```

Some real numbers represented by non-repeating decimals have symbolic names that can be approximated at the time of evaluation to any desired degree of precision (within the hardware limitations of the computers used).

- The real number *e*

The real number *e* is very different from the number 1 and other integers: it is irrational (cannot be represented by a terminating or repeating decimal expansion) and transcendental (is not the root of a polynomial with integer coefficients).

```
Exp[1]
```

```
e
```

Depending on its use, the transcendental number *e* can be approximated with the **N** function to any desired number of decimals:

```
N[Exp[1]]
```

```
2.71828
```

Mathematica assigns an infinite precision to the symbolic number *e*:

```
Precision[Exp[1]]
```

```
∞
```

- The real number π

The real number π has properties similar to the real number *e*: it is irrational and transcendental.

```
π
```

```
π
```

The transcendental number π can also be approximated with the **N** function to any desired number of decimals:

```
N[π]
```

```
3.14159
```

Mathematica also assigns an infinite precision of the symbolic number π:

```
Precision[π]
```

```
∞
```

On the other hand, *Mathematica* assigns what is called the MachinePrecision to the numerical expressions *e*.0 and π.0.

```
Precision[Exp[1] .0]
```

```
MachinePrecision
```

```
Precision[π .0]
```

```
MachinePrecision
```

However, these distinctions are neglected in most parts of this book. They do come up implicitly in the use of random matrices.

- An array of random real scalars

```
RandomReal[{0, 9}, {3, 2}]
```

```
{{7.27419, 0.948809}, {4.06732, 8.69741}, {3.22765, 2.77941}}
```

If we cut and paste this matrix into a new input field, their MachinePrecision appears.

```
MatrixForm[A = {{6.31429, 0.681646}, {7.38745, 5.25242}, {2.27482, 6.07207}}]
```

$$\begin{pmatrix} 6.31429 & 0.681646 \\ 7.38745 & 5.25242 \\ 2.27482 & 6.07207 \end{pmatrix}$$

Mathematica computes the sum, difference, product, and quotient of two real numbers as follows:

```
x = 3.999; y = -12.12356;
{x + y, x - y, x * y, x / y}
```

```
{-8.12456, 16.1226, -48.4821, -0.329854}
```

Complex scalars

Complex numbers require the use of the symbol *i* denoting the square root of minus one. In calculations, the symbols *i* is represented by the capital letter "I" in input statements, or it can be written by typing Esc ii Esc.

- Complex scalars

A complex scalar is an expression of the form (x + y *i*), where x and y are real scalars. The letter *i* denotes the square root of -1. In *Mathematica* calculations, the capital letter "I" is interpreted as *i*.

```
I == i
```

```
True
```

Mathematica computes the sum, difference, product, and quotient of two complex numbers as follows:

- Complex addition, subtraction, multiplication, and division

```
x = 3 + 4 I; y = -5 + 9 I;
Simplify[{x + y, x - y, x * y, x / y}]
```

$$\left\{-2 + 13\,i,\ 8 - 5\,i,\ -51 + 7\,i,\ \frac{21}{106} - \frac{47\,i}{106}\right\}$$

- Conjugate of a complex scalar

```
Conjugate[{3 - 4 I, -5 + 7 I}]
```

```
{3 + 4 i, -5 - 7 i}
```

- Complex exponential scalar

```
Exp[3 + 4 I] == e^(3+4 I)
```

True

- Power of a complex scalar

```
(3 + 4 I)^5
```

-237 - 3116 i

Manipulation

- Real arithmetic

```
Manipulate[{a + b, a - b, a b, a / b}, {a, -5, 5}, {b, 1, 5}]
```

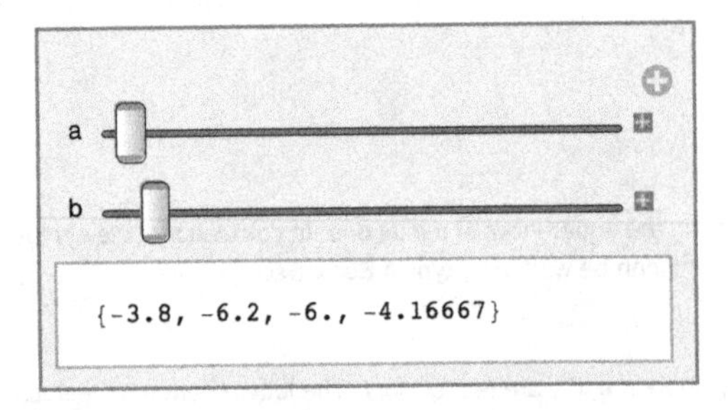

We use **Manipulate** to explore the sums, differences, products, and quotients of real numbers. For example, if a = -5 and b = 1.2, then

```
{a + b, a - b, a b, a / b} == {-3.8, -6.2, -6., -4.16667};
```

- Complex arithmetic

```
Manipulate[{(a + b I) + (c + d I), (a + b I) - (c + d I), (a + b I) * (c + d I), (a + b I) / (c + d I)},
 {a, 2, 5, 1}, {b, 1, 5, 1}, {c, 1, 3, 1}, {d, -1, 1, 1}]
```

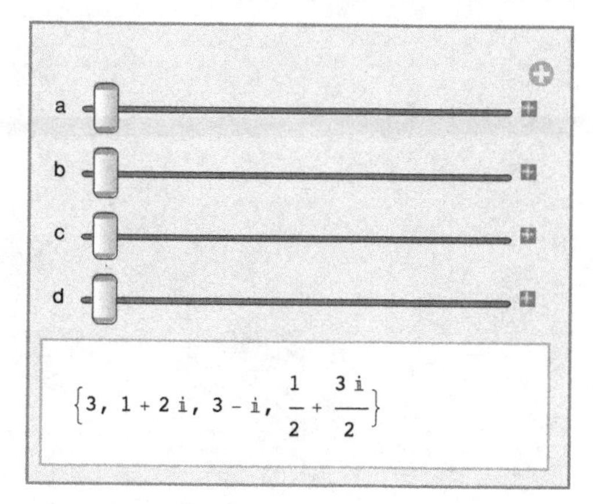

We use **Manipulate** to explore the sums, differences, products, and quotients of complex numbers. For example, if a = 2 and b = 1, c = 1 and d = -1, then

$$\{(a+b\,I)+(c+d\,I),\ (a+b\,I)-(c+d\,I),\ (a+b\,I)*(c+d\,I),\ (a+b\,I)/(c+d\,I)\} == \left\{3,\ 1+2\,i,\ 3-i,\ \frac{1}{2}+\frac{3\,i}{2}\right\}$$

For example,

```
(2 + I) * (1 - I) == 3 - I
```

```
True
```

Scalar multiple of a matrix

Matrices can be multiplied by scalars componentwise. The result of multiplying a matrix by a scalar is called a *scalar multiple* of the matrix. In *Mathematica*, placing a scalar to the left of a matrix with a space in between defines scalar multiplication.

Illustration

- Scalar multiple of a matrix

```
MatrixForm[A = {{1, 2, 3}, {4, 5, 6}}]
```

$$\begin{pmatrix} 1 & 2 & 3 \\ 4 & 5 & 6 \end{pmatrix}$$

```
MatrixForm[s A]
```

$$\begin{pmatrix} s & 2\,s & 3\,s \\ 4\,s & 5\,s & 6\,s \end{pmatrix}$$

Every element in the matrix A is multiplied by the scalar *s*.

Illustration

- Multiplication by the scalar 1

```
vector = Range[5]; scalar = 1;
```

```
1 vector
```

```
{1, 2, 3, 4, 5}
```

```
1 vector == vector
```

```
True
```

Manipulation

- Scalar multiple of a 3-by-2 matrix

```
Clear[a]
```

```
Manipulate[MatrixForm[a {{1, 2, 3}, {4, 5, 6}}], {a, -10, 10, 1}]
```

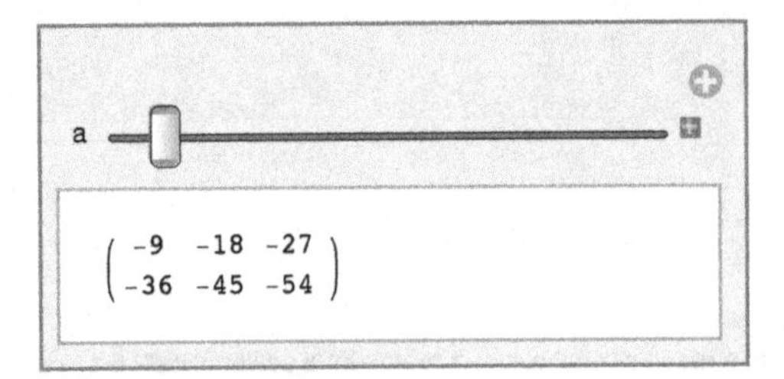

We use **Manipulate** and **MatrixForm** to explore the scalar products of a 2-by-3 matrix and display the result in two-dimensional form. For example, if a = -9, then the scalar product

```
a {{1, 2, 3}, {4, 5, 6}}
```

is the matrix

$$\begin{pmatrix} -9 & -18 & -27 \\ -36 & -45 & -54 \end{pmatrix}$$

Scalar multiplication

See Vector space

Scalar triple product

The *scalar triple product* of three vectors **u**, **v**, and **w** in $\mathbb{R}^3$ is the value of the function

```
Dot[u, Cross[v, w]]
```
(1)

The **ScalarTripleProduct** function is also available in the ClassroomUtilities package and can be activated by either loading the package using the command

```
<< ClassroomUtilities`
```

or by invoking the package using the **Needs** command:

```
Needs["ClassroomUtilities`"]
```

Illustration

- A scalar triple product in $\mathbb{R}^3$ calculated with the **Dot** and **Cross** functions

```
u = {1, 2, 3}; v = {-5, 3, 1}; w = {1, 0, 1};

Dot[u, Cross[v, w]] == Det[{u, v, w}]

True

Needs["ClassroomUtilities`"]

ScalarTripleProduct[u, v, w]

6
```

- Using the scalar triple product to find the equation of the plane passing through the points with position vectors **u**, **v**, **w**:

```
Clear[u, v, w, x, y, z, p]

u = {1, 2, 3}; v = {-2, 1, 1}; w = {4, 4, 4};

p = {x, y, z};

Det[{u - p, v - p, w - p}] == 0

-12 - 3 x + 3 y + 3 z == 0

Expand[Det[{u - p, v - p, w - p}] == Dot[u - p, Cross[v - p, w - p]]]

True
```

```
ContourPlot3D[Evaluate[-12 - 3 x + 3 y + 3 z == 0], {x, -10, 10}, {y, -10, 10}, {z, -10, 10}]
```

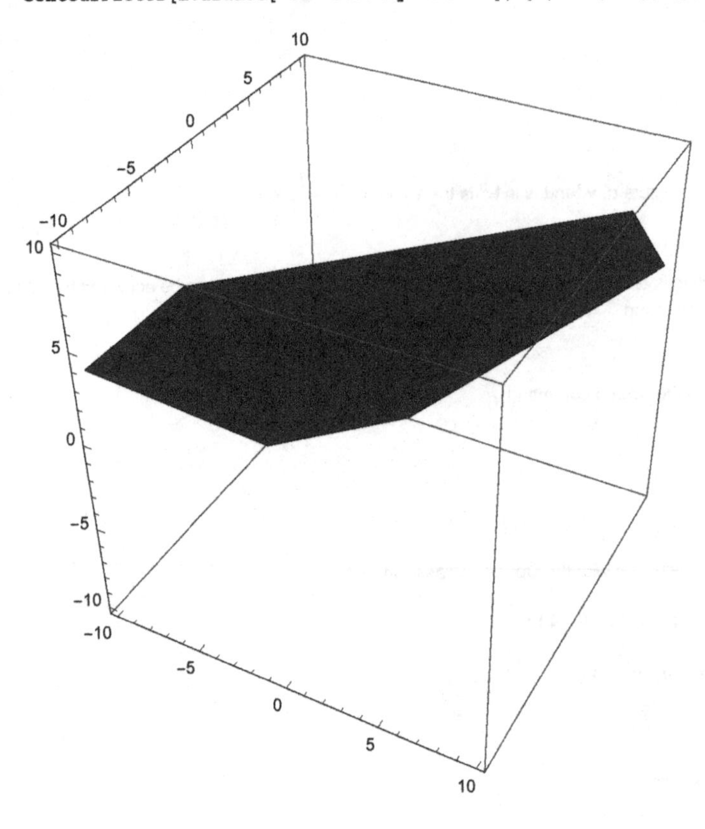

Manipulation

- A scalar triple product in $\mathbb{R}^3$

```
<< ClassroomUtilities`
```

```
Manipulate[ScalarTripleProduct[{1, 2, 3}, {-5 a, 3 b, 1}, {a, 0, 4 c}],
 {a, -3, 3, 1}, {b, -2, 2, 1}, {c, -1, 16, 1/4}]
```

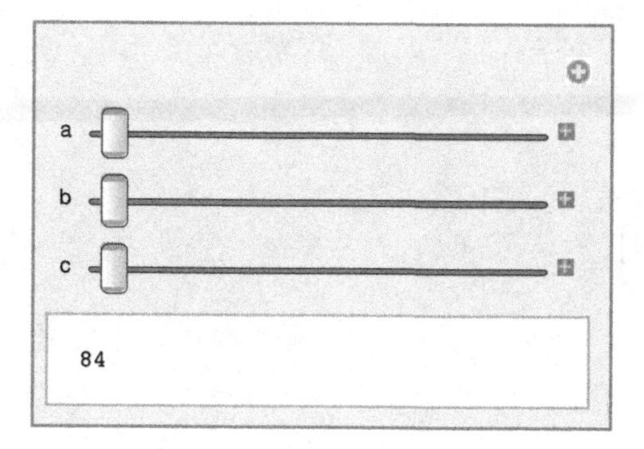

We use **Manipulate** in combination with the **ScalarTripleProduct** function in the ClassroomUtilities package to explore the scalar triple product of vectors in $\mathbb{R}^3$. For example, the scalar triple product of the vectors {1,2,3}, {15, -6, 1}, and {-3, 0, -4}, obtained by letting a = -3, b = -2, and c = -1, is 84.

Scaling

A *scaling* is a linear transformation on an *n*-dimensional Euclidean space represented by an *n*-by-*n* diagonal matrix all of whose diagonal elements are positive. The scaling is uniform if all diagonal elements are equal. In that case, the transformation enlarges or shrinks an object by a scale factor that is the same in all directions.

Illustration

- A uniform scaling on $\mathbb{R}^2$

```
Clear[a, b]
```

```
dM = {{2, 0}, {0, 2}}; vector = {a, b};
```

```
dM.vector
```

```
{2 a, 2 b}
```

The vector {*a*, *b*} is scaled by a factor of 2.

```
Simplify[Norm[dM.vector] == 2 Norm[vector]]
```

```
True
```

- A uniform scaling on $\mathbb{R}^3$

```
dM = {{1/2, 0, 0}, {0, 1/2, 0}, {0, 0, 1/2}}; vector = {a, b, c};
```

```
dM.vector
```

$$\left\{\frac{a}{2}, \frac{b}{2}, \frac{c}{2}\right\}$$

The vector {*a*, *b*, *c*} is scaled by a factor of 1/2.

```
Simplify[Norm[dM.vector] == 1/2 Norm[vector]]
```

```
True
```

- A Nonuniform scaling on $\mathbb{R}^4$

```
Clear[a, b, c, d]
```

```
dM = {{2, 0, 0, 0}, {0, 3, 0, 0}, {0, 0, 1/3, 0}, {0, 0, 0, 1}};
```

```
vector = {a, b, c, d};
```

```
dM.vector
```

$$\left\{2\,a, 3\,b, \frac{c}{3}, d\right\}$$

Manipulation

- Scaling along the *x*-, *y*-, and *z*-axis

```
Clear[x, y, z, a, b, c]
```

```
Manipulate[{{a, 0, 0}, {0, b, 0}, {0, 0, c}}.{1, 2, 3}, {a, .1, 5}, {b, .1, 5}, {c, .1, 5}]
```

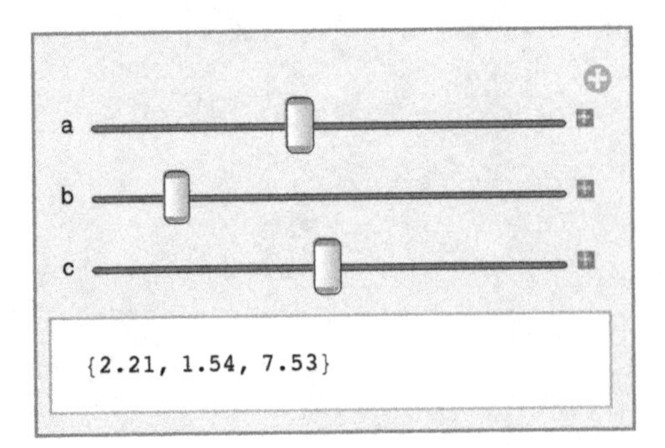

We use **Manipulate** to explore the scaling of vectors in $\mathbb{R}^3$ along the *x*-, *y*-, and *z*-axis. For example, if we let a = 2.21, b = 0.77, and c = 2.51, then the vector {1, 2, 3} becomes the vector {2.21, 1.54, 7.53}.

Schur decomposition

If A is an n-by-n complex matrix, then A can be decomposed into a product $Q\,U\,Q^{-1}$, where Q is a unitary matrix, Q^{-1} is the conjugate transpose of Q, and U is upper-triangular.

Illustration

- Schur decomposition of a 3-by-3 matrix

```
A = N[{{27, 48, 81 I}, {-6 I, 0, 0}, {1, 0, 3}}];
```

```
{q, u} = SchurDecomposition[A]
```

```
{{{-0.323801 + 0.923776 i, 0.0111915 + 0.203727 i, 0.0124449 + 0.00253488 i},
  {0.169698 + 0.107746 i, -0.951375 - 0.176081 i, 0.152471 + 0.0150415 i},
  {-0.0207515 + 0.0308893 i, -0.0198055 - 0.147845 i, -0.19754 - 0.968165 i}},
 {{28.4584 - 6.62034 i, 6.79676 + 32.1819 i, -42.0055 - 74.5259 i},
  {0. + 0. i, -2.41644 + 7.02789 i, -1.69622 - 17.6608 i},
  {0. + 0. i, 0. + 0. i, 3.95803 - 0.407549 i}}}
```

```
MatrixForm[q]
```

$$\begin{pmatrix} -0.323801 + 0.923776\,i & 0.0111915 + 0.203727\,i & 0.0124449 + 0.00253488\,i \\ 0.169698 + 0.107746\,i & -0.951375 - 0.176081\,i & 0.152471 + 0.0150415\,i \\ -0.0207515 + 0.0308893\,i & -0.0198055 - 0.147845\,i & -0.19754 - 0.968165\,i \end{pmatrix}$$

```
MatrixForm[u]
```

$$\begin{pmatrix} 28.4584 - 6.62034\,i & 6.79676 + 32.1819\,i & -42.0055 - 74.5259\,i \\ 0. + 0.\,i & -2.41644 + 7.02789\,i & -1.69622 - 17.6608\,i \\ 0. + 0.\,i & 0. + 0.\,i & 3.95803 - 0.407549\,i \end{pmatrix}$$

```
Chop[q.u.Inverse[q]] == A
```

```
True
```

```
ConjugateTranspose[q] == Inverse[q]
```

```
True
```

Self-adjoint transformation

If V is an inner product space, then a linear transformation $T : V \longrightarrow V$ is *self-adjoint* if <T[**u**], **v**> = <**u**, T[**v**]> for all vectors **u** and **v** in V.

If A is a matrix representing T in an orthonormal basis for V, then T is self-adjoint if and only if A is symmetric.

Illustration

- A self-adjoint linear transformation

```
T = {{Cos[π], -Sin[π]}, {Sin[π], Cos[π]}}
```

```
{{-1, 0}, {0, -1}}
```

```
T* = Transpose[T]
```

```
{{-1, 0}, {0, -1}}
```

```
T = {{1, 0, 2}, {4, 3, 0}, {0, 0, 3}};
```

```
u = {1, 2, 3}; v = {4, 5, 6};
```

```
Dot[T.u, v] == Dot[u, T.v]
```

```
True
```

For any linear transformation T on a real inner product space V, the transformation $T^* \circ T - I$ is self-adjoint (with I the appropriate identity transformation)

- Construction of a self-adjoint linear transformation

```
T = RandomInteger[{0, 9}, {4, 4}];
```

$$T = \begin{pmatrix} 9 & 7 & 7 & 0 \\ 8 & 3 & 4 & 6 \\ 4 & 7 & 3 & 1 \\ 4 & 1 & 8 & 8 \end{pmatrix};$$

```
MatrixForm[T* = Transpose[T]]
```

$$\begin{pmatrix} 9 & 8 & 4 & 4 \\ 7 & 3 & 7 & 1 \\ 7 & 4 & 3 & 8 \\ 0 & 6 & 1 & 8 \end{pmatrix}$$

```
MatrixForm[S = T*.T - IdentityMatrix[4]]
```

$$\begin{pmatrix} 176 & 119 & 139 & 84 \\ 119 & 107 & 90 & 33 \\ 139 & 90 & 137 & 91 \\ 84 & 33 & 91 & 100 \end{pmatrix}$$

The linear transformation S is self-adjoint:

```
u = {1, 2, 3, 4}; v = {5, 6, 7, 8};
```

```
Dot[S.u, v] == Dot[u, S.v]
```

```
True
```

Every self-adjoint transformation T on a finite-dimensional real inner product space can be represented by a diagonal matrix whose diagonal elements are the eigenvalues of T.

Manipulation

- Self-adjoint linear transformations defined by 3-by-3 symmetric matrices

```
MatrixForm[A = {{1, 2, 4 a}, {4 b, 4, 5}, {8, 2, 2}}]
```

$$\begin{pmatrix} 1 & 2 & 4a \\ 4b & 4 & 5 \\ 8 & 2 & 2 \end{pmatrix}$$

```
MatrixForm[S = Transpose[A].A - IdentityMatrix[3]]
```

$$\begin{pmatrix} 64 + 16 b^2 & 18 + 16 b & 16 + 4 a + 20 b \\ 18 + 16 b & 23 & 24 + 8 a \\ 16 + 4 a + 20 b & 24 + 8 a & 28 + 16 a^2 \end{pmatrix}$$

```
u = {1, 2, 3}; v = {5, 6, 7};
```

```
Manipulate[Evaluate[Dot[S.u, v] == Dot[u, S.v]], {a, -1, 1, 1}, {b, -1, 1, 1}]
```

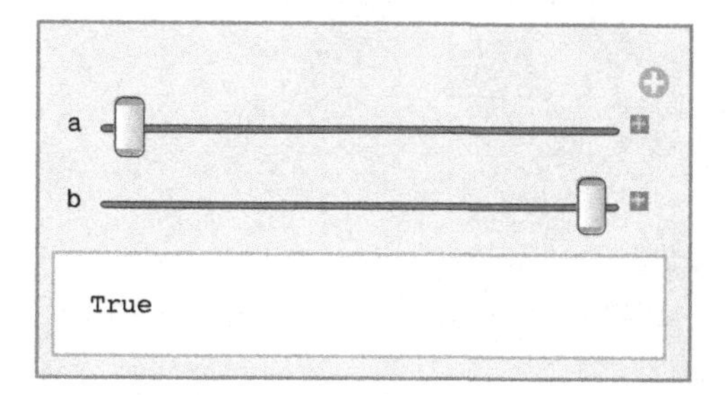

We use **Manipulate**, **Evaluate**, and **Dot** to confirm the self-adjointness of the linear transformation determined by a symmetric matrix. The manipulation shows that if *a* and *b* are any two integers between -1 and 1, for example, the linear operator S defined by the given symmetric matrix is self-adjoint.

Shear

1. For any real number *s*, a left-multiplication of a vector **v** in $\mathbb{R}^2$ by the matrix

```
MatrixForm[A = {{1, s}, {0, 1}}]
```

$$\begin{pmatrix} 1 & s \\ 0 & 1 \end{pmatrix}$$

is a shear along the *x*-axis.

2. For any real number *s*, a left-multiplication of a vector **v** in $\mathbb{R}^2$ by the matrix

```
MatrixForm[A = {{1, 0}, {s, 1}}]
```

$$\begin{pmatrix} 1 & 0 \\ s & 1 \end{pmatrix}$$

is a shear along the *y*-axis

Illustration

- A shear along the *x*-axis for s = 3

```
s = 3;

MatrixForm[A = {{1, s}, {0, 1}}]
```

$$\begin{pmatrix} 1 & 3 \\ 0 & 1 \end{pmatrix}$$

```
v = {4, 5}

{4, 5}

xshearedvector = A.v

{19, 5}

arrow1 = Arrow[{{0, 0}, {4, 5}}];

arrow2 = Arrow[{{0, 0}, {19, 5}}];

Graphics[{arrow1, arrow2}, Axes → True]
```

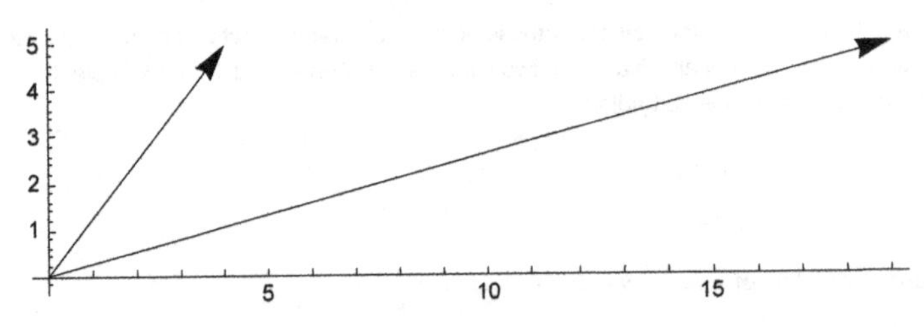

The matrix *A* sheared the vector {4, 5} along the *x*-axis to {19, 5}.

- A shear along the *y*-axis for *s* = 3

```
s = 3;

MatrixForm[A = {{1, 0}, {s, 1}}]
```

$$\begin{pmatrix} 1 & 0 \\ 3 & 1 \end{pmatrix}$$

```
v = {1, -1}
```

```
{1, -1}
```

```
yshearedvector = A.v
```

```
{1, 2}
```

```
arrow3 = Arrow[{{0, 0}, {1, -1}}];
```

```
arrow4 = Arrow[{{0, 0}, {1, 2}}];
```

```
Graphics[{arrow3, arrow4}, Axes → True]
```

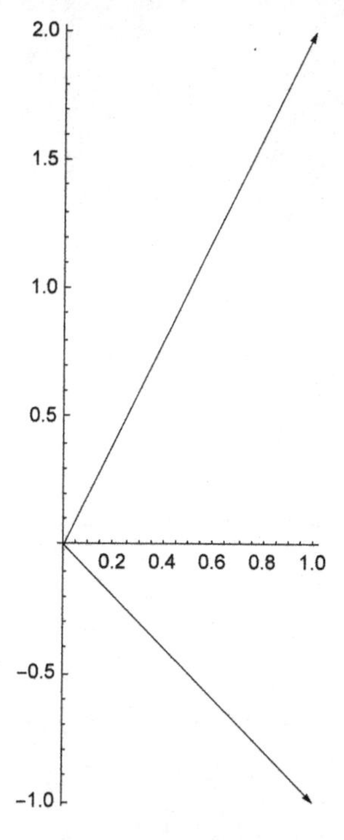

The matrix A sheared the vector {1, -1} along the *y*-axis to {1, 2}.

The **ShearingTransform**[θ, **v**, **n**] represents a shear by θ radians along the direction of the vector **v**, normal to the vector **n**, keeping the origin fixed.

- A 30° shear along the *x*-axis applied to the unit rectangle:

```
Graphics[GeometricTransformation[Rectangle[],
  ShearingTransform[30 Degree, {1, 0}, {0, 1}]], Frame → True]
```

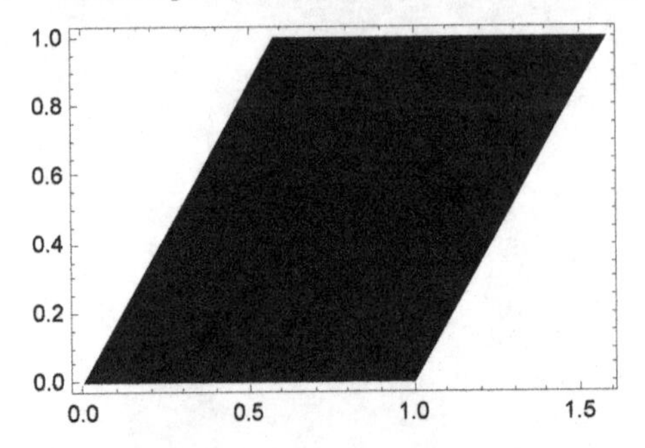

- A 30° shear along the *y*-axis applied to the unit rectangle:

```
Graphics[GeometricTransformation[Rectangle[],
  ShearingTransform[30 Degree, {0, 1}, {1, 0}]], Frame → True]
```

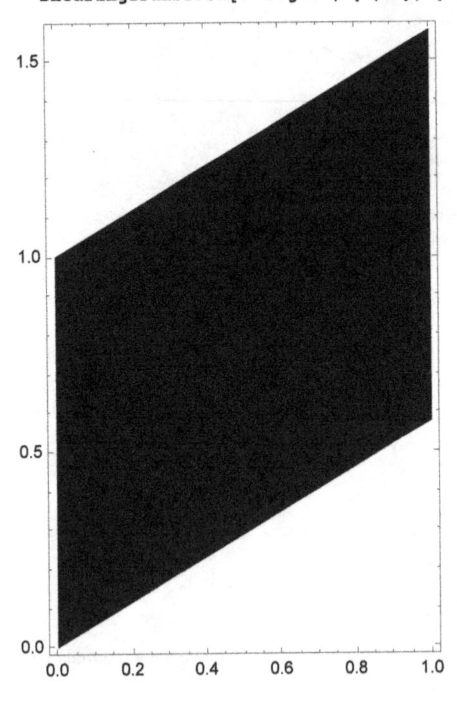

Sigma notation

Sums of indexed scalars, vectors, and linear combinations of scalars and vectors can be written compactly in *sigma notation*. Sigma is the capital Greek letter S and can be written by typing Esc Sigma Esc.

Illustration

- A sum of three indexed scalars in sigma notation

```
{s1, s2, s3} = {4, -3, 7};
```

$$\sum_{n=1}^{3} s_n$$

8

This sum can also be written using the **Sum** function:

```
Sum[sn, {n, 1, 3}]
```

8

- A sum of four indexed vectors in sigma notation

```
A = RandomInteger[{0, 9}, {4, 5}]
```

{{4, 6, 1, 8, 4}, {9, 8, 4, 7, 4}, {8, 0, 9, 0, 0}, {5, 7, 5, 6, 4}}

```
{v1, v2, v3, v4} = {A[[1]], A[[2]], A[[3]], A[[4]]}
```

{{4, 6, 1, 8, 4}, {9, 8, 4, 7, 4}, {8, 0, 9, 0, 0}, {5, 7, 5, 6, 4}}

$$\sum_{k=1}^{4} v_k$$

{26, 21, 19, 21, 12}

- A linear combination of six indexed scalars and vectors in sigma notation

```
{s1, s2, s3, s4, s5, s6} = {2, 3, 1, 5, 6, -2};
A = {{3, 6, 7}, {3, 5, 6}, {1, 6, 6}, {5, 2, 5}, {0, 8, 4}, {6, 8, 4}};
{v1, v2, v3, v4, v5, v6} = {A[[1]], A[[2]], A[[3]], A[[4]], A[[5]], A[[6]]};
```

$$\sum_{k=1}^{6} s_k\, v_k$$

{29, 75, 79}

Similar matrices

Two matrices A and B are *similar* if there exists an invertible matrix P for which $A = PBP^{-1}$.

Illustration

- Two similar matrices

```
A = RandomInteger[{0, 9}, {3, 3}];
```

$$A = \begin{pmatrix} 9 & 5 & 4 \\ 6 & 6 & 1 \\ 8 & 1 & 3 \end{pmatrix};$$

```
P = RandomInteger[{0, 9}, {3, 3}];
```

$$P = \begin{pmatrix} 7 & 2 & 7 \\ 5 & 4 & 1 \\ 9 & 7 & 8 \end{pmatrix};$$

```
Det[P]
```

```
106
```

Since the determinant of P is nonzero, the matrix P is invertible and we can use it to construct a matrix B similar to A.

```
MatrixForm[B = Inverse[P].A.P]
```

$$\begin{pmatrix} \frac{3485}{106} & \frac{2003}{106} & \frac{1121}{53} \\ -\frac{1947}{106} & -\frac{1199}{106} & -\frac{612}{53} \\ -\frac{1051}{106} & -\frac{661}{106} & -\frac{189}{53} \end{pmatrix}$$

By construction, the matrices A and B are similar.

```
A == P.B.Inverse[P]
```

```
True
```

Similarity matrix

If two matrices A and B are connected by an invertible matrix S for which $A = SBS^{-1}$, then the matrix S is called a *similarity matrix*.

Illustration

- A 2-by-2 similarity matrix

```
MatrixForm[A = {{1, 2}, {3, 4}}]
```

$$\begin{pmatrix} 1 & 2 \\ 3 & 4 \end{pmatrix}$$

```
MatrixForm[B = {{-2, 3}, {-4, 7}}]
```

$$\begin{pmatrix} -2 & 3 \\ -4 & 7 \end{pmatrix}$$

```
MatrixForm[S = {{0, 2}, {2, 4}}]
```

$$\begin{pmatrix} 0 & 2 \\ 2 & 4 \end{pmatrix}$$

```
B == S.A.Inverse[S]
```

```
True
```

The matrix S is a similarity matrix.

- A 3-by-3 similarity matrix

```
A = RandomInteger[{0, 9}, {3, 3}];
```

$$A = \begin{pmatrix} 9 & 6 & 4 \\ 9 & 2 & 9 \\ 5 & 0 & 7 \end{pmatrix};$$

```
{values, vectors} = N[Eigensystem[A]]
```

```
{{16.6603, 2., -0.660254}, {{1.93205, 1.8, 1.}, {-2., 1., 2.}, {-1.53205, 1.8, 1.}}}
```

```
A == Chop[Transpose[vectors].DiagonalMatrix[values].Inverse[Transpose[vectors]]]
```

```
True
```

The transpose of the matrix of eigenvectors is a similarity matrix.

Manipulation

- Similarity matrices

```
Clear[a, b]

A = {{1, 2, 3}, {4, 5, 6}, {1, 1, 1}};

S = {{a, 3, 2}, {1, 5, b}, {a, 4, 4}};

Reduce[Det[S] == 0, {a, b}]
```

$$a \neq 0 \,\&\&\, b == \frac{2\,(-2 + 5\,a)}{a}$$

```
B = S.A.Inverse[S];
```

This calculation shows that if *a* is not zero and *b* is $\frac{2(-2+5a)}{a}$, the matrix *S* can be used to convert the matrix *A* to the similar matrix $B = SAS^{-1}$.

```
Manipulate[Evaluate[B == S.A.Inverse[S]], {a, 1, 2, 1}, {b, -3, 5}]
```

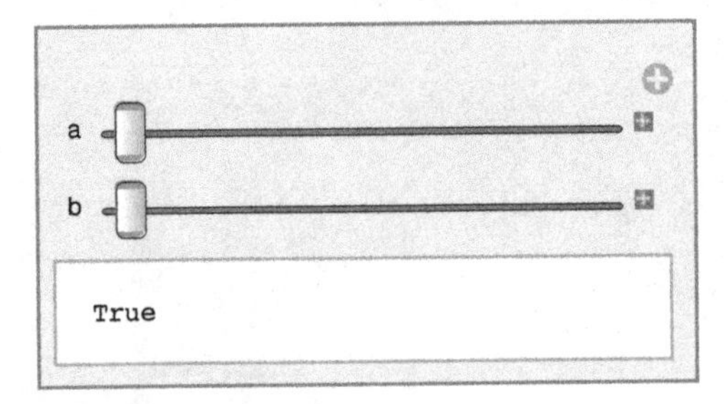

We use **Manipulate**, **Evaluate**, and **Inverse** to construct similar matrices. The manipulation shows that the constructed matrices are similar.

Similarity transformation

The multiplication $A \longrightarrow PAP^{-1}$ of a matrix *A* by invertible matrix *P* is called a *similarity transformation*.

Illustration

- A similarity transformation of a 2-by-2 matrix

```
P = RandomInteger[{0, 9}, {2, 2}];
```

$$P = \begin{pmatrix} 4 & 6 \\ 8 & 5 \end{pmatrix};$$

```
Det[P]
```

-28

Since *P* is invertible, we can use it to define a similarity transformation.

```
simtrans[A_] := P.A.Inverse[P]
```

```
MatrixForm[simtrans[{{1, 2}, {3, 4}}]]
```

$$\begin{pmatrix} \frac{73}{14} & \frac{1}{7} \\ \frac{173}{28} & -\frac{3}{14} \end{pmatrix}$$

Manipulation

- Exploring similarity transformation

```
A = {{1, 2}, {3, 4}};
```

```
S = {{a, 3}, {1, 5}};
```

```
Reduce[Det[S] == 0, a]
```

$a == \frac{3}{5}$

```
Det[{{3/5, 3}, {1, 5}}]
```

0

The transformation $A \longrightarrow SAS^{-1}$ is a similarity transformation for all a ≠ 3/5.

```
Manipulate[Evaluate[{A, S.A.Inverse[S]}], {a, -5, 0}]
```

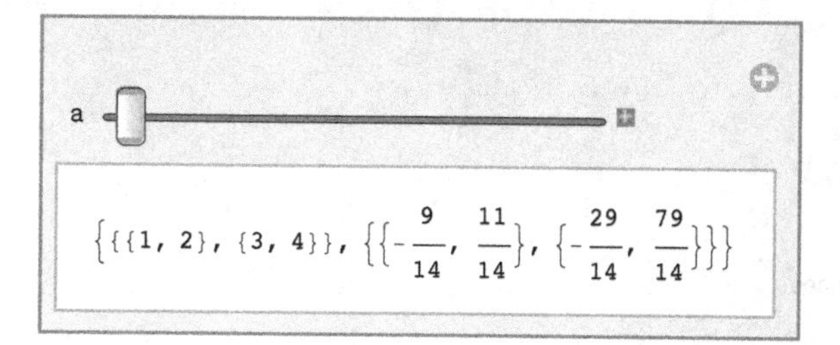

a
{{{1, 2}, {3, 4}}, {{-9/14, 11/14}, {-29/14, 79/14}}}

We use **Manipulate**, **Evaluate**, and **Inverse** to construct similar transformations. The manipulation shows that the constructed transformations are similarity transformations.

Singular matrix

A square numerical matrix is *singular* if it is not invertible. A numerical matrix is singular if and only if its determinant is 0.

Illustration

- A singular diagonal matrix with a zero element in the diagonal

```
v = {1, 0, 3, 4};
```

```
MatrixForm[A = DiagonalMatrix[v]]
```

$$\begin{pmatrix} 1 & 0 & 0 & 0 \\ 0 & 0 & 0 & 0 \\ 0 & 0 & 3 & 0 \\ 0 & 0 & 0 & 4 \end{pmatrix}$$

```
Det[A]
```

0

Since the determinant of A is zero, the matrix is singular.

```
Inverse[A];
```

Inverse::sing : Matrix {{1, 0, 0, 0}, {0, 0, 0, 0}, {0, 0, 3, 0}, {0, 0, 0, 4}} is singular. ≫

- A singular sparse matrix

```
MatrixForm[S = Normal[SparseArray[{{2, 3} -> a, {3, 2} -> b}, {3, 3}]]]
```

$$\begin{pmatrix} 0 & 0 & 0 \\ 0 & 0 & a \\ 0 & b & 0 \end{pmatrix}$$

```
Det[S]
```

0

Manipulation

- Exploring the singularity of 2-by-2 matrices

```
Clear[a, b]
```

```
Det[{{a, 2}, {2, b}}]
```

-4 + a b

```
Manipulate[Det[{{a, 2}, {2, b}}], {a, -10, 10, 1}, {b, -10, 10, 1}]
```

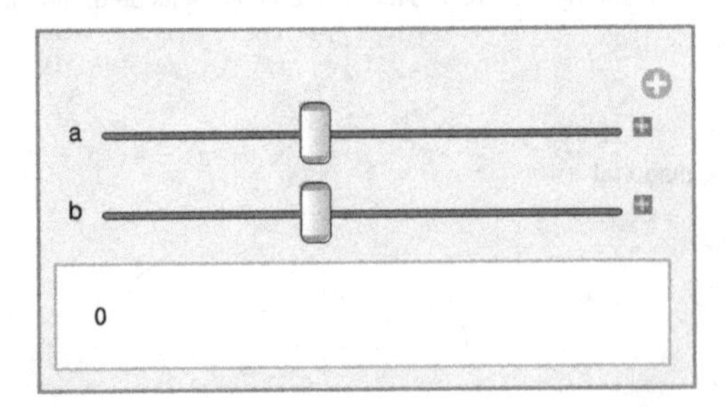

We use **Manipulate** and **Det** to explore the singularity of matrices. The manipulation shows, for example, that the generated matrix is singular if $a = b = -2$ and nonsingular otherwise since for two nonzero parameters c and d,

```
Solve[Det[{{c, 2}, {2, d}}] == 0]
```

$$\left\{\left\{d \to \frac{4}{c}\right\}\right\}$$

if and only if d == 4/c.

Singular value

A *singular value* of a real matrix *A* is the positive square root of an eigenvalue of the symmetric matrix AA^T or $A^T A$.

Illustration

- The singular values of $A^T A$ and AA^T of a 2-by-3 matrix *A*

```
A = {{1, 2, 3}, {4, 5, 6}};

SingularValueList[A]
```

$$\left\{\sqrt{\frac{1}{2}\left(91 + \sqrt{8065}\right)}, \sqrt{\frac{1}{2}\left(91 - \sqrt{8065}\right)}\right\}$$

```
MatrixForm[AtA = Transpose[A].A]
```

$$\begin{pmatrix} 17 & 22 & 27 \\ 22 & 29 & 36 \\ 27 & 36 & 45 \end{pmatrix}$$

```
Sqrt[Eigenvalues[AtA]]
```

$$\left\{\sqrt{\frac{1}{2}\left(91 + \sqrt{8065}\right)}, \sqrt{\frac{1}{2}\left(91 - \sqrt{8065}\right)}, 0\right\}$$

```
MatrixForm[AAt = A.Transpose[A]]
```

$$\begin{pmatrix} 14 & 32 \\ 32 & 77 \end{pmatrix}$$

```
Sqrt[Eigenvalues[AAt]]
```

$$\left\{\sqrt{\frac{1}{2}\left(91 + \sqrt{8065}\right)}, \sqrt{\frac{1}{2}\left(91 - \sqrt{8065}\right)}\right\}$$

```
Sqrt[Eigenvalues[AAt]] == SingularValueList[A]

True
```

The difference between computing singular values using AA^T and $A^T A$ is that in one case 0 may come up as a singular value whereas it may not do so in the other case.

- The singular values of a matrix A coinciding with the square roots of the eigenvalues of both of $A^T A$ and AA^T

$$A = \begin{pmatrix} 1 & 0 & 1 \\ 0 & 3 & 0 \\ 1 & 0 & 3 \end{pmatrix};$$

```
sv1 = Sqrt[Eigenvalues[Transpose[A].A]]
```

$$\left\{\sqrt{2\left(3+2\sqrt{2}\right)},\ 3,\ \sqrt{2\left(3-2\sqrt{2}\right)}\right\}$$

```
sv2 = Sqrt[Eigenvalues[A.Transpose[A]]]
```

$$\left\{\sqrt{2\left(3+2\sqrt{2}\right)},\ 3,\ \sqrt{2\left(3-2\sqrt{2}\right)}\right\}$$

```
SingularValueList[A] == sv1 == sv2
```

```
True
```

Manipulation

- Exploring the singular values of 3-by-2 matrices

```
Manipulate[SingularValueList[{{4., 1}, {2, b}, {3, a}}], {a, -3, 3}, {b, -3, 3}]
```

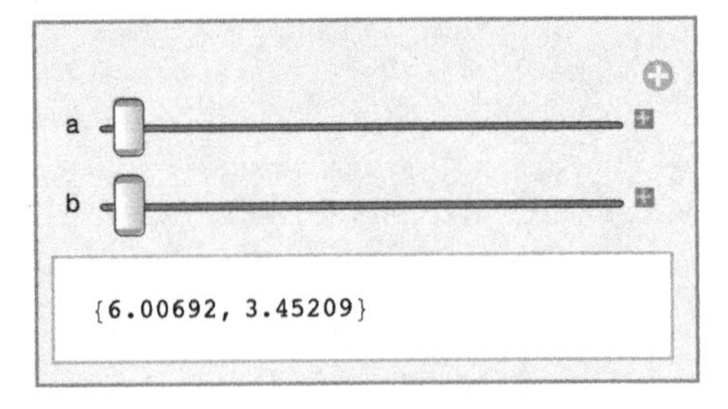

We use **Manipulate** and **SingularValueList** to explore the singular values of 3-by-2 matrix. The manipulation shows, for example, that the singular values of the matrix

```
MatrixForm[{{4., 1}, {2, -3}, {3, -3}}]
```

$$\begin{pmatrix} 4. & 1 \\ 2 & -3 \\ 3 & -3 \end{pmatrix}$$

are 6.00692 and 3.45209.

Singular value decomposition

The *singular value decomposition* of a matrix allows us to write any rectangular real matrix *A* as a product **u.w.Transpose[v]** of two orthogonal matrices **u** and **v** and a diagonal matrix **w**. If A is not square, the matrix **w** is padded with rows or columns of zeros. (*Mathematica* uses the lower-case letters **u** and **v** for the orthogonal matrices and the letter **w** for the diagonal matrix calculated by the **SingularValueDecomposition** function.)

Illustration

- A singular value decomposition of a 3-by-3 real matrix

The computing cost of powers of *A* is reduced by decomposing *A* as a product of the form **u.w.Transpose[v]** and then computing the power of the decomposed product.

```
MatrixForm[A = {{3, 1, -1}, {1, -1, 1}, {2, -1, -1}}]
```

$$\begin{pmatrix} 3 & 1 & -1 \\ 1 & -1 & 1 \\ 2 & -1 & -1 \end{pmatrix}$$

```
{u, w, v} = N[SingularValueDecomposition[A], 5];
```

```
MatrixForm[u]
```

$$\begin{pmatrix} -0.81441 & -0.43925 & 0.37920 \\ -0.15694 & 0.79585 & 0.58481 \\ -0.55866 & 0.41676 & -0.71708 \end{pmatrix}$$

```
Inverse[u] == Transpose[u]
```

True

```
MatrixForm[v]
```

$$\begin{pmatrix} -0.95013 & 0.16667 & 0.26357 \\ -0.025253 & -0.88353 & 0.46769 \\ 0.31082 & 0.43771 & 0.84368 \end{pmatrix}$$

```
Inverse[v] == Transpose[v]
```

True

```
MatrixForm[w]
```

$$\begin{pmatrix} 3.9126 & 0 & 0 \\ 0 & 1.8696 & 0 \\ 0 & 0 & 1.0936 \end{pmatrix}$$

```
A == u.w.Transpose[v]
```

True

```
Timing[MatrixPower[A, 100]]
```

```
{0.000919, {{73 278 692 513 007 179 808 797 058 918 974 595 206 140 665 801,
    21 889 768 781 166 777 101 194 626 797 160 254 862 590 538 775,
    -13 032 913 129 465 196 384 521 495 342 374 527 860 094 984 975},
   {26 318 196 607 017 567 459 531 192 524 553 118 363 838 315 675,
    7 861 756 517 594 023 195 701 380 367 297 893 262 017 395 201,
    -4 680 798 173 937 965 048 824 767 567 196 926 144 896 122 625},
   {30 494 254 084 781 183 127 379 556 412 141 919 221 437 746 850,
    9 109 225 999 788 755 407 161 333 294 589 789 646 143 899 525,
    -5 423 526 959 958 347 879 308 316 814 880 697 241 725 935 499}}}
```

```
Timing[MatrixPower[u.w.Transpose[v], 100]]
```

$\{0.000286, \{\{7.\times 10^{46}, 2.2\times 10^{46}, -1.3\times 10^{46}\}, \{2.6\times 10^{46}, 8.\times 10^{45}, -5.\times 10^{45}\}, \{3.0\times 10^{46}, 9.\times 10^{45}, -5.\times 10^{45}\}\}\}$

This calculation shows that the computing cost of the matrix A^{100} measured in time is 0.000919 seconds, whereas the corresponding computing cost of (u.w.**Transpose[v]**)^{100} is 0.000286 seconds. These values are hardware dependent and may differ from computer to computer.

- A singular value decomposition of a 2-by-3 matrix

$$A = \begin{pmatrix} 3 & 1 & -1 \\ 1 & -1 & 1 \end{pmatrix};$$

```
{u, w, v} = SingularValueDecomposition[A];
```

```
MatrixForm[N[u]]
```

$$\begin{pmatrix} 0.992508 & -0.122183 \\ 0.122183 & 0.992508 \end{pmatrix}$$

```
N[Inverse[u] == Transpose[u]]
```

True

```
N[MatrixForm[v]]
```

$$\begin{pmatrix} 0.92941 & 0.369048 & 0. \\ 0.260956 & -0.657192 & 0.707107 \\ -0.260956 & 0.657192 & 0.707107 \end{pmatrix}$$

```
N[Chop[MatrixForm[Inverse[v] - Transpose[v]]] == MatrixForm[ConstantArray[0, {3, 3}]]]
```

$$\begin{pmatrix} 3.33067\times 10^{-16} & -5.55112\times 10^{-17} & 5.55112\times 10^{-17} \\ 1.11022\times 10^{-16} & 0. & 0. \\ 0. & 0. & 0. \end{pmatrix} == \begin{pmatrix} 0. & 0. & 0. \\ 0. & 0. & 0. \\ 0. & 0. & 0. \end{pmatrix}$$

```
N[MatrixForm[w]]
```

$$\begin{pmatrix} 3.33513 & 0. & 0. \\ 0. & 1.69614 & 0. \end{pmatrix}$$

```
N[A == u.w.Inverse[v]]
```

True

- A singular value decomposition of a 3-by-2 matrix

$$A = \begin{pmatrix} 3 & 1 \\ 1 & -1 \\ -1 & 1 \end{pmatrix};$$

```
{u, w, v} = SingularValueDecomposition[A];
```

```
MatrixForm[N[u]]
```

$$\begin{pmatrix} 0.92941 & 0.369048 & 0. \\ 0.260956 & -0.657192 & 0.707107 \\ -0.260956 & 0.657192 & 0.707107 \end{pmatrix}$$

The following display shows that the matrix w is diagonal:

```
MatrixForm[N[w]]
```

$$\begin{pmatrix} 3.33513 & 0. \\ 0. & 1.69614 \\ 0. & 0. \end{pmatrix}$$

```
MatrixForm[N[v]]
```

$$\begin{pmatrix} 0.992508 & -0.122183 \\ 0.122183 & 0.992508 \end{pmatrix}$$

The following calculation shows that the matrices u and v are orthogonal:

```
{N[Inverse[u] == Transpose[u]], N[Inverse[v] == Transpose[v]]}
```

{True, True}

Manipulation

- The singular value decomposition of 2-by-2 matrices

```
Manipulate[{u, w, v} = SingularValueDecomposition[{{2., a}, {3, 4}}], {a, -1, 1, 1}]
```

```
a

{{{0.0985376, 0.995133}, {0.995133, -0.0985376}},
 {{5.01976, 0.}, {0., 2.19134}}, {{0.633989, 0.773342}, {0.773342, -0.633989}}}
```

We use **Manipulate** and **SingularValueDecomposition** to explore the singular value decomposition of 2-by-2 matrices. If we let a = -1, for example, then the manipulation produces a list consisting of three matrices {u, w, v}, with the property that the matrix {{2., -1}, {3, 4}} is the product of u, w, and the transpose of v.

Singular vector

For any real or complex *m*-by-*n* matrix *A*, the left-singular vectors of *A* are the eigenvectors of AA^T. They are equal to the columns of the matrix **u** in the singular value decomposition {**u**, **w**, **v**} of *A*. The right-singular vectors of *A* are the eigenvectors of the matrix **v** in the singular value decomposition of *A*.

Illustration

- Left-singular vectors of a 2-by-3 matrix

```
MatrixForm[A = {{1, 2, 3}, {4, 5, 6}}]
```

$$\begin{pmatrix} 1 & 2 & 3 \\ 4 & 5 & 6 \end{pmatrix}$$

```
{u, w, v} = SingularValueDecomposition[A];
```

```
MatrixForm[N[Transpose[u]]]
```

$$\begin{pmatrix} 0.386318 & 0.922366 \\ -0.922366 & 0.386318 \end{pmatrix}$$

```
MatrixForm[AAt = A.Transpose[A]]
```

$$\begin{pmatrix} 14 & 32 \\ 32 & 77 \end{pmatrix}$$

```
MatrixForm[Eigenvectors[N[AAt]]]
```

$$\begin{pmatrix} 0.386318 & 0.922366 \\ -0.922366 & 0.386318 \end{pmatrix}$$

- Right-singular vectors of a 2-by-3 matrix

```
MatrixForm[A = {{1, 2, 3}, {4, 5, 6}}]
```

$$\begin{pmatrix} 1 & 2 & 3 \\ 4 & 5 & 6 \end{pmatrix}$$

```
{u, w, v} = SingularValueDecomposition[A];
```

```
MatrixForm[N[Transpose[v]]]
```

$$\begin{pmatrix} 0.428667 & 0.566307 & 0.703947 \\ 0.805964 & 0.112382 & -0.581199 \\ 0.408248 & -0.816497 & 0.408248 \end{pmatrix}$$

```
MatrixForm[AtA = Transpose[A].A]
```

$$\begin{pmatrix} 17 & 22 & 27 \\ 22 & 29 & 36 \\ 27 & 36 & 45 \end{pmatrix}$$

```
MatrixForm[Eigenvectors[N[AtA]]]
```

$$\begin{pmatrix} 0.428667 & 0.566307 & 0.703947 \\ 0.805964 & 0.112382 & -0.581199 \\ 0.408248 & -0.816497 & 0.408248 \end{pmatrix}$$

Skew symmetric matrix

A square matrix is *skew symmetric* if its transpose is equal to the matrix multiplied by -1, that is $A^T = -A$.

Illustration

- A skew symmetric 4-by-4 matrix

```
MatrixForm[A = {{2, 1, 5, 4}, {5, 7, 7, 1}, {2, 5, 8, 6}, {4, 2, 1, 6}}]
```

$$\begin{pmatrix} 2 & 1 & 5 & 4 \\ 5 & 7 & 7 & 1 \\ 2 & 5 & 8 & 6 \\ 4 & 2 & 1 & 6 \end{pmatrix}$$

```
SymmetricMatrixQ[A + Transpose[A]]

True
```

$$A == \frac{1}{2}(A + \text{Transpose}[A]) + \frac{1}{2}(A - \text{Transpose}[A])$$

```
True
```

For every square matrix A, the matrix (A - **Transpose**[A]) is skew symmetric.

```
MatrixForm[B = A - Transpose[A]]
```

$$\begin{pmatrix} 0 & -4 & 3 & 0 \\ 4 & 0 & 2 & -1 \\ -3 & -2 & 0 & 5 \\ 0 & 1 & -5 & 0 \end{pmatrix}$$

```
(-1) B == Transpose[B]
```

True

Every real n-by-n matrix B is skew symmetric if and only if **Dot**[B.**x**, **y**] = - **Dot**[**x**, B.**y**].

- A skew symmetric matrix characterized by the dot product

```
MatrixForm[A = RandomInteger[{0, 9}, {3, 3}]]
```

$$\begin{pmatrix} 6 & 1 & 6 \\ 7 & 9 & 5 \\ 0 & 2 & 6 \end{pmatrix}$$

```
MatrixForm[B = A - Transpose[A]]
```

$$\begin{pmatrix} 0 & -6 & 6 \\ 6 & 0 & 3 \\ -6 & -3 & 0 \end{pmatrix}$$

```
Dot[B.{1, 2, 3}, {a, b, c}] == Simplify[-Dot[{1, 2, 3}, B.{a, b, c}]]
```

6 a + 15 b - 12 c == 3 (2 a + 5 b - 4 c)

Solution of a linear system

The solutions of a linear system in n equations and m variables $\{x_1, \ldots, x_m\}$ are lists of scalars $\{a_1, \ldots, a_m\}$ satisfying each of the n equations.

Illustration

Linear systems can be solved in several ways.

- Solving a linear system using Gaussian elimination

```
Clear[x, y]

system = {6 x + 2 y == 5, 3 x + 4 y == 9};

TraditionalForm[matrixequation = {{6, 2}, {3, 4}}.{{x}, {y}} == {{5}, {9}}]
```

$$\begin{pmatrix} 6x+2y \\ 3x+4y \end{pmatrix} = \begin{pmatrix} 5 \\ 9 \end{pmatrix}$$

```
lhs1 = {{1, -2}, {0, 1}}.{{6, 2}, {3, 4}};
rhs1 = {{1, -2}, {0, 1}}.{{5}, {9}};

TraditionalForm[lhs1.{{x}, {y}} == rhs1]
```

$$\begin{pmatrix} -6y \\ 3x+4y \end{pmatrix} = \begin{pmatrix} -13 \\ 9 \end{pmatrix}$$

```
lhs2 = {{0, 1}, {1, 0}}.lhs1;
rhs2 = {{0, 1}, {1, 0}}.rhs1;

TraditionalForm[lhs2.{{x}, {y}} == rhs2]
```

$$\begin{pmatrix} 3x+4y \\ -6y \end{pmatrix} = \begin{pmatrix} 9 \\ -13 \end{pmatrix}$$

```
lhs3 = {{1, 4/6}, {0, 1}}.lhs2;
rhs3 = {{1, 4/6}, {0, 1}}.rhs2;

TraditionalForm[lhs3.{{x}, {y}} == rhs3]
```

$$\begin{pmatrix} 3x \\ -6y \end{pmatrix} = \begin{pmatrix} \frac{1}{3} \\ -13 \end{pmatrix}$$

```
lhs4 = {{1/3, 0}, {0, 1}}.lhs3;
rhs4 = {{1/3, 0}, {0, 1}}.rhs3;

TraditionalForm[lhs4.{{x}, {y}} == rhs4]
```

$$\begin{pmatrix} x \\ -6y \end{pmatrix} = \begin{pmatrix} \frac{1}{9} \\ -13 \end{pmatrix}$$

```
lhs5 = {{1, 0}, {0, -1/6}}.lhs4;
rhs5 = {{1, 0}, {0, -1/6}}.rhs4;

TraditionalForm[lhs5.{{x}, {y}} == rhs5]
```

$$\begin{pmatrix} x \\ y \end{pmatrix} = \begin{pmatrix} \frac{1}{9} \\ \frac{13}{6} \end{pmatrix}$$

The values $x = 1/9$ and $y = 13/6$ are a solution of the given linear system:

```
{6 x + 2 y == 5, 3 x + 4 y == 9} /. {x → 1/9, y → 13/6}

{True, True}
```

- Another method of applying Gaussian elimination to solve a linear system

```
system = {6 x + 2 y == 5, 3 x + 4 y == 9};
```

  - Step 1

```
A = {{6, 2, 5}, {3, 4, 9}}
```

```
{{6, 2, 5}, {3, 4, 9}}
```

- Step 2

```
A = {A[[2]], A[[1]]}
```

```
{{3, 4, 9}, {6, 2, 5}}
```

- Step 3

```
A[[2]] = A[[2]] - 2 A[[1]];
A
```

```
{{3, 4, 9}, {0, -6, -13}}
```

- Step 4

$$A[[2]] = \frac{-1}{6} A[[2]];$$

A

$$\left\{\{3, 4, 9\}, \left\{0, 1, \frac{13}{6}\right\}\right\}$$

```
A[[1]] = A[[1]] - 4 A[[2]];
A
```

$$\left\{\left\{3, 0, \frac{1}{3}\right\}, \left\{0, 1, \frac{13}{6}\right\}\right\}$$

$$A[[1]] = \frac{1}{3} A[[1]];$$

A

$$\left\{\left\{1, 0, \frac{1}{9}\right\}, \left\{0, 1, \frac{13}{6}\right\}\right\}$$

$$\text{system} /. \left\{x \to \frac{1}{9}, y \to \frac{13}{6}\right\}$$

```
{True, True}
```

- Solving a linear system in two equations and two variables using matrix inversion

```
system = {6 x + 2 y == 5, 3 x + 4 y == 9};
```

```
A = {{6, 2}, {3, 4}}; b = {5, 9};
```

```
solution = Inverse[A].b
```

$\left\{\frac{1}{9}, \frac{13}{6}\right\}$

- Solving a linear equation in two equations and three variables using the **LinearSolve** command

```
system = {6 x + 2 y + z == 5, 3 x + 4 y - z == 9};

A = {{6, 2, 1}, {3, 4, -1}}; b = {5, 9};

solution = LinearSolve[A, b]
```

$\left\{\frac{1}{9}, \frac{13}{6}, 0\right\}$

$\texttt{system /. } \left\{x \to \frac{1}{9}, y \to \frac{13}{6}, z \to 0\right\}$

```
{True, True}
```

- Solving a linear system using an upper and lower matrix decomposition

```
A = {{1, -3, 2, -2}, {3, -2, 0, -1}, {2, 36, -28, 27}, {1, -3, 22, 5}};
b = {-11, -4, 155, 10};

{lu, p, c} = LUDecomposition[A]

{{{1, -3, 2, -2}, {3, 7, -6, 5}, {2, 6, 4, 1}, {1, 0, 5, 2}}, {1, 2, 3, 4}, 0}

MatrixForm[lower = {{1, 0, 0, 0}, {3, 1, 0, 0}, {2, 6, 1, 0}, {1, 0, 5, 1}}]
```

$$\begin{pmatrix} 1 & 0 & 0 & 0 \\ 3 & 1 & 0 & 0 \\ 2 & 6 & 1 & 0 \\ 1 & 0 & 5 & 1 \end{pmatrix}$$

```
MatrixForm[upper = {{1, -3, 2, -2}, {0, 7, -6, 5}, {0, 0, 4, 1}, {0, 0, 0, 2}}]
```

$$\begin{pmatrix} 1 & -3 & 2 & -2 \\ 0 & 7 & -6 & 5 \\ 0 & 0 & 4 & 1 \\ 0 & 0 & 0 & 2 \end{pmatrix}$$

```
A == lower.upper

True

LinearSolve[A, b] == LinearSolve[upper, LinearSolve[lower, b]]

True
```

Span of a list of vectors

The span of a list of vectors is the set of all linear combinations of the vectors.

Illustration

- Span of two vectors in $\mathbb{R}^2$

```
Clear[a, b]
```

```
vectors = {{1, 2}, {3, 4}};
```

```
span = {a {1, 2} + b {3, 4}}
```

```
{{a + 3 b, 2 a + 4 b}}
```

Thus the span of the vectors {1, 2} and {3, 4} is the set of all vectors of the form {a + 3 b, 2 a + 4 b} for all real numbers *a* and *b*. All vectors in $\mathbb{R}^2$ lie in the span of the two vectors.

Let {*x*, *y*} be a given vector in $\mathbb{R}^2$. Then the following calculation shows that {*x*, *y*} can be written as {a + 3 b, 2 a + 4 b} for suitable scalars *a* and *b*:

```
Clear[x, y, a, b]
```

```
{x, y} = {1, 2};
```

```
Solve[{a + 3 b, 2 a + 4 b} == {x, y}, {a, b}]
```

```
{{a → 1, b → 0}}
```

```
{1, 2} == {1 + 3 × 0, 2 × 1 + 4 × 0}
```

```
True
```

- Spans and linear independence

If every vector within that span has a unique expression as a linear combination of the vectors on the left, then any solution is unique. In that case, the span also forms a basis of the space generated by the given vectors. The number of vectors in this basis is called the dimension of the space.

- Four linearly independent vectors in $\mathbb{R}^4$ span $\mathbb{R}^4$

```
Clear[w, x, y, z]
```

$$\text{equation} = w\begin{pmatrix}1\\2\\3\\1\end{pmatrix} + x\begin{pmatrix}4\\0\\6\\3\end{pmatrix} + y\begin{pmatrix}3\\3\\9\\8\end{pmatrix} + z\begin{pmatrix}5\\4\\3\\1\end{pmatrix} == \begin{pmatrix}0\\0\\0\\0\end{pmatrix};$$

```
Solve[equation, {w, x, y, z}]
```

```
{{w → 0, x → 0, y → 0, z → 0}}
```

This shows that the four given vectors are linearly independent.

The same conclusion can be drawn from the fact that the matrix

```
MatrixForm[A = {{1, 4, 3, 5}, {2, 0, 3, 4}, {3, 6, 9, 3}, {1, 3, 8, 1}}]
```

$$\begin{pmatrix} 1 & 4 & 3 & 5 \\ 2 & 0 & 3 & 4 \\ 3 & 6 & 9 & 3 \\ 1 & 3 & 8 & 1 \end{pmatrix}$$

is invertible.

```
Det[A]
```

```
-264
```

This means that every vector {*w*, *x*, *y*, *z*} in $\mathbb{R}^4$ can be written as a unique linear combination of the four columns or rows of the matrix *A*.

```
Clear[a, b, c, d]
```

```
Reduce[{w, x, y, z} == a A[[All,1]] + b A[[All,2]] + c A[[All,3]] + d A[[All,4]], {a, b, c, d}]
```

$$a == -\frac{31 w}{88} + \frac{9 x}{44} + \frac{125 y}{264} - \frac{21 z}{44} \;\&\&\; b == \frac{5 w}{44} - \frac{5 x}{22} + \frac{7 y}{44} - \frac{3 z}{22} \;\&\&$$
$$c == -\frac{w}{44} + \frac{x}{22} - \frac{13 y}{132} + \frac{5 z}{22} \;\&\&\; d == \frac{17 w}{88} + \frac{5 x}{44} - \frac{43 y}{264} + \frac{3 z}{44}$$

Manipulation

- Linear combinations of the columns of a matrix

```
Clear[A, a, b, c, d, w, x, y, z]
```

$$A = \begin{pmatrix} 1 & 4 & 3 & 5 \\ 2 & 0 & 3 & 4 \\ 3 & 6 & 9 & 3 \\ 1 & 3 & 8 & 1 \end{pmatrix};$$

```
Manipulate[
 Reduce[{w, x, y, z} == a A[[All,1]] + b A[[All,2]] + c A[[All,3]] + d A[[All,4]], {a, b, c, d}],
 {w, -2, 2, 1}, {x, -2, 2, 1}, {y, -2, 2, 1}, {z, -2, 2, 1}]
```

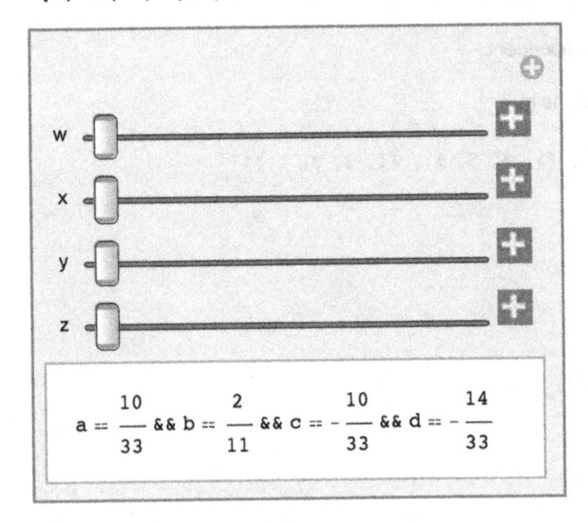

We use **Manipulate** and **Reduce** to explore the linear combinations of the columns of a generated matrix. The manipulation shows, for example, that the vector {-2, -2, -2} can be represented as a linear combination of the columns of a given 4-by-4 matrix.

Sparse matrix

Mathematica has special *sparse array* technology for efficiently handling arrays with literally astronomical numbers of elements when only a small fraction of the elements are nonzero.

Illustration

- A sparse 3-by-3 matrix

```
S = SparseArray[{{1, 1} → 1, {2, 2} → 2, {3, 3} → 3, {1, 3} → 4}]
```

```
MatrixForm[Normal[S]]
```

$$\begin{pmatrix} 1 & 0 & 4 \\ 0 & 2 & 0 \\ 0 & 0 & 3 \end{pmatrix}$$

The **SparseArray**[<4>, {3,3}] output tells us that the matrix has three rows and three columns and four zero elements. The *Normal* command reveals all or part of the matrix in a *Mathematica* window, depending on the dimensions of the matrix.

- A sparse matrix generated with four nonzero elements

```
SparseArray[Table[{2^i} → 1, {i, 4}]]
```

```
MatrixForm[Partition[Normal[%], 4]]
```

$$\begin{pmatrix} 0 & 1 & 0 & 1 \\ 0 & 0 & 0 & 1 \\ 0 & 0 & 0 & 0 \\ 0 & 0 & 0 & 1 \end{pmatrix}$$

- A large sparse array

```
S = SparseArray[Table[{2^i, 3^i + i} → 1, {i, 10}]]
```

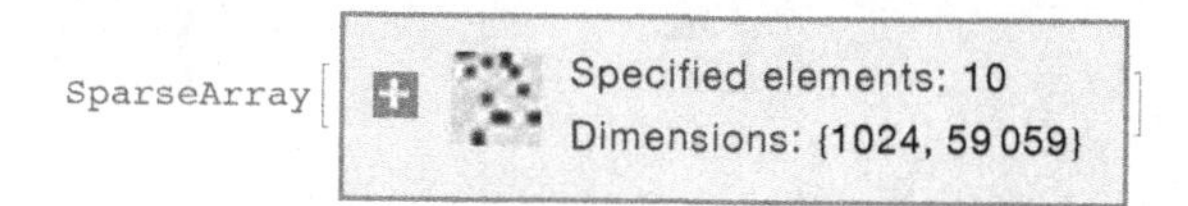

The matrix has 10 zero elements and the pair {1024,59059} specifies is dimension: 1024 rows and 59059 columns.

```
Dimensions[S]
```

```
{1024, 59059}
```

- A sparse array with 13 nonzero elements

```
S = SparseArray[{{i_, i_} → -2, {i_, j_} /; Abs[i - j] == 1 → 1}, {5, 5}]
```

```
SparseArray[ Specified elements: 13
             Dimensions: {5, 5} ]
```

```
MatrixForm[Normal[S]]
```

$$\begin{pmatrix} -2 & 1 & 0 & 0 & 0 \\ 1 & -2 & 1 & 0 & 0 \\ 0 & 1 & -2 & 1 & 0 \\ 0 & 0 & 1 & -2 & 1 \\ 0 & 0 & 0 & 1 & -2 \end{pmatrix}$$

Spectral decomposition

For every real symmetric matrix A there exists an orthogonal matrix Q and a diagonal matrix dM such that $A = (Q^T\, dM\, Q)$. This decomposition is called a spectral decomposition of A since Q consists of the eigenvectors of A and the diagonal elements of dM are corresponding eigenvalues. The terminology derives from the fact that the set of eigenvalues of a matrix is also called the "spectrum" of the matrix.

Illustration

- Spectral decomposition of a 2-by-2 real symmetric matrix

```
MatrixForm[A = {{3, -4}, {-4, -3}}];
```

```
MatrixForm[Q = Orthogonalize[Eigenvectors[A]]]
```

$$\begin{pmatrix} \frac{1}{\sqrt{5}} & \frac{2}{\sqrt{5}} \\ -\frac{2}{\sqrt{5}} & \frac{1}{\sqrt{5}} \end{pmatrix}$$

```
MatrixForm[dM = DiagonalMatrix[Eigenvalues[A]]]
```

$$\begin{pmatrix} -5 & 0 \\ 0 & 5 \end{pmatrix}$$

```
A == Transpose[Q].dM.Q
```

```
True
```

- Spectral decomposition of a 4-by-4 real symmetric matrix

```
MatrixForm[A = {{1, 0, 0, 0}, {0, 1, 1, 0}, {0, 1, 1, 0}, {0, 0, 0, 1}}];
```

```
MatrixForm[Q = Orthogonalize[Eigenvectors[A]]]
```

$$\begin{pmatrix} 0 & -\frac{1}{\sqrt{2}} & -\frac{1}{\sqrt{2}} & 0 \\ 0 & 0 & 0 & 1 \\ 1 & 0 & 0 & 0 \\ 0 & -\frac{1}{\sqrt{2}} & \frac{1}{\sqrt{2}} & 0 \end{pmatrix}$$

```
MatrixForm[dM = DiagonalMatrix[Eigenvalues[A]]]
```

$$\begin{pmatrix} 2 & 0 & 0 & 0 \\ 0 & 1 & 0 & 0 \\ 0 & 0 & 1 & 0 \\ 0 & 0 & 0 & 0 \end{pmatrix}$$

```
A == Transpose[Q].dM.Q
```

```
True
```

Spectral theorem

The complex spectral theorem for a linear operator $T : V \longrightarrow V$ on a finite-dimensional unitary vector space V says that V has an orthonormal basis consisting of eigenvectors of T if and only if T is normal.

The real spectral theorem for a linear operator $S : V \longrightarrow V$ on a finite-dimensional inner product space V says that if S is a self-adjoint linear operator on V, then V has an orthonormal basis consisting of eigenvectors of T if and only if T is self-adjoint.

The *spectral theorem* for real symmetric matrices A says that there exists an orthogonal matrix Q and a diagonal matrix dM whose diagonal elements are the eigenvalues of A such that $A = (Q\, dM\, Q^T)$. Numerous other spectral theorems provide conditions under which matrices can be diagonalized.

Illustration

- The spectral decomposition of A as a linear combination of eigenvalues and eigenvectors

For the orthogonal decomposition $(Q\, dM\, Q^T)$ of the matrix A,

$$\begin{pmatrix} 0 & -\frac{1}{\sqrt{2}} & -\frac{1}{\sqrt{2}} & 0 \\ 0 & 0 & 0 & 1 \\ 1 & 0 & 0 & 0 \\ 0 & -\frac{1}{\sqrt{2}} & \frac{1}{\sqrt{2}} & 0 \end{pmatrix} . \begin{pmatrix} 2 & 0 & 0 & 0 \\ 0 & 1 & 0 & 0 \\ 0 & 0 & 1 & 0 \\ 0 & 0 & 0 & 0 \end{pmatrix} . \begin{pmatrix} 0 & 0 & 1 & 0 \\ -\frac{1}{\sqrt{2}} & 0 & 0 & -\frac{1}{\sqrt{2}} \\ -\frac{1}{\sqrt{2}} & 0 & 0 & \frac{1}{\sqrt{2}} \\ 0 & 1 & 0 & 0 \end{pmatrix}$$

we can rewrite this matrix product as,

$$A = \lambda_1\, u_1 . u_1{}^T + \lambda_2\, u_2 . u_2{}^T + \lambda_3\, u_2 . u_2{}^T + \lambda_4\, u_4 . u_4{}^T \tag{1}$$

where $\lambda_1 = 2$, $\lambda_2 = \lambda_3 = 1$, and $\lambda_4 = 0$. The vectors $u_i = Q_{[[i]]}$ and $u_i{}^T = Q_{[[All,i]]}$ are the rows and columns of Q, respectively. This linear combination is also referred to as the "spectral decomposition" of A. If **u** is a 4-by-1 matrix, then u^T is a 1-by-4 matrix. So that $u.u^T$ *is a* 4-by-1-by-4 = 4-by-4 matrix. This explains the equation

$$A = \lambda_1\, u_1\, u_1{}^T + \lambda_2\, u_2\, u_2{}^T + \lambda_3\, u_3\, u_3{}^T + \lambda_3\, u_4\, u_4{}^T$$

as a sum of 4-by-4 matrices.

$$\textbf{A} = \begin{pmatrix} 1 & 0 & 0 & 0 \\ 0 & 1 & 1 & 0 \\ 0 & 1 & 1 & 0 \\ 0 & 0 & 0 & 1 \end{pmatrix};$$

$$\textbf{MatrixForm}\left[\begin{pmatrix} 0 & 0 & 1 & 0 \\ -\frac{1}{\sqrt{2}} & 0 & 0 & -\frac{1}{\sqrt{2}} \\ -\frac{1}{\sqrt{2}} & 0 & 0 & \frac{1}{\sqrt{2}} \\ 0 & 1 & 0 & 0 \end{pmatrix} . \begin{pmatrix} 2 & 0 & 0 & 0 \\ 0 & 1 & 0 & 0 \\ 0 & 0 & 1 & 0 \\ 0 & 0 & 0 & 0 \end{pmatrix} . \begin{pmatrix} 0 & -\frac{1}{\sqrt{2}} & -\frac{1}{\sqrt{2}} & 0 \\ 0 & 0 & 0 & 1 \\ 1 & 0 & 0 & 0 \\ 0 & -\frac{1}{\sqrt{2}} & \frac{1}{\sqrt{2}} & 0 \end{pmatrix}\right] == \textbf{MatrixForm[A]}$$

```
True
```

We can rewrite this matrix product as

$$\mathbf{A1} = 2\begin{pmatrix} 0 \\ -\frac{1}{\sqrt{2}} \\ -\frac{1}{\sqrt{2}} \\ 0 \end{pmatrix}.\begin{pmatrix} 0 & -\frac{1}{\sqrt{2}} & -\frac{1}{\sqrt{2}} & 0 \end{pmatrix}; \mathbf{A2} = 1\begin{pmatrix} 0 \\ 0 \\ 0 \\ 1 \end{pmatrix}.(0\ 0\ 0\ 1);$$

$$\mathbf{A3} = 1\begin{pmatrix} 1 \\ 0 \\ 0 \\ 0 \end{pmatrix}.(1\ 0\ 0\ 0); \mathbf{A4} = 0\begin{pmatrix} 0 \\ -\frac{1}{\sqrt{2}} \\ \frac{1}{\sqrt{2}} \\ 0 \end{pmatrix}.\begin{pmatrix} 0 & -\frac{1}{\sqrt{2}} & \frac{1}{\sqrt{2}} & 0 \end{pmatrix};$$

```
A1 + A2 + A3 + A4 == A
```

```
True
```

Square matrix

A matrix is square if it has the same number of rows and columns. The **Dimensions** function calculates the number of rows and columns of a matrix and can therefore be used to test whether or not a matrix is square. Only square matrices have eigenvalues and eigenvectors.

Illustration

- A 4-by-4 square matrix

```
A = RandomInteger[{0, 9}, {4, 4}];
```

$$\mathbf{A} = \begin{pmatrix} 7 & 3 & 6 & 1 \\ 5 & 9 & 5 & 9 \\ 6 & 2 & 9 & 3 \\ 0 & 4 & 1 & 3 \end{pmatrix};$$

```
Dimensions[A]
```

```
{4, 4}
```

- A non-square (rectangular) matrix

The following two matrices are not square:

$$\mathbf{B1} = \begin{pmatrix} 5 & 4 & 7 & 3 \\ 9 & 5 & 3 & 9 \\ 1 & 2 & 6 & 2 \end{pmatrix};$$

```
Dimensions[B1]
```

```
{3, 4}
```

$$B2 = \begin{pmatrix} 4 & 7 & 3 \\ 5 & 3 & 9 \\ 2 & 6 & 2 \\ 2 & 0 & 4 \end{pmatrix};$$

```
Dimensions[B2]
```

```
{4, 3}
```

Standard basis

A standard basis for $\mathbb{R}^n$ is a basis consisting of vectors all of whose coordinates are 0 except for a single entry equal to 1. A standard basis for the space $\mathbb{R}$[t,n] of polynomials consists of powers of *t*.

Illustration

- The standard basis of $\mathbb{R}^3$

```
Clear[sB]
```

```
sB = {{1, 0, 0}, {0, 1, 0}, {0, 0, 1}};
```

- The standard basis of the polynomial space $\mathbb{R}$[t,4] of polynomials in *t* of degree less than or equal to 4

```
Clear[sB]
```

$$sB = \{1, t, t^2, t^3, t^4\}$$

- The standard basis of $\mathbb{R}^{2\times 3}$

```
Clear[aB, a, b]
```

```
sB = {B11, B12, B13, B21, B22, B23};
```

where

```
MatrixForm[B11 = {{1, 0, 0}, {0, 0, 0}}]
```

$$\begin{pmatrix} 1 & 0 & 0 \\ 0 & 0 & 0 \end{pmatrix}$$

```
MatrixForm[B12 = {{0, 1, 0}, {0, 0, 0}}]
```

$$\begin{pmatrix} 0 & 1 & 0 \\ 0 & 0 & 0 \end{pmatrix}$$

```
MatrixForm[B13 = {{0, 0, 1}, {0, 0, 0}}]
```

$$\begin{pmatrix} 0 & 0 & 1 \\ 0 & 0 & 0 \end{pmatrix}$$

```
MatrixForm[B21 = {{0, 0, 0}, {1, 0, 0}}]
```

$$\begin{pmatrix} 0 & 0 & 0 \\ 1 & 0 & 0 \end{pmatrix}$$

```
MatrixForm[B22 = {{0, 0, 0}, {0, 1, 0}}]
```

$$\begin{pmatrix} 0 & 0 & 0 \\ 0 & 1 & 0 \end{pmatrix}$$

```
MatrixForm[B23 = {{0, 0, 0}, {0, 0, 1}}]
```

$$\begin{pmatrix} 0 & 0 & 0 \\ 0 & 0 & 1 \end{pmatrix}$$

Standard deviation of a numerical vector

The population standard deviation σ measures the spread of a vector in $\mathbb{R}^n$. It is defined to be the square root of the population variance of the vector. The sample standard deviation s is defined to the square root of the sample variance of the vector.

Illustration

- The population standard deviation of a vector in $\mathbb{R}^6$

```
x = {2, 6, 3, 1, 8, 9}; mx = Mean[x];
```

$$\sigma = \text{Sqrt}\left[\text{Total}\left[\frac{1}{6}\text{Table}\left[(x[[i]] - mx)^2, \{i, 1, 6\}\right]\right]\right]$$

$$\frac{\sqrt{329}}{6}$$

- The sample standard deviation of a vector in $\mathbb{R}^6$

```
x = {2, 6, 3, 1, 8, 9}; mx = Mean[x];
```

$$s = \text{Sqrt}\left[\text{Total}\left[\frac{1}{5}\text{Table}\left[(x[[i]] - mx)^2, \{i, 1, 6\}\right]\right]\right]$$

$$\sqrt{\frac{329}{30}}$$

The *Mathematica* **StandardDeviation** function computes the sample standard deviation of a vector.

```
s == StandardDeviation[x]
```

```
True
```

Stochastic matrix

A stochastic matrix is a square matrix whose columns are probability vectors. A probability vector is a numerical vector whose entries are real numbers between 0 and 1 whose sum is 1.

1. A *stochastic matrix* is a matrix describing the transitions of a Markov chain. It is also called a *Markov matrix*.
2. A *right stochastic matrix* is a square matrix of nonnegative real numbers whose rows add up to 1.
3. A *left stochastic matrix* is a square matrix of nonnegative real numbers whose columns add up to 1.
4. A *doubly stochastic matrix* is a square matrix of nonnegative real numbers with each row and column adding up to 1.

Illustration

- A right stochastic matrix

```
MatrixForm[RM = {{.5, 0, .5}, {.5, .25, .25}, {1, 0, 0}}]
```

$$\begin{pmatrix} 0.5 & 0 & 0.5 \\ 0.5 & 0.25 & 0.25 \\ 1 & 0 & 0 \end{pmatrix}$$

- A left stochastic matrix

```
MatrixForm[LM = Transpose[{{.5, 0, .5}, {.5, .25, .25}, {1, 0, 0}}]]
```

$$\begin{pmatrix} 0.5 & 0.5 & 1 \\ 0 & 0.25 & 0 \\ 0.5 & 0.25 & 0 \end{pmatrix}$$

- A doubly stochastic matrix

```
MatrixForm[DM = {{.5, 0, .5}, {.5, .25, .25}, {0, .75, .25}}]
```

$$\begin{pmatrix} 0.5 & 0 & 0.5 \\ 0.5 & 0.25 & 0.25 \\ 0 & 0.75 & 0.25 \end{pmatrix}$$

- A finite Markov process

A *finite Markov process* is a random process on a graph, where from each state you specify the probability of selecting each available transition to a new state.

Finite Markov processes are used to model a variety of decision processes in areas such as games, weather, manufacturing, business, and biology.

The **DiscreteMarkovProcess** function generates a Markov chain from a stochastic matrix *A* and an initial probability vector **v**.

```
A = {{0, 1/2, 1/2}, {1/2, 0, 1/2}, {1/2, 1/2, 0}};
```

```
v = {1/3, 0, 2/3};
```

```
P = DiscreteMarkovProcess[v, A];

data1 = Normal[RandomFunction[P, {0, 10}]];

data2 = Normal[RandomFunction[P, {0, 10}]];

plot1 = ListLinePlot[data1, PlotStyle → Directive[Red, Thick]];

plot2 = ListLinePlot[data2, PlotStyle → Directive[Green, Thickness[0.02]]];

Show[plot1, plot2]
```

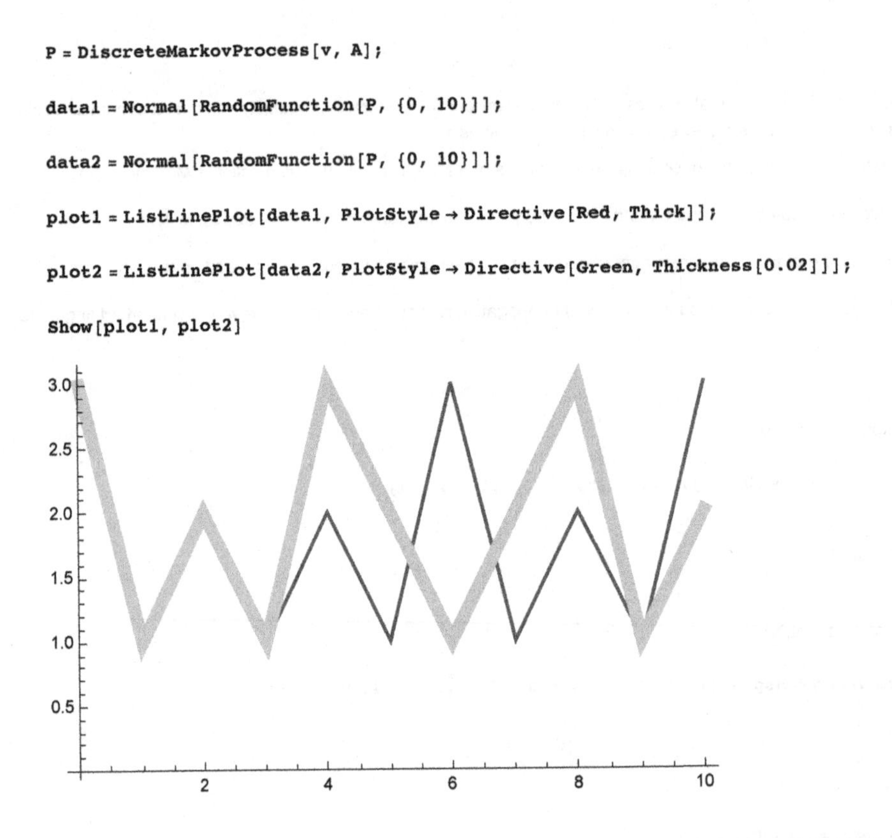

Subdiagonal of a matrix

The subdiagonal of an *n*-by-*m* matrix *A* is the list of all elements $A_{[[i+1,j]]}$ of the matrix *A*.

Illustration

- The subdiagonal of a 4-by-3 matrix

```
A = RandomInteger[{0, 9}, {4, 3}];
```

$$A = \begin{pmatrix} 5 & 9 & 3 \\ 0 & 9 & 0 \\ 8 & 1 & 7 \\ 4 & 6 & 0 \end{pmatrix};$$

```
Diagonal[A, -1]

{0, 1, 0}
```

```
Diagonal[A, -1] == {A[[2,1]], A[[3,2]], A[[4,3]]}
```

```
True
```

Submatrix

A *submatrix* of a matrix A is a matrix obtained from A by deleting some rows and/or columns of A.

Illustration

```
A = RandomInteger[{0, 9}, {4, 6}];
```

$$A = \begin{pmatrix} 1 & 5 & 0 & 1 & 8 & 2 \\ 6 & 8 & 4 & 9 & 3 & 3 \\ 1 & 3 & 7 & 7 & 8 & 1 \\ 3 & 9 & 2 & 9 & 3 & 7 \end{pmatrix};$$

- The submatrix of A obtained by deleting the first row and first column of A

$$S_{11} = \begin{pmatrix} 8 & 4 & 9 & 3 & 3 \\ 3 & 7 & 7 & 8 & 1 \\ 9 & 2 & 9 & 3 & 7 \end{pmatrix};$$

- The submatrix of A obtained by deleting the second and third rows and the sixth column

$$S_{23/6} = \begin{pmatrix} 1 & 5 & 0 & 1 & 8 \\ 3 & 9 & 2 & 9 & 3 \end{pmatrix};$$

Subspace

A *subspace* S of a vector space V is a subset of all vectors in V, with the same vector addition and scalar multiplication as those on V. Strictly speaking, the operations on S are the restrictions of the operations on V to the subset S. For this to make sense, however, the set S has to be closed under the given vector addition and scalar multiplication. That is, the sum of two vectors in S must also be a vector in S.

Every vector space is a subspace of itself. But most spaces have proper subspaces whose sets of vectors are only part of the entire set of vectors of the ambient space. However, all subspaces must have the same set of scalars of the given space.

Every vector space has a *zero subspace*, a subspace whose only vector is the zero vector of the original space. By definition, the empty set is a basis of a zero space and the space therefore has *dimension zero*.

Illustration

- Subspaces of $\mathbb{R}^2$

Every line through the point {0, 0} defines a subspace of $\mathbb{R}^2$.

- A one-dimensional subspace of $\mathbb{R}^2$

```
line = y == 3 x;
```

```
subspace = {{x, y} : y == 3 x ∧ x, y ∈ Reals}
```

Every point in $\mathbb{R}^2$ other than the origin {0, 0} determines a subspace of $\mathbb{R}^2$

- A basis for a one-dimensional subspace of $\mathbb{R}^2$

```
point = {1, 3};
```

```
basis = {point}
```

```
{{1, 3}}
```

```
subspace = {x, y} == a {1, 3}
```

```
{{2, 6, 3, 1, 8, 9}, y} == {a, 3 a}
```

Two intersecting lines through the origin {0, 0} determine a subspace of $\mathbb{R}^2$. If the lines do not overlap, they generate a two-dimensional subspace of $\mathbb{R}^2$. In that case, the subspace is actually all of $\mathbb{R}^2$.

- Two disjoint subspaces of $\mathbb{R}^2$

```
Clear[x, y]
```

```
line1 = y == 3 x;
```

```
line2 = y == -2 x;
```

```
subspace1 = {{x, y} : y == 3 x ∧ x, y ∈ Reals}
```

```
subspace2 = {{x, y} : y == -2 x ∧ x, y ∈ Reals}
```

The two subspaces have only the zero vector in common:

```
Solve[3 x == -2 x, x]
```

```
{{x → 0}}
```

- A basis for a subspace of $\mathbb{R}^2$

```
basis = {{3, 1}, {1, -2}};
```

This set forms a basis for a subspace of dimension 2. In other words, it forms a basis for the ambient space $\mathbb{R}^2$ since all vectors {*x*, *y*} in $\mathbb{R}^2$ are linear combinations of the vectors {3, 1} and {1, -2}:

```
Solve[{x, y} == a {3, 1} + b {1, -2}, {a, b}]
```

$$\left\{\left\{a \to \frac{1}{7}(2x + y),\ b \to \frac{1}{7}(x - 3y)\right\}\right\}$$

- Subspaces of $\mathbb{R}^3$

Lines through the origin {0,0,0}, planes through the origin, the zero subspace, and $\mathbb{R}^3$ itself are all subspaces of $\mathbb{R}^3$.

- A plane through the origin of $\mathbb{R}^3$ forms a two-dimensional subspace of $\mathbb{R}^3$

```
Clear[x, y, z]
```

```
plane = z == 3 x - 2 y
```

```
z == 3 x - 2 y
```

```
Reduce[plane, {x, y, z}]
```

```
z == 3 x - 2 y
```

```
ContourPlot3D[Evaluate[plane], {x, -5, 5}, {y, -5, 5}, {z, -5, 5}, Axes → True]
```

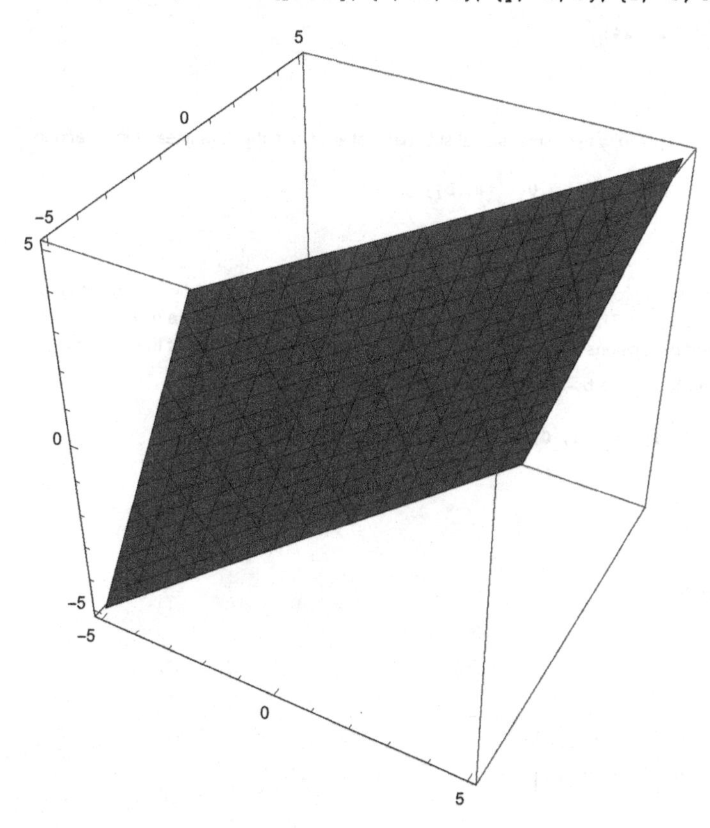


```
Reduce[3 x - 2 y == 0, {x, y}]
```

$$y == \frac{3 x}{2}$$

```
Reduce[3 x - 2 y == 0, {x, y}]
```

```
     3 x
y == ---
      2
```

If we assign two appropriate values to x, we get two linearly independent vectors for different values of z lying in the given plane. They form a basis for the space.

- A basis for a two-dimensional subspace of $\mathbb{R}^3$

```
basis = {{2, 3, 0}, {6, -3, 24}};
```

```
z == 3 x - 2 y /. {x → 2, y → 3, z → 0}
```

```
True
```

```
z == 3 x - 2 y /. {x → 6, y → -3, z → 24}
```

```
True
```

The two vectors form a basis for the plane, a two-dimensional subspace of $\mathbb{R}^3$, since they are linearly independent:

```
Solve[a {2, 3, 0} + b {6, -3, 24} == {0, 0, 0}, {a, b}]
```

```
{{a → 0, b → 0}}
```

Many important vector spaces arise as proper parts of larger spaces, coordinate spaces in particular. Among them are four spaces associated with every real matrix. These are called the *row space*, the *column space*, the *null space*, and the *left null space* of a given matrix. Their dimensions do not exceed the number of rows and columns of the given matrix.

- The four vector spaces generated by a 3-by-5 real matrix

```
MatrixForm[A = {{3, 1, 0, 2, 4}, {1, 1, 0, 0, 2}, {5, 2, 0, 3, 7}}]
```

$$\begin{pmatrix} 3 & 1 & 0 & 2 & 4 \\ 1 & 1 & 0 & 0 & 2 \\ 5 & 2 & 0 & 3 & 7 \end{pmatrix}$$

```
MatrixForm[ns = NullSpace[A]]
```

$$\begin{pmatrix} -1 & -1 & 0 & 0 & 1 \\ -1 & 1 & 0 & 1 & 0 \\ 0 & 0 & 1 & 0 & 0 \end{pmatrix}$$

$$\text{LinearSolve}\left[\begin{pmatrix} -1 & -1 & 0 & 0 & 1 \\ -1 & 1 & 0 & 1 & 0 \\ 0 & 0 & 1 & 0 & 0 \end{pmatrix}, \{0, 0, 0\}\right]$$

```
{0, 0, 0, 0, 0}
```

Therefore the null space of A has dimension 0 since only the zero vector is in its basis.

```
lns = NullSpace[Transpose[A]]
```

```
{{-3, -1, 2}}
```

Therefore the left null space of *A* has dimension 1. Every vector in the left null space is a multiple of the vector {-3,-1,2}.

```
{-3, -1, 2}.A
```

```
{0, 0, 0, 0, 0}
```

```
MatrixForm[rs = RowReduce[A]]
```

$$\begin{pmatrix} 1 & 0 & 0 & 1 & 1 \\ 0 & 1 & 0 & -1 & 1 \\ 0 & 0 & 0 & 0 & 0 \end{pmatrix}$$

Therefore the row space of *A* has dimension 2. Every vector in the row space of *A* is a linear combination of the vectors {1, 0, 0, 1,1 } and {0, 1, 0, -1, 1}.

```
a {1, 0, 0, 1, 1} + b {0, 1, 0, -1, 1}
```

```
{a, b, 0, a - b, a + b}
```

```
MatrixForm[cs = Transpose[RowReduce[Transpose[A]]]]
```

$$\begin{pmatrix} 1 & 0 & 0 & 0 & 0 \\ 0 & 1 & 0 & 0 & 0 \\ \frac{3}{2} & \frac{1}{2} & 0 & 0 & 0 \end{pmatrix}$$

$$a \begin{pmatrix} 1 \\ 0 \\ \frac{3}{2} \end{pmatrix} + b \begin{pmatrix} 0 \\ 1 \\ \frac{1}{2} \end{pmatrix}$$

$$\left\{\{a\}, \{b\}, \left\{\frac{3a}{2} + \frac{b}{2}\right\}\right\}$$

This tells us that the column space of *A* has dimension 2. Every vector in the column space is a linear combination of the vectors {1, 0, 3/2} and {0, 1, 1/2}.

The four spaces are connected by the direct sum operation:

nullSpace[A] ⊕ rowSpace[A] (1)

leftnullSpace[A] ⊕ columnSpace[A] (2)

The direct sum of two vector spaces is the union of two subspaces of a given space, provided that they only have the zero vector in common. Every vector in the ambient space is a unique sum of vectors from the two subspaces.

All vector spaces have a zero-dimensional subspace whose only vector is the zero vector of the space. It is convenient to consider the empty set { } to be the basis of the zero subspace. All subspaces of a given vector space have the zero vector in common. If this is the only common vector, the subspaces are said to be *disjoint*.

Manipulation

- Straight lines as subspaces of $\mathbb{R}^2$

```
Clear[a, b, x, y]
```

```
Manipulate[ContourPlot[{y == a x, y == b x}, {x, -5, 5}, {y, -5, 5}, Axes → True],
 {a, -20, 20}, {b, -20, 20}]
```

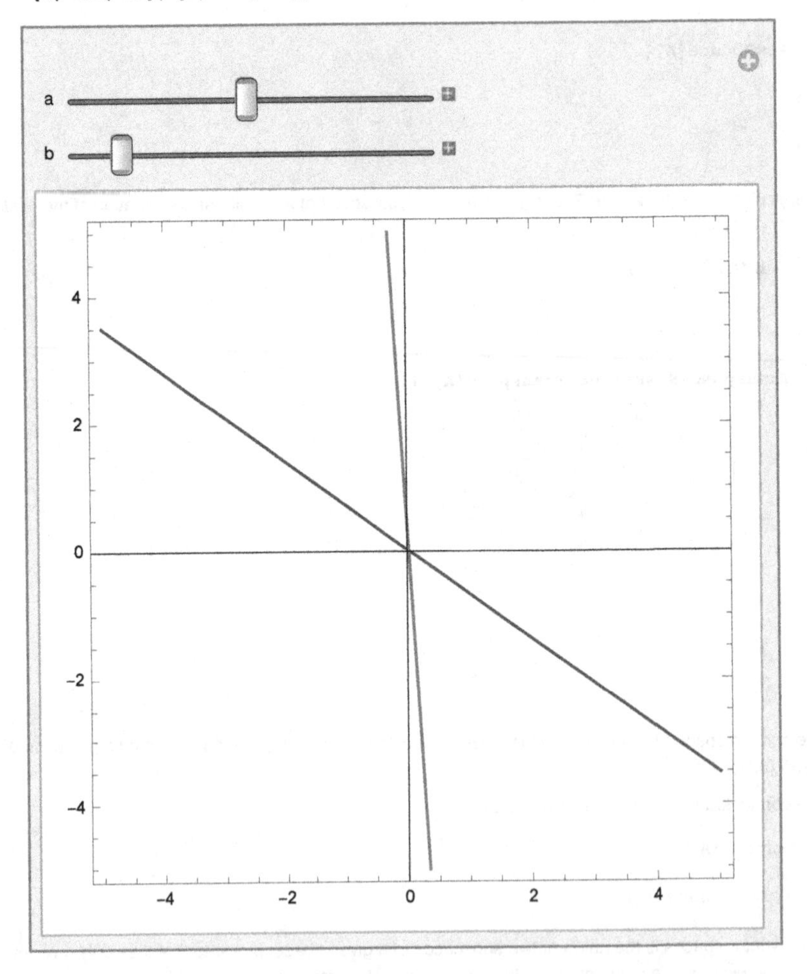

We use **Manipulate** and **ContourPlot** to visualize lines through the origin as subspaces of $\mathbb{R}^2$. For example, the manipulation displays the subspace defined by the equations y = -0.7x and y = -16x.

Sum of subspaces

If U and V are two subspaces of a vector space W, the sum $(U + V)$ of U and V is the span of the set-theoretical union of U and V.

Illustration

- The sum of two subspaces of $\mathbb{R}^4$

```
U = {{a, a, 0, b} : a, b ∈ ℝ}
V = {{0, b, b, c} : b, c ∈ ℝ}
```

```
U + V = span[{a {1, 1, 0, 0} + c {0, 0, 0, 1} + b {0, 1, 1, 0} + c {0, 0, 0, 1}} : a, b, c ∈ ℝ]
```

- The sum of two disjoint subspaces of $\mathbb{R}^3$

```
U = {{a, a, 0} : a ∈ ℝ}
V = {{0, 0, b} : bℝ}
```

```
U + V = span[{a {1, 1, 0} + b {0, 0, 1}} : a, b ∈ ℝ]
```

Superdiagonal of a matrix

The superdiagonal of an n-by-m matrix A is the list of all elements $A_{[[i,i+1]]}$ of the matrix A.

Illustration

- The superdiagonal of a 3-by-4 matrix

```
A = RandomInteger[{0, 9}, {3, 4}];
```

$$A = \begin{pmatrix} 9 & 9 & 8 & 8 \\ 0 & 9 & 7 & 2 \\ 5 & 9 & 7 & 8 \end{pmatrix};$$

```
Diagonal[A, 1]
```

```
{9, 7, 8}
```

```
Diagonal[A, 1] == {A[[1,2]], A[[2,3]], A[[3,4]]}
```

```
True
```

Surjective linear transformation

A linear transformation $T : V \longrightarrow W$ from a vector space V to a vector space W is *surjective* (onto) if for every vector **w** in W there exists a vector **v** in V for which $\mathbf{w} = T[\mathbf{v}]$.

Illustration

- A surjective linear transformation $T : \mathbb{R}^4 \longrightarrow \mathbb{R}^{2\times 2}$

```
Clear[a, b, c, d, T, A, S]

T[{a_, b_, c_, d_}] := {{a, b}, {c, d}}

A = {{1, 2}, {3, 4}};

Solve[A == T[{a, b, c, d}], {a, b, c, d}]

{{a → 1, b → 2, c → 3, d → 4}}
```

The transformation T is also injective. Its inverse S is defined by

```
S[{{a_, b_}, {c_, d_}}] := {a, b, c, d}

S[A]

SparseArray[ Specified elements: 13  Dimensions: {5, 5} ][{{1, 2}, {3, 4}}]

T[S[{{a, b}, {c, d}}]] == {{a, b}, {c, d}}

{{9, 7, 7, 0}, {8, 3, 4, 6}, {4, 7, 3, 1}, {4, 1, 8, 8}}[
  SparseArray[ Specified elements: 13  Dimensions: {5, 5} ][{{a, b}, {c, d}}]] == {{a, b}, {c, d}}

S[T[{a, b, c, d}]] == {a, b, c, d}

SparseArray[ Specified elements: 13  Dimensions: {5, 5} ][
  {{9, 7, 7, 0}, {8, 3, 4, 6}, {4, 7, 3, 1}, {4, 1, 8, 8}}[{a, b, c, d}]] == {a, b, c, d}
```

- A surjective linear transformation T: $\mathbb{R}^3 \longrightarrow \mathbb{R}^2$

```
Clear[T, a, b, c]

T[{a_, b_, c_}] := {a, c}
```

```
T[{1, 2, 3}]
```

```
{1, 3}
```

The transformation T is surjective since every vector {a, b, c} gets mapped to the vector {a, c} for all real numbers a and c. However, it is not injective. The vectors {a, 0, c} and {a, 1, c}, for example, get mapped to the same vector {a, c}.

Sylvester's theorem

If q is a real quadratic form on $\mathbb{R}^n$, then there exist integers r and s, with $s \leq r \leq n$, depending uniquely on q, such that

$$q[x_1, \ldots, x_n] = x_1^2 + \cdots + x_s^2 - x_{s+1}^2 - \cdots - x_r^2 \quad (1)$$

in some orthonormal basis for $\mathbb{R}^n$.

Sylvester's theorem tells us, for example, that by choosing a suitable basis, we can make an ellipse look like a circle, but we can never choose a basis, for example, that makes an ellipse look like a hyperbola.

Illustration

- Ellipses and circles

```
q1[x_, y_] := 2 x^2 + 5 y^2;
```

```
plot1 = ContourPlot[q1[x, y], {x, -4, 4}, {y, -4, 4}]
```

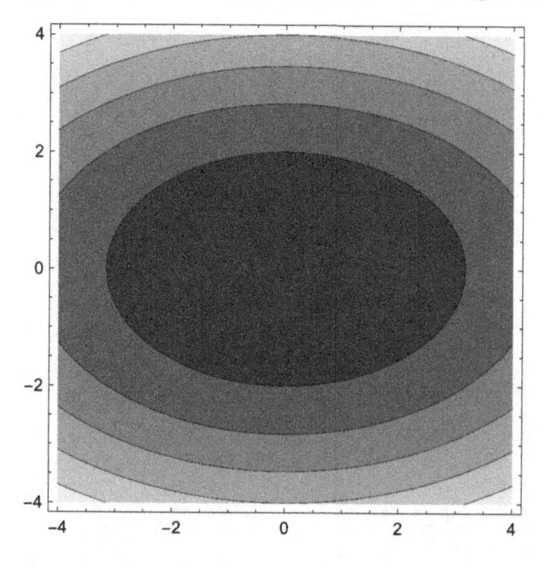


```
q2[x_, y_] := x^2 + y^2;
```

```
plot2 = ContourPlot[q2[x, y], {x, -4, 4}, {y, -4, 4}]
```

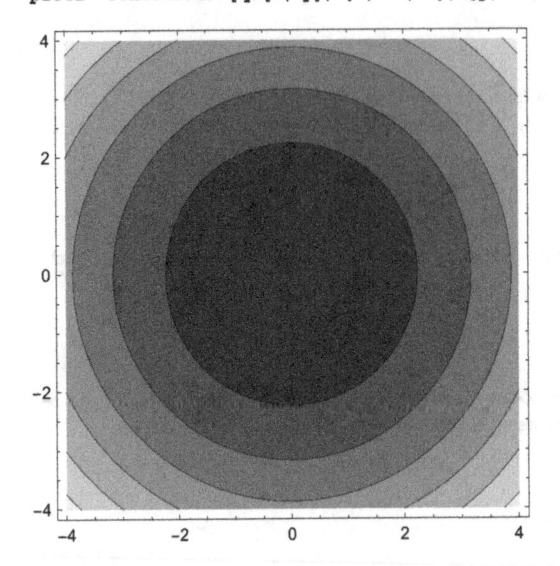

The number of pluses and minuses in the canonical representation of a quadratic form provided by Sylvester's theorem is sometimes called the *signature* of the form and can be used to give a classification of these forms:

- Classifying quadratic forms

$$q1[x, y] = x^2 + y^2 \tag{1}$$

$$q2[x, y] = -x^2 + y^2 \tag{2}$$

$$q3[x, y] = x^2 - y^2 \tag{3}$$

$$q4[x, y] = -x^2 - y^2 \tag{4}$$

The signatures of these quadratic forms are {1, 1}, {-1, 1}, {1, -1}, and {-1, -1}.

More generally,

$$q1[x, y] = \lambda x^2 + \mu y^2 \text{ and } \lambda > 0, \mu > 0 \tag{1}$$

$$q2[x, y] = \lambda x^2 + \mu y^2 \text{ and } \lambda < 0, \mu > 0 \tag{2}$$

$$q3[x, y] = \lambda x^2 + \mu y^2 \text{ and } \lambda > 0, \mu < 0 \tag{3}$$

$$q4[x, y] = \lambda x^2 + \mu y^2 \text{ and } \lambda < 0, \mu < 0 \tag{4}$$

where λ and μ are the eigenvalues of q.

Manipulation

- Plotting quadratic forms

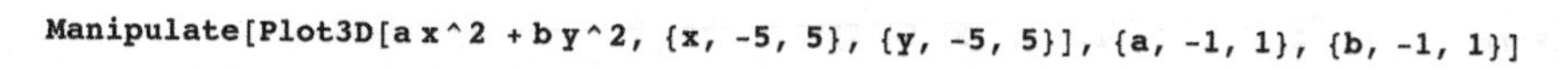

```
Manipulate[Plot3D[a x^2 + b y^2, {x, -5, 5}, {y, -5, 5}], {a, -1, 1}, {b, -1, 1}]
```

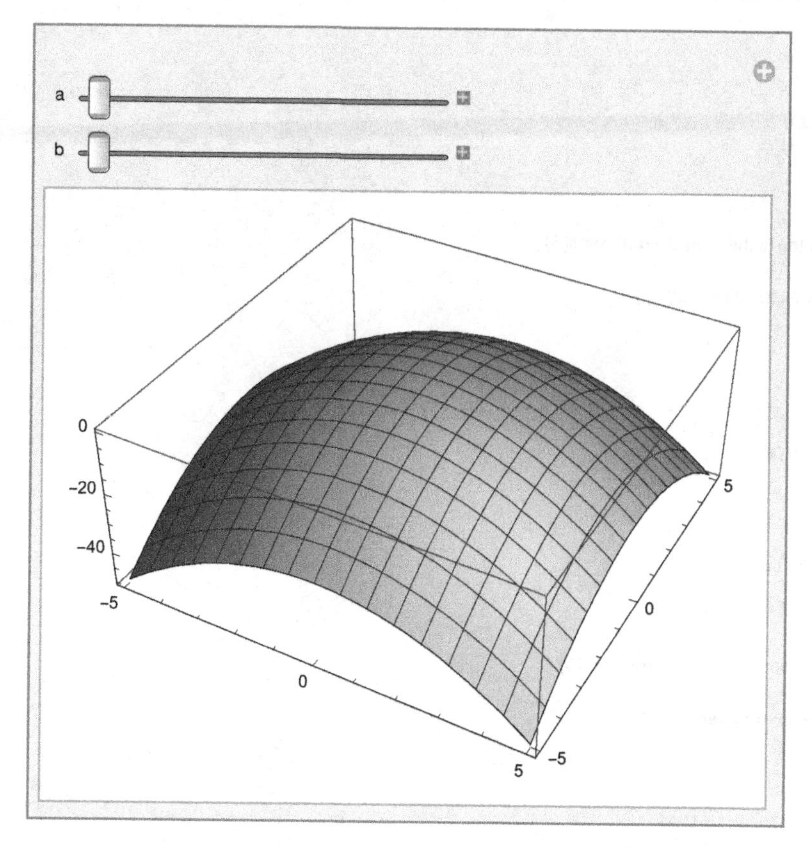

We use **Manipulate** and **Plot3D** to visualize the quadratic forms determined by the equations $ax^2 + by^2$. The image in the manipulation window displays the graph of the quadratic form $-x^2 - y^2$.

Symmetric matrix

A matrix S is *symmetric* if it equals its transpose. Real matrices of the form AA^T and $A^T A$ are symmetric and have nonnegative eigenvalues.

Illustration

- A symmetric matrix

```
MatrixForm[S = {{1, 2, 3}, {2, 4, 5}, {3, 5, 6}}]
```

$$\begin{pmatrix} 1 & 2 & 3 \\ 2 & 4 & 5 \\ 3 & 5 & 6 \end{pmatrix}$$

```
SymmetricMatrixQ[S]
```

```
True
```

- A symmetric matrix of the form S.**Transpose**[S]

```
MatrixForm[R1 = S.Transpose[S]]
```

$$\begin{pmatrix} 14 & 25 & 31 \\ 25 & 45 & 56 \\ 31 & 56 & 70 \end{pmatrix}$$

```
SymmetricMatrixQ[R1]
```

```
True
```

```
N[Eigenvalues[R1]]
```

```
{128.705, 0.265977, 0.029212}
```

- A symmetric matrix of the form **Transpose**[S].*S*

```
MatrixForm[R2 = Transpose[S].S]
```

$$\begin{pmatrix} 14 & 25 & 31 \\ 25 & 45 & 56 \\ 31 & 56 & 70 \end{pmatrix}$$

```
SymmetricMatrixQ[R2]
```

```
True
```

```
N[Eigenvalues[R2]]
```

```
{128.705, 0.265977, 0.029212}
```

- A symmetric matrix of the form $(A + A^T)$

For every square matrix A, the matrix $(A + A^T)$ is symmetric

$$\mathbf{A} = \begin{pmatrix} 8 & 3 & 6 & 7 \\ 5 & 0 & 6 & 7 \\ 5 & 5 & 0 & 0 \\ 0 & 4 & 5 & 1 \end{pmatrix};$$

```
SymmetricMatrixQ[A + Transpose[A]]
```

```
True
```

System of linear equations

See Linear system

T

Toeplitz matrix

A n-by-n matrix is a *Toeplitz matrix* if the elements in the first row and the first column are successive integers.

The **ToeplitzMatrix** function built into *Mathematica* can be used to create Toeplitz matrices.

Illustration

- A 4-by-4 Toeplitz matrix

```
MatrixForm[T = ToeplitzMatrix[4]]
```

$$\begin{pmatrix} 1 & 2 & 3 & 4 \\ 2 & 1 & 2 & 3 \\ 3 & 2 & 1 & 2 \\ 4 & 3 & 2 & 1 \end{pmatrix}$$

The elements in row one and column one are the successive integers from 1 to 4.

- Building Toeplitz matrices

```
Table[MatrixForm[ToeplitzMatrix[n]], {n, 2, 5, 1}]
```

$$\left\{\begin{pmatrix} 1 & 2 \\ 2 & 1 \end{pmatrix}, \begin{pmatrix} 1 & 2 & 3 \\ 2 & 1 & 2 \\ 3 & 2 & 1 \end{pmatrix}, \begin{pmatrix} 1 & 2 & 3 & 4 \\ 2 & 1 & 2 & 3 \\ 3 & 2 & 1 & 2 \\ 4 & 3 & 2 & 1 \end{pmatrix}, \begin{pmatrix} 1 & 2 & 3 & 4 & 5 \\ 2 & 1 & 2 & 3 & 4 \\ 3 & 2 & 1 & 2 & 3 \\ 4 & 3 & 2 & 1 & 2 \\ 5 & 4 & 3 & 2 & 1 \end{pmatrix}\right\}$$

Trace

The *trace* of a square matrix A is the sum of the diagonal elements of the matrix A. The trace of a matrix can be computed with the built-in **Tr** function. The trace of A is also the (approximate) sum of the eigenvalues of A. The trace function satisfies the following properties:

Properties of the trace function

```
Tr[A] = Tr[Transpose[A]]          (1)

Tr[A.B] = Tr[B.A]                 (2)

Tr[B.A.Inverse[B]] = Tr[A]        (3)
```

Illustration

- The trace of a general 3-by-3 matrix

```
Clear[A, a, b, c, d, e, f, g, h, i]
```

```
MatrixForm[A = {{a, b, c}, {d, e, f}, {g, h, i}}]
```

$$\begin{pmatrix} a & b & c \\ d & e & f \\ g & h & i \end{pmatrix}$$

```
Tr[A]
```

a + e + i

- The trace of a 4-by-4 matrix

```
MatrixForm[A = RandomInteger[{0, 9}, {4, 4}]]
```

$$\begin{pmatrix} 1 & 2 & 4 & 1 \\ 2 & 5 & 7 & 8 \\ 2 & 5 & 3 & 7 \\ 4 & 5 & 3 & 6 \end{pmatrix}$$

```
Tr[A]
```

15

```
Total[N[Eigenvalues[A]]]
```

15. + 0. i

- The trace of a 2-by-2 real matrix

```
A = RandomReal[{0, 9}, {2, 2}];
```

$$A = \begin{pmatrix} 6.073493017009433` & 2.5207312136769584` \\ 7.104724528204983` & 0.09315320872764765` \end{pmatrix};$$

```
Tr[A]
```

6.16665

- The trace of a 4-by-4 matrix as the sum of the eigenvalues of the matrix

```
A = RandomInteger[{0, 9}, {4, 4}];
```

$$A = \begin{pmatrix} 1 & 0 & 8 & 6 \\ 8 & 7 & 5 & 1 \\ 2 & 3 & 9 & 8 \\ 0 & 5 & 1 & 9 \end{pmatrix};$$

```
Tr[A] == N[Total[Eigenvalues[A]]]
```

```
True
```

Manipulation

- Exploring the trace of 2-by-2 matrices

```
Manipulate[{A = {{a, b}, {4, 4}}, Tr[A]}, {a, -5, 5, 1}, {b, -5, 5, 1}]
```

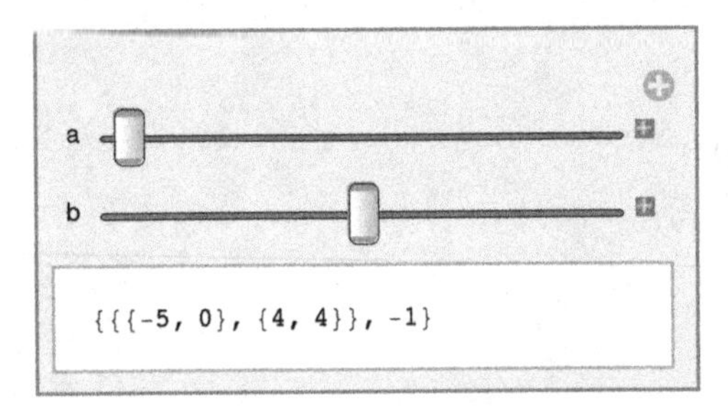

We use **Manipulate** and **Tr** to explore the trace of 2-by-2 matrices with integer elements. If we let a = -5 and b = 0, for example, the manipulation produces the matrix

```
MatrixForm[{{-5, 0}, {4, 4}}]
```

$$\begin{pmatrix} -5 & 0 \\ 4 & 4 \end{pmatrix}$$

and its trace -1.

- Calculating the trace of a 3-by-3 matrix

```
Clear[A, a, b, c]
```

```
MatrixForm[A = {{a, 1, 2}, {3, b, 4}, {5, 6, c}}]
```

$$\begin{pmatrix} a & 1 & 2 \\ 3 & b & 4 \\ 5 & 6 & c \end{pmatrix}$$

```
Manipulate[Evaluate[Tr[A]], {a, -5, 5, 1}, {b, -3, 3, 1}, {c, -4, 4, 1}]
```

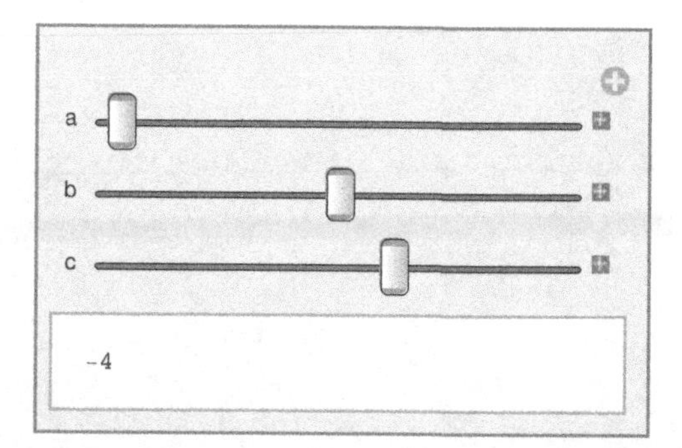

We use **Manipulate**, **Evaluate,** and **Tr** to display the traces of selected matrices. The manipulation produces that the trace of the matrix obtain by letting $a = -5$, $b = 0$, and $c = 1$ is - 4.

Transformation

See Affine transformation, linear transformation

Transformational geometry

The name transformation, as used in linear algebra for matrix-based functions, comes from *transformational geometry*, where rotations, shears, reflections, and translations are studied. All of these transformations can be represented by invertible matrices, except for translations.

Linear transformations, as the name suggests, transform vectors from $\{a, b\}$ to a new location $\{a', b'\}$; this is the image of the vector $\{a, b\}$ under the transformation. However, they do not transform the origin $\{0, 0\}$. So any geometric figure such as a triangle with one vertex at the origin cannot be moved by matrix multiplication.

This problem has led to the development of *affine geometry*, a combination of matrix multiplication and the subsequent addition of translation vectors. It is intriguing that these two types of operations can nevertheless be combined in special 3-by-3 matrices using *homogeneous coordinates.*

Transition matrix

See Stochastic matrix

Translation

A *translation* is a displacement of a vector by the addition of a nonzero vector.

Illustration

- Translation of vectors in $\mathbb{R}^2$

```
translation[vector_] := vector + {3, 4}
```

```
translation[{5, 6}]
```

```
{8, 10}
```

- Translation of vectors in $\mathbb{R}^3$

```
translation[vector_] := vector + {a, b, c}
```

```
translation[{5, 6, 7}]
```

```
{5 + a, 6 + b, 7 + c}
```

Manipulation

- Translation of vectors in $\mathbb{R}^4$

```
Manipulate[{1, 2, 3, 4} + {a, b, c, d}, {a, -5, 5}, {b, -4, 4}, {c, 0, 9}, {d, -8, -3}]
```

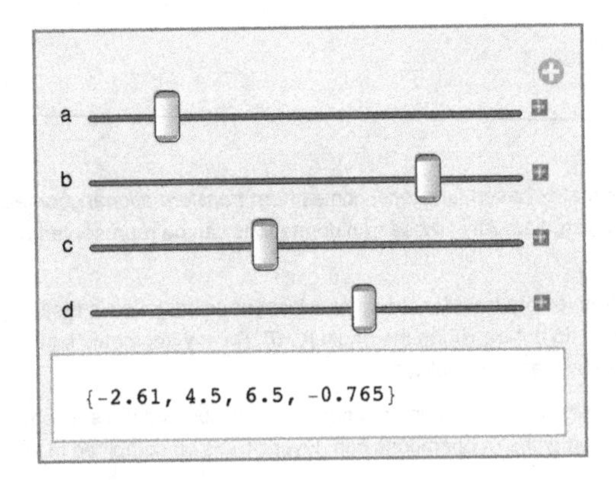

We use **Manipulate** to translate vectors in $\mathbb{R}^4$. If we let a = -3.61, b = 2.5, c = 3.5, and d = -4.765, then the manipulation shows that the sum of the vectors {1, 2, 3, 4} and {a, b, c, d} is {-2.61, 4.5, 6.5, -0.765}.

Transpose of a matrix

Matrices can be converted into other matrices by interchanging their rows and columns. This operation is called matrix transposition and produces a matrix called the *transpose* of the given matrix. All matrices can be transposed. In addition to the **Transpose** function, *Mathematica* supports the superscript notation A^T, also denoting the transpose of A. The transpose function satisfies the following properties:

Properties of the transpose function

```
Transpose[Transpose[A]] = A                                    (1)

Transpose[A.B] = Transpose[B].Transpose[A]                     (2)
```

```
Transpose[A + B] = Transpose[A] + Transpose[B]
```
(3)

```
Transpose[a A] = a Transpose[A]
```
(4)

Illustration

- Transposition of a matrix

```
MatrixForm[A = {{1, 2, 3}, {4, 5, 6}}]
```

$$\begin{pmatrix} 1 & 2 & 3 \\ 4 & 5 & 6 \end{pmatrix}$$

```
MatrixForm[Transpose[A]]
```

$$\begin{pmatrix} 1 & 4 \\ 2 & 5 \\ 3 & 6 \end{pmatrix}$$

As expected, the transpose of the transpose of a matrix returns the original matrix.

```
Transpose[Transpose[A]] == A
True
```

Triangle inequality

In a normed vector space, the norms $\|\mathbf{u}\|$, $\|\mathbf{v}\|$, and $\|\mathbf{u} + \mathbf{v}\|$ of any three vectors **u**, **v**, and (**u** + **v**) are related by the inequality

$$\|\mathbf{u} + \mathbf{v}\| \leq \|\mathbf{u}\| + \|\mathbf{v}\| \tag{1}$$

The inequality is an example of a *triangle inequality* and corresponds to the relationship between the lengths of the three sides of a triangle.

Illustration

- The triangle inequality for two real numbers *x* and *y*,

```
Clear[x, y]

Abs[x + y] ≤ Abs[x] + Abs[y];

x = 5; y = -7;

Abs[x + y] ≤ Abs[x] + Abs[y]

True
```

- The triangle inequality for two complex numbers ($a + b\,i$) and ($c + d\,i$),

```
Clear[a, b, c, d]

Abs[(a + b i) + (c + d i)] ≤ Abs[(a + b i)] + Abs[(c + d i)];

a = 5; b = 2; c = -4; d = -1;

Abs[(a + b i) + (c + d i)] ≤ Abs[(a + b i)] + Abs[(c + d i)]

True
```

- The triangle inequality for two vectors in the Euclidean space $\mathbb{E}^3$

```
x = {1, 2, 3}; y = {4, 5, 6};

Norm[x + y, 2] ≤ Norm[x, 2] + Norm[y, 2]

True
```

Triangular matrix

A *triangular matrix* is a square matrix whose elements above or below the main diagonal are zero. Diagonal matrices are automatically triangular.

The **UpperTriangularize** and **LowerTriangularize** built into *Mathematica* can be used to convert given matrices to triangular ones.

Illustration

- A triangular matrix with integer elements

```
A = {{6, 1, 4, 3, 2}, {7, 0, 1, 4, 2}, {0, 3, 1, 3, 4}, {6, 2, 4, 0, 4}, {2, 4, 5, 8, 4}};

MatrixForm[UpperTriangularize[A]]
```

$$\begin{pmatrix} 6 & 1 & 4 & 3 & 2 \\ 0 & 0 & 1 & 4 & 2 \\ 0 & 0 & 1 & 3 & 4 \\ 0 & 0 & 0 & 0 & 4 \\ 0 & 0 & 0 & 0 & 4 \end{pmatrix}$$

```
MatrixForm[LowerTriangularize[A]]
```

$$\begin{pmatrix} 6 & 0 & 0 & 0 & 0 \\ 7 & 0 & 0 & 0 & 0 \\ 0 & 3 & 1 & 0 & 0 \\ 6 & 2 & 4 & 0 & 0 \\ 2 & 4 & 5 & 8 & 4 \end{pmatrix}$$

U

Underdetermined linear system

A linear system is *underdetermined* if it has more unknowns than equations. When the system is expressed as matrix-vector equation, the matrix of coefficients will have more columns than rows. The solutions of consistent underdetermined systems may contain a free variable.

Illustration

- An underdetermined consistent linear system in two equations and three variables

```
system = {3 x + 4 y - z == 5, x - y + 7 z == 4};

solutions = Solve[system, {x, y, z}]
```

Solve::svars : Equations may not give solutions for all "solve" variables. ≫

$$\left\{\left\{y \to \frac{13}{9} - \frac{22\,x}{27},\ z \to \frac{7}{9} - \frac{7\,x}{27}\right\}\right\}$$

The variable x is said to be free. We can assign to it any real number value, and find a corresponding value for y and z.

```
Simplify[system /. solutions]

{{True, True}}

solutions /. {x → 1}
```

$$\left\{\left\{y \to \frac{17}{27},\ z \to \frac{14}{27}\right\}\right\}$$

- An underdetermined linear system in three equations and four variables

```
Clear[w, x, y, z]

system = {x + y - z == 1, 2 x - y == 4, x + 5 y - 5 w == 3};

Reduce[system, {w, x, y, z}]
```

$$x == \frac{23}{11} + \frac{5\,w}{11}\ \&\&\ y == \frac{2}{11} + \frac{10\,w}{11}\ \&\&\ z == \frac{14}{11} + \frac{15\,w}{11}$$

- An underdetermined inconsistent linear system

```
A = {{1, 2, 3}, {1, 2, 3}}; b = {1, 2};

LinearSolve[A, b]
```

LinearSolve::nosol : Linear equation encountered that has no solution. ≫

```
LinearSolve[{{1, 2, 3}, {1, 2, 3}}, {1, 2}]
```

Unit circle

The *unit circle* in a normed vector space *V* consists of all unit vectors of *V*.

Illustration

- The unit circle of $\mathbb{R}^2$ relative to the Euclidean norm

```
Clear[a, b]
```

```
v = {a, b}; u = Normalize[v];
```

```
Simplify[Norm[u]]
```

1

The unit circle is thus the set of all vectors whose components are of the form

$$\left\{\frac{a}{\sqrt{|a|^2+|b|^2}}, \frac{b}{\sqrt{|a|^2+|b|^2}}\right\}$$

- The unit circle of $\mathbb{R}^2$ relative to the one-norm

$$\text{v = \{a, b\}; u = } \frac{1}{\text{Norm[v, 1]}}\text{ v}$$

$$\left\{\frac{a}{\text{Abs}[a]+\text{Abs}[b]}, \frac{b}{\text{Abs}[a]+\text{Abs}[b]}\right\}$$

```
Simplify[Norm[u, 1]]
```

1

- The unit circle of $\mathbb{R}^2$ relative to the infinity-norm

$$\text{v = \{a, b\}; a = 3; b = -5; u = } \frac{1}{\text{Norm[v, Infinity]}}\text{ v}$$

$$\left\{\frac{3}{5}, -1\right\}$$

```
Norm[u, Infinity]
```

1

- Plotting the three unit circles relative to different norms

```
ContourPlot[
 {Norm[{x, y}, 1] == 1, Norm[{x, y}, 2] == 1, Norm[{x, y}, Infinity] == 1},
 {x, -1.1, 1.1}, {y, -1.1, 1.1}, Axes → True]
```

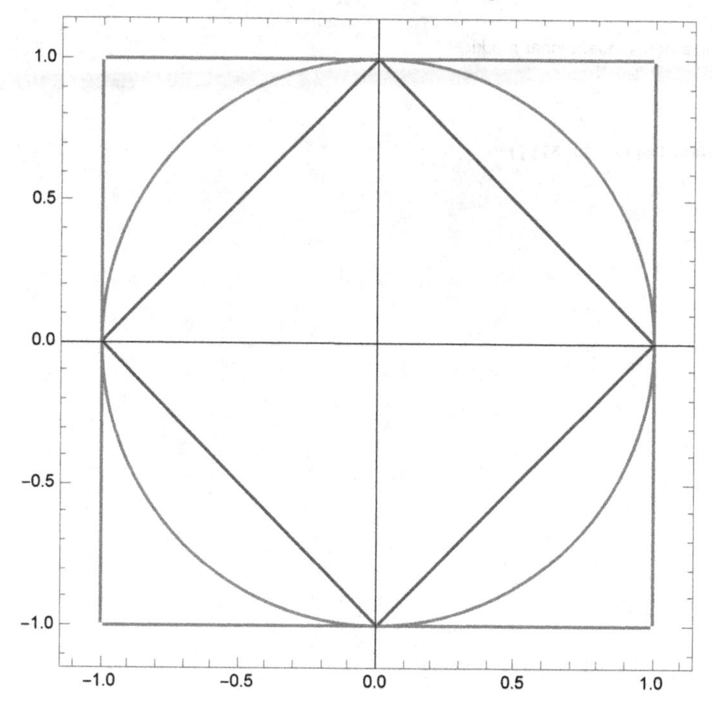

Unit vector

A *unit vector* in a normed vector space is a vector for which $\|\mathbf{v}\| = 1$ in the norm of the space.

Illustration

- A unit vector in the Euclidean space $\mathbb{R}^2$

```
v = {3, 7};
```

```
u = 1/Norm[v] v
```

$$\left\{\frac{3}{\sqrt{58}}, \frac{7}{\sqrt{58}}\right\}$$

```
Norm[u] == 1
```

```
True
```

- A unit vector in $\mathbb{R}^3$ relative to a nonstandard inner product

```
Clear[x, y, z]

MatrixForm[A = DiagonalMatrix[{1, 2, 3}]];

w = {x, y, z};

⟨u_, v_⟩ := u.A.v

‖w_‖ := Sqrt[⟨w, w⟩]

‖w‖
```

$$\sqrt{x^2 + 2\,y^2 + 3\,z^2}$$

```
w123 = w /. {x → 1, y → 2, z → 3}
```

```
{1, 2, 3}
```

```
‖w123‖
```

```
6
```

$$u = \frac{1}{\|w123\|}\,w123$$

$$\left\{\frac{1}{6}, \frac{1}{3}, \frac{1}{2}\right\}$$

```
‖u‖
```

```
1
```

Unitary matrix

A *unitary matrix* is a matrix whose inverse equals it conjugate transpose. Unitary matrices are the complex analog of real orthogonal matrices. If U is a square, complex matrix, then the following conditions are equivalent :

- U is unitary.
- The conjugate transpose U^* of U is unitary.
- U is invertible and $U^{-1} = U^*$.

- The columns of *U* form an orthonormal basis with respect to the inner product determined by *U*.
- The rows of *U* form an orthonormal basis with respect to the inner product determined by *U*.
- *U* is an isometry with respect to the inner product determined by *U*.
- U is a normal matrix with eigenvalues lying on the unit circle.

Illustration

- A 2-by-2 unitary matrix

```
MatrixForm[A = {{0, I}, {I, 0}}]
```

$$\begin{pmatrix} 0 & i \\ i & 0 \end{pmatrix}$$

```
Inverse[A] == ConjugateTranspose[A]
```

True

```
MatrixForm[ConjugateTranspose[A]]
```

$$\begin{pmatrix} 0 & -i \\ -i & 0 \end{pmatrix}$$

```
MatrixForm[A.ConjugateTranspose[A]]
```

$$\begin{pmatrix} 1 & 0 \\ 0 & 1 \end{pmatrix}$$

- A unitary matrix of the form *i*.IdentityMatrix[2]

```
MatrixForm[B = I IdentityMatrix[2]]
```

$$\begin{pmatrix} i & 0 \\ 0 & i \end{pmatrix}$$

```
Inverse[B] == ConjugateTranspose[B]
```

True

- A more general unitary matrix

$$\text{MatrixForm}\left[M = \left(\frac{1}{2}\ \{\{1 - I,\ 1 + I\},\ \{1 + I,\ 1 - I\}\}\right)\right]$$

$$\begin{pmatrix} \frac{1}{2} - \frac{i}{2} & \frac{1}{2} + \frac{i}{2} \\ \frac{1}{2} + \frac{i}{2} & \frac{1}{2} - \frac{i}{2} \end{pmatrix}$$

```
Inverse[M] == ConjugateTranspose[M]
```

True

- A 3-by-3 real unitary matrix

```
MatrixForm[A = 1/√2 {{1, 0, 1}, {0, √2, 0}, {-1, 0, 1}}]
```

$$\begin{pmatrix} \frac{1}{\sqrt{2}} & 0 & \frac{1}{\sqrt{2}} \\ 0 & 1 & 0 \\ -\frac{1}{\sqrt{2}} & 0 & \frac{1}{\sqrt{2}} \end{pmatrix}$$

```
UnitaryMatrixQ[A]
```

True

```
A.ConjugateTranspose[A] == IdentityMatrix[3]
```

True

- A 2-by-2 complex unitary matrix

```
A = 1/√2 {{1, i}, {i, 1}};
UnitaryMatrixQ[A]
```

True

```
A.ConjugateTranspose[A] == A.Inverse[A]
```

True

Upper-triangular matrix

An n-by-n matrix $A = A_{[[i,j]]}$ is upper-triangular if $A_{[[i,j]]} = 0$ for all $i > j$. That is, if all entries below the main diagonal are 0.

Illustration

- An upper-triangular matrix

```
MatrixForm[A = {{1, 2, 3, 4}, {0, 5, 6, 7}, {0, 0, 8, 9}, {0, 0, 0, 10}}]
```

$$\begin{pmatrix} 1 & 2 & 3 & 4 \\ 0 & 5 & 6 & 7 \\ 0 & 0 & 8 & 9 \\ 0 & 0 & 0 & 10 \end{pmatrix}$$

- An upper-triangular matrix obtained by triangularization

```
A = RandomInteger[{0, 9}, {5, 5}];
```

$$A = \begin{pmatrix} 8 & 5 & 5 & 1 & 0 \\ 7 & 4 & 6 & 3 & 1 \\ 2 & 3 & 0 & 8 & 2 \\ 8 & 9 & 7 & 4 & 1 \\ 5 & 2 & 4 & 1 & 3 \end{pmatrix};$$

```
MatrixForm[B = UpperTriangularize[A]]
```

$$\begin{pmatrix} 8 & 5 & 5 & 1 & 0 \\ 0 & 4 & 6 & 3 & 1 \\ 0 & 0 & 0 & 8 & 2 \\ 0 & 0 & 0 & 4 & 1 \\ 0 & 0 & 0 & 0 & 3 \end{pmatrix}$$

Manipulation

- Exploring upper-triangular matrices

```
Manipulate[MatrixForm[UpperTriangularize[{{1, 2 a, 3}, {4, 5, 6 b}, {7, 8 c, 9}}]],
  {a, -2, 2, 1}, {b, -3, 3, 1}, {c, -5, 5, 1}]
```

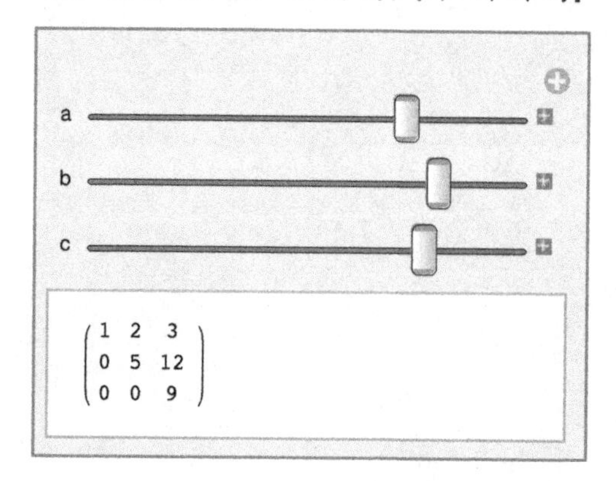

We use **Manipulate**, **MatrixForm**, and **UpperTriangularize** to construct and explore upper-triangular matrices. If we let a = 1, b = 2, and c = 3, then the **UpperTriangualize** function converts the matrix

```
MatrixForm[{{1, 2, 3}, {4, 5, 12}, {7, 24, 9}}]
```

$$\begin{pmatrix} 1 & 2 & 3 \\ 4 & 5 & 12 \\ 7 & 24 & 9 \end{pmatrix}$$

to the upper-triangular matrix

$$\begin{pmatrix} 1 & 2 & 3 \\ 0 & 5 & 12 \\ 0 & 0 & 9 \end{pmatrix}$$

V

Vandermonde matrix

A *Vandermonde matrix* arises in the solutions of linear systems, the calculation of the coefficients of Lagrange interpolating polynomials, As expectedand similar situations.

Illustration

- Vandermonde matrix of order 4 in the variables x_1, x_2, x_3 and x_4

```
MatrixForm[V = Table[{1, xi, xi^2, xi^3}, {i, 1, 4}]]
```

$$\begin{pmatrix} 1 & x_1 & x_1^2 & x_1^3 \\ 1 & x_2 & x_2^2 & x_2^3 \\ 1 & x_3 & x_3^2 & x_3^3 \\ 1 & x_4 & x_4^2 & x_4^3 \end{pmatrix}$$

In many applications, $\{x_1, x_2, x_3, x_4\} = \{1, 2, 3, 4\}$. In the case of n variables, the list of variables is often **Range**[*n*].

The following defined function produced Vandermonde matrices:

```
Vandermonde[x_] := Table[Range[x]^n, {n, 0, x - 1}]
```

- A Vandermonde matrix of order 4 for $\{x_1, x_2, x_3, x_4\} = \{1, 2, 3, 4\}$

```
MatrixForm[Vandermonde[4]]
```

$$\begin{pmatrix} 1 & 1 & 1 & 1 \\ 1 & 2 & 3 & 4 \\ 1 & 4 & 9 & 16 \\ 1 & 8 & 27 & 64 \end{pmatrix}$$

- A Vandermonde matrix of order 4 in the variables *w, x, y, z*

```
MatrixForm[Outer[Power, {w, x, y, z}, Range[0, 3]]]
```

$$\begin{pmatrix} 1 & w & w^2 & w^3 \\ 1 & x & x^2 & x^3 \\ 1 & y & y^2 & y^3 \\ 1 & z & z^2 & z^3 \end{pmatrix}$$

- The **InterpolatingPolynomial** function and its Vandermonde matrix counterpart

```
InterpolatingPolynomial[{1, 4, 9, 16, 25}, x]
```

$(x-1)(x+1)+1$

Consider the list of five points

```
points = {{1, 3}, {2, 1}, {3, 6}, {4, 2}, {5, 4}};
```

then the Lagrange polynomial is a polynomial of degree four passing through the five points. The given data determine a linear system and the coefficient matrix is an example of a Vandermonde matrix.

```
Expand[InterpolatingPolynomial[points, x]]
```

$$\frac{31x^4}{24} - \frac{187x^3}{12} + \frac{1553x^2}{24} - \frac{1277x}{12} + 59$$

We can find this polynomial by finding its list of coefficients as the solution of a linear system.

```
p[x_] := a0 + a1 x + a2 x^2 + a3 x^3 + a4 x^4
```

```
p[1] == 3
```

$$a_0 + a_1 + a_2 + a_3 + a_4 = 3$$

```
p[2] == 1
```

$$a_0 + 2a_1 + 4a_2 + 8a_3 + 16a_4 = 1$$

```
p[3] == 6
```

$$a_0 + 3a_1 + 9a_2 + 27a_3 + 81a_4 = 6$$

```
p[4] == 2
```

$$a_0 + 4a_1 + 16a_2 + 64a_3 + 256a_4 = 2$$

```
p[5] == 4
```

$$a_0 + 5a_1 + 25a_2 + 125a_3 + 625a_4 = 4$$

```
sys = {p[1] == 3, p[2] == 1, p[3] == 6, p[4] == 2, p[5] == 4};
```

```
Solve[sys, {a0, a1, a2, a3, a4}]
```

$$\left\{\left\{a_0 \to 59, a_1 \to -\frac{1277}{12}, a_2 \to \frac{1553}{24}, a_3 \to -\frac{187}{12}, a_4 \to \frac{31}{24}\right\}\right\}$$

- Solving a linear system involving a 5-by-5 Vandermonde matrix

```
MatrixForm[Vandermonde[5]]
```

$$\begin{pmatrix} 1 & 1 & 1 & 1 & 1 \\ 1 & 2 & 4 & 8 & 16 \\ 1 & 3 & 9 & 27 & 81 \\ 1 & 4 & 16 & 64 & 256 \\ 1 & 5 & 25 & 125 & 625 \end{pmatrix}$$

```
LinearSolve[Vandermonde[5], {3, 1, 6, 2, 4}]
```

$$\left\{59, -\frac{1277}{12}, \frac{1553}{24}, -\frac{187}{12}, \frac{31}{24}\right\}$$

```
fp = {3, 1, 6, 2, 4}
```

{3, 1, 6, 2, 4}

- Constructing the associated approximating polynomial and comparing it with the polynomial obtained by using the built-in **Fit** function.

$$N\left[\frac{31\,x^4}{24} - \frac{187\,x^3}{12} + \frac{1553\,x^2}{24} - \frac{1277\,x}{12} + 59\right]$$

$$59. - 106.417\,x + 64.7083\,x^2 - 15.5833\,x^3 + 1.29167\,x^4$$

$$\text{Fit}\left[\text{fp}, \left\{1, x, x^2, x^3, x^4\right\}, x\right]$$

$$59. - 106.417\,x + 64.7083\,x^2 - 15.5833\,x^3 + 1.29167\,x^4$$

Manipulation

- Exploring Vandermonde matrices

We can manipulate the defined **Vandermonde** function

```
Clear[x, n]
```

```
Vandermonde[x_] := Transpose[Table[Range[x]^n, {n, 0, x - 1}]]
```

to explore Vandermonde matrices of any dimension. For example, the following command generated all Vandermonde matrices *n*-by-*n* matrices for all *n* between 5 and 20:

```
Manipulate[MatrixForm[Vandermonde[x]], {x, 5, 20, 1}]
```

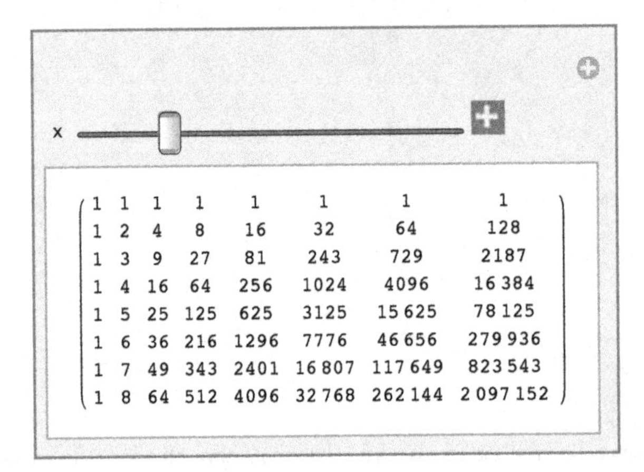

$$\begin{pmatrix} 1 & 1 & 1 & 1 & 1 & 1 & 1 & 1 \\ 1 & 2 & 4 & 8 & 16 & 32 & 64 & 128 \\ 1 & 3 & 9 & 27 & 81 & 243 & 729 & 2187 \\ 1 & 4 & 16 & 64 & 256 & 1024 & 4096 & 16\,384 \\ 1 & 5 & 25 & 125 & 625 & 3125 & 15\,625 & 78\,125 \\ 1 & 6 & 36 & 216 & 1296 & 7776 & 46\,656 & 279\,936 \\ 1 & 7 & 49 & 343 & 2401 & 16\,807 & 117\,649 & 823\,543 \\ 1 & 8 & 64 & 512 & 4096 & 32\,768 & 262\,144 & 2\,097\,152 \end{pmatrix}$$

We use **Manipulate**, **MatrixForm,** and the defined function **Vandermonde** to explore the Vandermonde matrix generated by letting x = 8.

Variance of a vector

The *population variance* of a vector $x = \{x_1, \ldots, x_n\}$ in $\mathbb{R}^n$ is $\frac{1}{n}$ times the sum of squared differences from the mean. The *sample variance* of the vector x is $\frac{1}{n-1}$ times the squared differences from the mean.

Illustration

- The population variance of a vector in $\mathbb{R}^6$

```
v = {2, 6, 3, 1, 8, 9}; mx = Mean[v];
```

$$\text{pv} = \text{Total}\left[\frac{1}{6}\text{Table}\left[(\text{x[[i]]} - \text{mx})^2, \{\text{i}, 1, 6\}\right]\right]$$

$$\frac{329}{36}$$

- The sample variance of the vector **v**

$$\text{sv} = \frac{6}{5}\text{pv}$$

$$\frac{329}{30}$$

The *Mathematica* variance function computes the sample variance of the vector.

Vector

A vector is any element of a vector space. The name *vector* is a collective name for mathematical objects studied in linear algebra. The terminology stems from physics, where vectors represent forces and are often depicted by arrows.

The direction and length of the arrow starting at point *P* and ending at point *Q* have physical interpretations. In Latin, the word *vector* means "carrier," an appropriate term for a force that carries an object from *P* to *Q*. Some of the properties of the calculus of arrows have given rise to the more general notion of *vector* in linear algebra.

The *Mathematica* vector testing function **VectorQ** applied to an expression **v** returns **True** if **v** is a list or a one-dimensional **SparseArray** object where none of the elements are lists and returns **False** otherwise. This means that objects in certain vector spaces such as polynomials and matrices are not classified as vectors by the **VectorQ** function.

Illustration

- Vector test

The *Mathematica* function **VectorQ** returns *True* if the argument **v** in **VectorQ[v]** is a list or a sparse array and false otherwise.

```
x = Range[5]
```

```
{1, 2, 3, 4, 5}
```

```
VectorQ[x]
```

```
True
```

```
x = RandomInteger[{0, 9}, {2, 2}];
```

```
VectorQ[x]
```

```
False
```

- Vectors in the coordinate space $\mathbb{R}^2$

```
space = ℝ²; vector = {1, 1.2};
```

```
Graphics[Arrow[{{0, 0}, {1, 1.2}}], Axes → True]
```

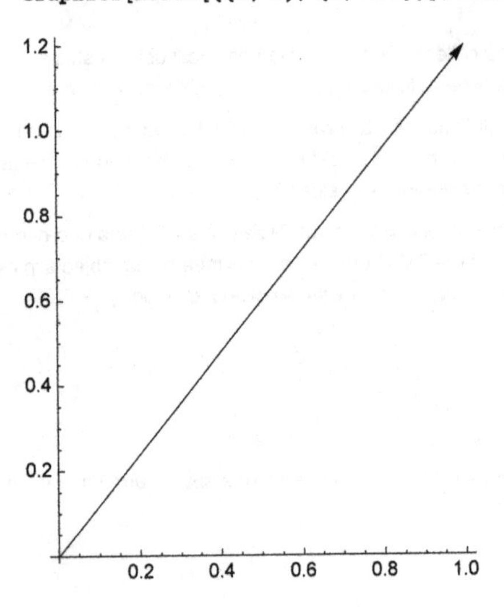

- Vectors in the coordinate space $\mathbb{R}^3$

```
space = ℝ³; vector = {1, 2, 3};
```

```
VectorQ[vector]
```

```
True
```

```
Graphics3D[Arrow[{{0, 0, 0}, {1, 2, 3}}], Axes → True]
```

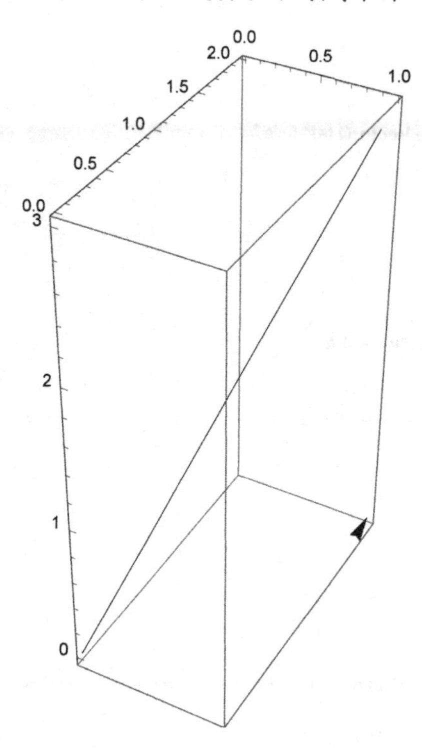

- Vectors in the polynomial space ℝ[t,3]

```
space = ℝ[t, 3] (real polynomials of degree at most 3);
```

```
vector = 2 + 4 t - 5 t^2 + t^3;
```

Mathematica does not consider polynomials to be vectors. Hence the **VectorQ** test fails.

```
VectorQ[vector]
```

```
False
```

```
vector3 = {vector /. {t → 1}, vector /. {t → 2}, vector /. {t → 3}}
```

```
{2, -2, -4}
```

However, the list of integers obtained from the polynomial by assigning numerical values to *t* qualifies again, as expected, as a vector in the Mathematica sense.

```
VectorQ[vector3]
```

```
True
```

- Vectors in the matrix space $\mathbb{R}^{2\times 3}$

```
space = ℝ^{2×3} (real 2 × 3 matrices);
```

```
vector = {{1, 2, 3}, {4, 5, 6}};
```

Mathematica does not consider matrices to be vectors. Hence the **VectorQ** test fails.

```
VectorQ[vectors]
```

```
False
```

- Test for a vector of real-valued numeric quantities

```
VectorQ[{1, Pi, Sin[1], Sqrt[2]}, NumericQ[#] && Im[#] == 0 &]
```

```
True
```

Vector addition

See Vector space

Vector component

The arrows corresponding to the vectors *{x, 0, 0}*, *{0, y, 0}*, and *{0, 0, z}* determined by the arrow corresponding to the vector **v** = *{x, y, z}* in $\mathbb{R}^3$ are often called the *components* of **v**.

Illustration

- The components of the vector {1, 1} in $\mathbb{R}^2$

```
Graphics[
 {Arrow[{{0, 0}, {1, 1}}], Arrow[{{0, 0}, {1, 0}}], Arrow[{{0, 0}, {0, 1}}]}, Axes -> True]
```

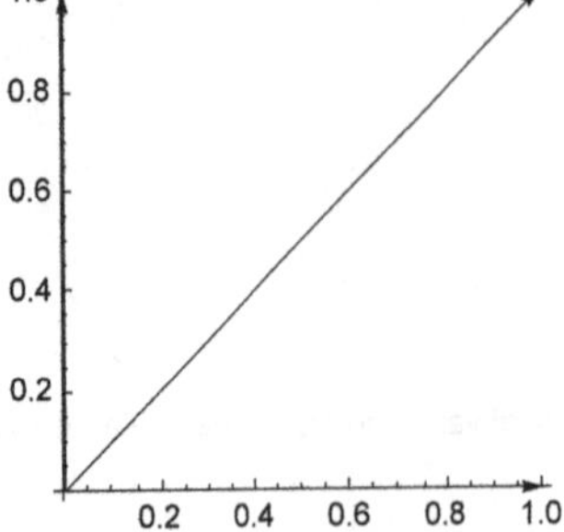

```
{1, 1} == {1, 0} + {0, 1}
```

True


```
{1, 1} == {1, 0} + {0, 1}
```

True

- The components of the vector {1, 1, 1} in $\mathbb{R}^3$

```
Graphics3D[{Arrow[{{0, 0, 0}, {1, 1, 1}}], Arrow[{{0, 0, 0}, {1, 0, 0}}],
  Arrow[{{0, 0, 0}, {0, 1, 0}}], Arrow[{{0, 0, 0}, {0, 0, 1}}]}]
```

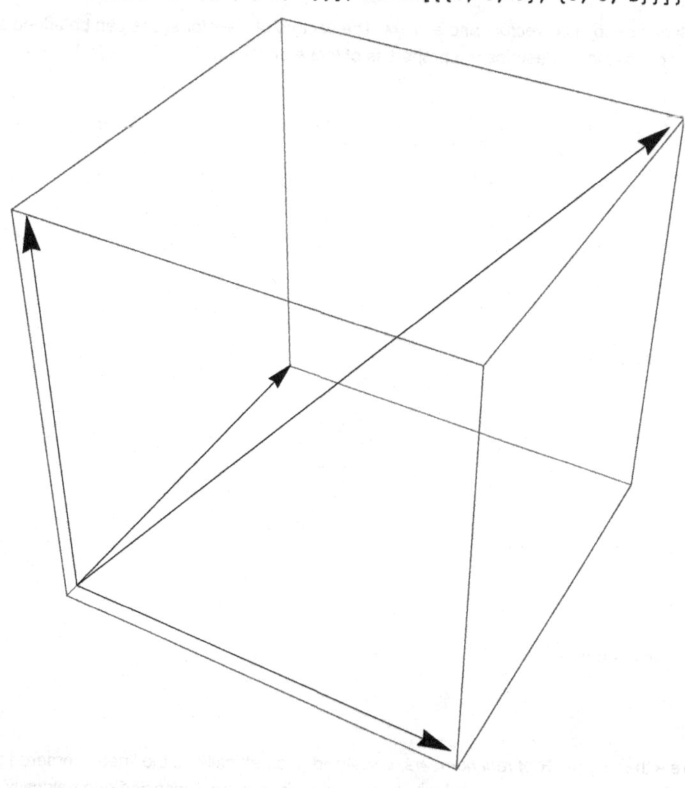

```
{1, 1, 1} == {1, 0, 0} + {0, 1, 0} + {0, 0, 1}
```

True

Vector cross product

See Cross product

Vector norm

See Norm

Vector spaces

Vector spaces are made of two types of objects, *vectors* and *scalars.* The vectors of a vector space can be added and multiplied by scalars. The vector space axioms describe the properties of these operations.

Properties of vector addition

$$(\mathbf{u} + \mathbf{v}) + \mathbf{w} = \mathbf{u} + (\mathbf{v} + \mathbf{w}) \quad (1)$$

$$\mathbf{u} + \mathbf{v} = \mathbf{v} + \mathbf{u} \quad (2)$$

$$\mathbf{u} + \mathbf{0} = \mathbf{u} \quad (3)$$

$$\mathbf{u} + (-\mathbf{u}) = \mathbf{0} \quad (4)$$

for all vectors u, v, w and the zero vector 0

Properties of scalar multiplication

$$(a + b)\,\mathbf{u} = a\,\mathbf{u} + b\,\mathbf{u} \quad (1)$$

$$a\,(\mathbf{u} + \mathbf{v}) = a\,\mathbf{u} + a\,\mathbf{v} \quad (2)$$

$$1\,\mathbf{u} = \mathbf{u} \quad (3)$$

$$a\,(b\,\mathbf{u}) = (a\,b)\,\mathbf{u} \quad (4)$$

for all scalars *a, b,* and 1 and all vectors **u** and **v**.

Scalars of a vector space

The *scalars* of vector spaces are either the set ℝ of *real numbers*, visualized geometrically as the linearly ordered points on a line, or the set of *complex numbers* ℂ, visualized geometrically as the points in a plane, organized geometrically on concentric circles centered at the point *{0, 0}*.

Vector spaces over real numbers are known as *real vector spaces* and vector spaces with complex scalars are known as *complex vector spaces.* All real vector spaces have all real numbers as scalars and all complex vector spaces have all complex numbers as scalars. The difference between the spaces arises from the nature of their vectors and from the algebraic differences of the real and complex numbers.

Real scalars

```
realscalars = Table[{n, 0}, {n, -20, 20}];
```

```
ListPlot[realscalars]
```

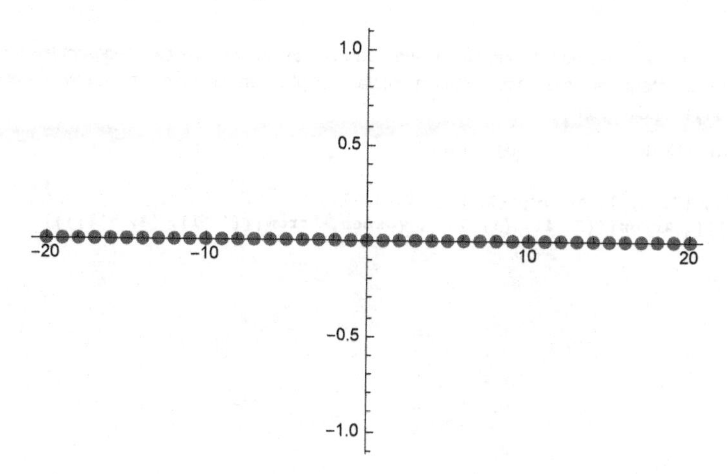

Complex scalars

```
complexscalars =
  Table[{{Cos[n], Sin[n]}, {1.5 Cos[n], 1.5 Sin[n]}, {2 Cos[n], 2 Sin[n]}}, {n, -20, 20}];
```

```
ListPlot[complexscalars, AspectRatio → 1]
```

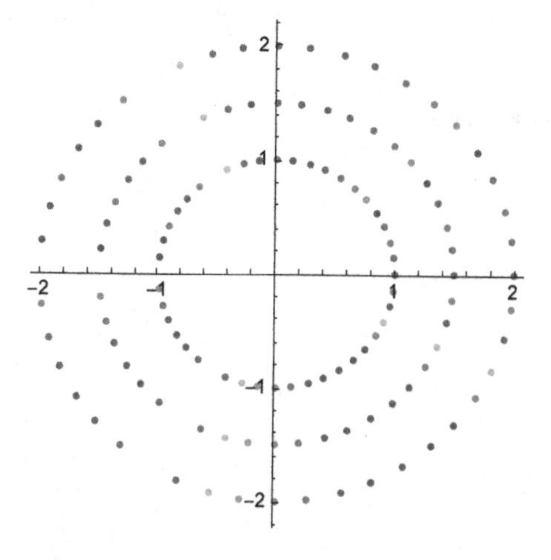

Vectors of a vector space

The *vectors* of a vector space come in all shapes and kinds. They can be numbers, lists of numbers, matrices, functions, polynomials, solutions of certain types of equations, and so on. However, in order to count as vectors, it must be possible to add the objects involved in a consistent way and it must also be possible to multiply them by scalars subject to certain compatibility conditions.

The parallelogram law

In *Mathematica*, vectors are often represented as lists and arrays and visualized as arrows. The *parallelogram law* for arrows can be used to give a visual interpretation of vector addition. Scalar multiplication can then depicted by stretching or shrinking arrows and by inverting their directions.

- Using **Graphics** and **Arrow** to illustrate the parallelogram law

```
Graphics[{Arrow[{{1, 1}, {3, 2}}], Arrow[{{1, 1}, {2, 4}}],
  Arrow[{{3, 2}, {4, 5}}], Arrow[{{2, 4}, {4, 5}}], {Green, Arrow[{{1, 1}, {4, 5}}]}}]
```

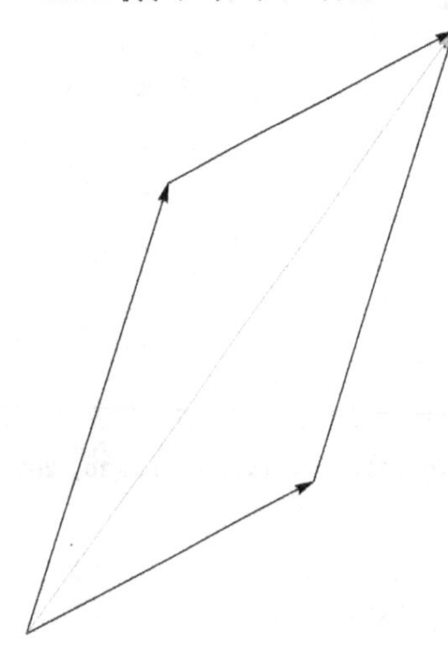

- Using the **ClassroomUtilities** package to illustrate parallelogram law

```
Needs["ClassroomUtilities`"]
```

```
With[{a = {0, 1}, b = {2, 1}},
 VectorDiagram[{a, b, a + b}, {{a, a + b}, {b, a + b}}]]
```

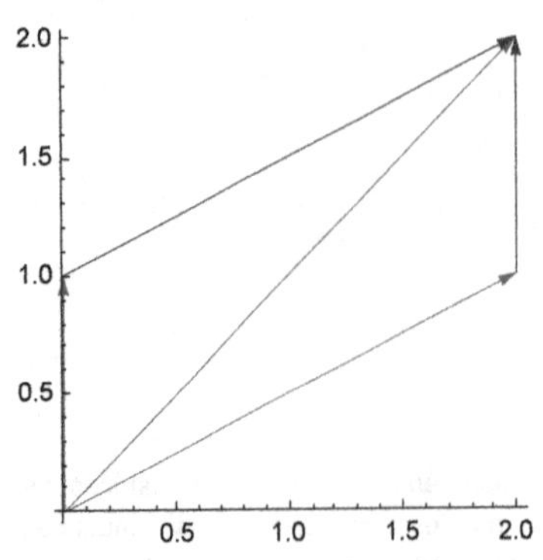

Coordinate spaces

The sets $\mathbb{R}^n$ of columns of real numbers of height n form vector spaces for all n = 1,2,3,... We call these spaces *real coordinate spaces*. The numbers *n* are called the *dimensions* of the spaces.

- The real coordinate space $V = \mathbb{R}^1 = \mathbb{R}$ of dimension 1
  - The vectors of *V* are the real numbers.
  - The scalars of *V* are the real numbers.
  - Vector addition and scalar multiplication are the addition and multiplication of real numbers.
  - The set {1} is the standard basis of V.
- The real coordinate space V = $\mathbb{R}^2$ of dimension 2
  - The vectors of V are the pairs {x, y} of real numbers x and y.
  - The scalars of V are the real numbers.
  - Vector addition is the addition of pairs of real numbers.
    - $\{u_1, u_2\} + \{v_1, v_2\} = \{u_1 + v_1, u_2 + v_2\}$
  - Scalar multiplication is the componentwise multiplication of real numbers.
    - s {u, v} = {s u, s v}
  - The set {{1, 0}, {0, 1}} is the standard basis of V.
- The real coordinate space V = $\mathbb{R}^3$ of dimension 3
  - The vectors of V are the triples {x, y, z} of real numbers x, y, z.
  - The scalars of V are the real numbers.
  - Vector addition is the addition of triples of real numbers.
    - $\{u_1, u_2, u_3\} + \{v_1, v_2, v_3\} = \{u_1 + v_1, u_2 + v_2, u_3 + v_3\}$
  - Scalar multiplication is the componentwise multiplication of real numbers.
    - s {u, v, w} = {s u, s v, s w}
  - The set { {1, 0, 0}, {0, 1, 0}, {0, 0, 1} } is the standard basis of V.
- The complex coordinate space $V = \mathbb{C}^1 = \mathbb{C}$ of dimension 1
  - The vectors of *V* are the complex numbers.
  - The scalars of *V* are the complex numbers.
  - Vector addition and scalar multiplication are the addition and multiplication of complex numbers.
  - The set {1} is the standard basis of V.

Polynomial spaces

Polynomials over the real or complex numbers can be added and modified by multiplying them by real or complex numbers. The operations satisfy the axioms of a vector space. Limits on the degrees of the polynomials can be imposed to obtain finite-dimensional spaces.

- The set ℝ[t] of real polynomials in the variable *t* forms a vector space. We call this space an *infinite-dimensional polynomial space* since it has no finite basis.
  - The vectors of V = ℝ[t] are real polynomials.
  - The scalars of V are the real numbers.
  - Vector addition is the addition of real polynomials.
    - p = Dot[{1, 2, 0, 3}, {1, t, t^2, t^3}]
    - $1 + 2t + 3t^3$
    - q = Dot[{0, 0, 1, 1, 0, 1}, {1, t, t^2, t^3, t^4, t^5}]
    - $t^2 + t^3 + t^5$
    - p + q
    - $1 + 2t + t^2 + 4t^3 + t^5$
  - Scalar multiplication is the multiplication of real polynomials by real numbers.
    - Expand[s (1 + 2 t + t^2 + 4 t^3 + t^5)] = $s + 2st + st^2 + 4st^3 + st^5$
  - The standard basis of V is the infinite set of all powers of the variable t.
- The polynomial space of real polynomials of degree ≤ 3
  - The vectors of V are the polynomials $a_0 + a_1 t + a_2 t^2 + a_3 t^3$ of dimension less than or equal to 3 over the real numbers.
  - The scalars of V are the real numbers.
  - Vector addition is the addition of real polynomials.
    - $(a_0 + a_1 t + a_2 t^2 + a_3 t^3) + (b_0 + b_1 t + b_2 t^2 + b_3 t^3) = (a_0 + b_0) + (a_1 + b_1) t + (a_2 + b_2) t^2 + (a_3 + b_3) t^3$
  - Scalar multiplication is the multiplication of polynomials by real numbers.
    - $s(a_0 + a_1 t + a_2 t^2 + a_3 t^3) = (s a_0) + (s a_1) t + (s a_2) t^2 + (s a_3) t^3$
  - The set $\{1, t, t^2, t^3\}$ of powers of t is the standard basis of V.

Matrix spaces

Matrices over the real or complex number can be added and modified by multiplying them by real or complex numbers. The operations require that the matrices to be added or multiplied by scalars have the same dimension. All matrix spaces are finite-dimensional.

- The real matrix space $V = \mathbb{R}^{2\times 3}$ of dimension 6
  - The vectors of V are 2-by-3 matrices with real elements.
  - The scalars of V are the real numbers.
  - Vector addition is the addition of 2-by-3 matrices.
    - $\begin{pmatrix} 4 & 1 & 9 \\ 9 & 0 & 7 \end{pmatrix} + \begin{pmatrix} 1 & 7 & 2 \\ 8 & 6 & 8 \end{pmatrix} = \begin{pmatrix} 5 & 8 & 11 \\ 17 & 6 & 15 \end{pmatrix}$
  - Scalar multiplication is the componentwise multiplication of the matrix elements by fixed scalars.
    - $s \begin{pmatrix} 4 & 1 & 9 \\ 9 & 0 & 7 \end{pmatrix} = \begin{pmatrix} 4\,s & s & 9\,s \\ 9\,s & 0 & 7\,s \end{pmatrix}$
  - The set of the following six 2-by-3 matrices is a basis for V:
    - $\left\{\begin{pmatrix} 1 & 0 & 0 \\ 0 & 0 & 0 \end{pmatrix}, \begin{pmatrix} 0 & 1 & 0 \\ 0 & 0 & 0 \end{pmatrix}, \begin{pmatrix} 0 & 0 & 1 \\ 0 & 0 & 0 \end{pmatrix}, \begin{pmatrix} 0 & 0 & 0 \\ 1 & 0 & 0 \end{pmatrix}, \begin{pmatrix} 0 & 0 & 0 \\ 0 & 1 & 0 \end{pmatrix}, \begin{pmatrix} 0 & 0 & 0 \\ 0 & 0 & 1 \end{pmatrix}\right\}$

A nonstandard four-dimensional real vector space

A vector space whose vectors form a subset of another vector space may fail to be a subspace of the given space because its vector space operations may fail to agree with the operations on the ambient space. Here is an example. It comes from *affine geometry*. The ideas involved in this example arise in the study of affine transformations.

- The vectors of the space *V* are real 3-by-3 matrices of the form

$$\begin{pmatrix} a & b & 0 \\ c & d & 0 \\ 0 & 0 & 1 \end{pmatrix}$$

- The scalars of the space *V* are real numbers.
- Vector addition is defined by

$$\begin{pmatrix} a & b & 0 \\ c & d & 0 \\ 0 & 0 & 1 \end{pmatrix} + \begin{pmatrix} e & f & 0 \\ g & h & 0 \\ 0 & 0 & 1 \end{pmatrix} = \begin{pmatrix} a+e & b+f & 0 \\ c+g & d+h & 0 \\ 0 & 0 & 1 \end{pmatrix}$$

- Scalar addition is defined by

$$s \begin{pmatrix} a & b & 0 \\ c & d & 0 \\ 0 & 0 & 1 \end{pmatrix} = \begin{pmatrix} s\,a & sb & 0 \\ s\,c & s\,d & 0 \\ 0 & 0 & 1 \end{pmatrix}$$

The defined operations satisfy the axioms of a vector space. However, they are not the operations in $\mathbb{R}^3$.

- The set of the following four matrices is a basis of *V*:

$$\left\{\begin{pmatrix} 1 & 0 & 0 \\ 0 & 0 & 0 \\ 0 & 0 & 1 \end{pmatrix}, \begin{pmatrix} 0 & 1 & 0 \\ 0 & 0 & 0 \\ 0 & 0 & 1 \end{pmatrix}, \begin{pmatrix} 0 & 0 & 0 \\ 1 & 0 & 0 \\ 0 & 0 & 1 \end{pmatrix}, \begin{pmatrix} 0 & 0 & 0 \\ 0 & 1 & 0 \\ 0 & 0 & 1 \end{pmatrix}\right\}$$

Arrows representing vectors

```
Needs["ClassroomUtilities`"]
```

```
VectorDiagram3D[{{2, 0, 0}, {0, 2, 0}, {1, 0, 1}},
 {}, VectorStyles → {Directive[Red], Automatic}]
```

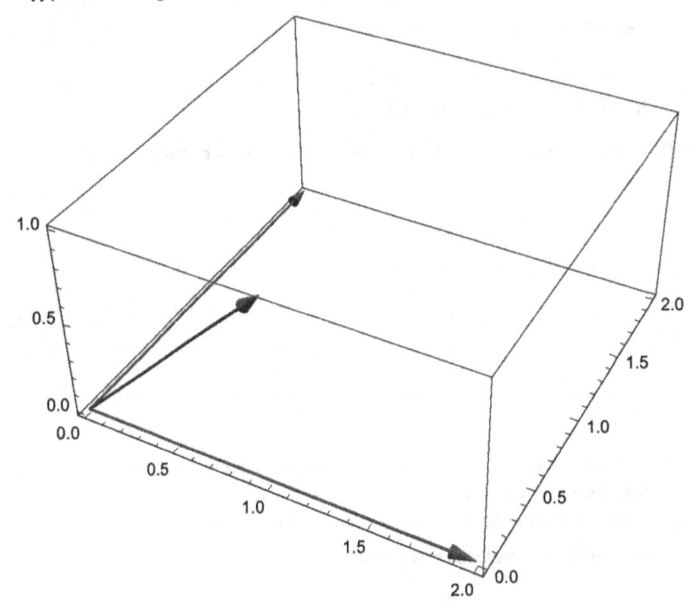

Vector triple product

The *vector triple product* of three vectors **u**, **v**, **w** in $\mathbb{R}^3$ is the vector **Cross[Cross[u, v], w]** .

Illustration

- The vector triple product of three vectors in $\mathbb{R}^3$

```
u = {1, 2, 3}; v = {4, 5, 6}; w = {7, 8, 9};
```

```
VectorTripleProduct[u_, v_, w_] := Cross[Cross[u, v], w]
```

```
VectorTripleProduct[u, v, w]
```

```
{78, 6, -66}
```

- A vector triple product identity

```
Cross[Cross[u, v], w] == -Cross[w, Cross[u, v]]
```

```
True
```

Manipulation

- Vector triple products

```
Manipulate[Cross[Cross[{1, 2, a}, {4, b, 6}], {7, 8, c}],
 {a, 0, 5, 1}, {b, 0, 5, 1}, {c, 0, 5, 1}]
```

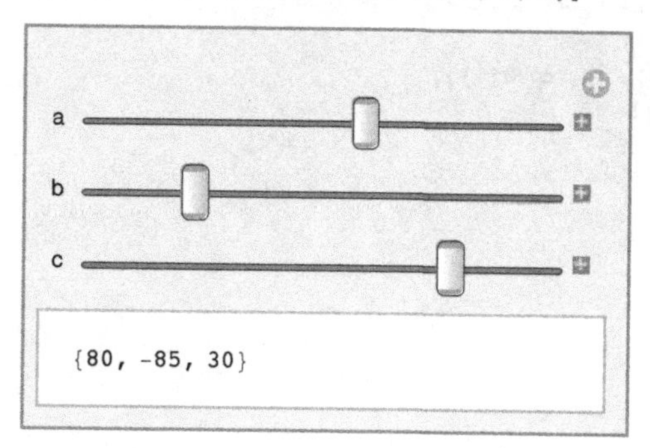

We use **Manipulate** and **Cross** to explore vector triple products. If a = 3, b = 1, and c = 4, for example, the manipulation shows that the vector triple product of the vectors {1, 2, 3}, {4, 1, 6}, and {7, 8, 4} is the vector {80, -85, 30}.

Volume of a parallelepiped

A parallelepiped in $\mathbb{R}^3$ is a prism whose faces are all parallelograms. If **u** = {*a, b, c*}, **v** = {*d, e, f*}, and **w** = {*g, h, i*} are three vectors defining the parallelepiped, then its *volume* is the absolute value of the determinant of the 3-by-3 matrix {**u**, **v**, **w**}.

Illustration

- The volume of a parallelepiped calculated using determinants

```
u = {1, 2, -3}; v = {-3, 4, 5}; w = {-2, 1, 8};
```

```
volumne = Abs[Det[{u, v, w}]]
```

```
40
```

The volume is also the absolute value of the *scalar triple product*.

- The volume of a parallelepiped calculated using scalar triple products

```
u = {1, 2, 3}; v = {4, 5, 6}; w = {-2, 1, 8};
```

```
Abs[Dot[u, Cross[v, w]]]
```

12

```
Abs[ScalarTripleProduct[u, v, w]]
```

12

Manipulation

- The volume of a parallelepiped

```
Manipulate[Abs[Det[{{1, 2, 3 a}, {3 b, 4, 5}, {2, c, 8}}]],
 {a, -2, 2, 1}, {b, -2, 2, 1}, {c, -2, 2, 1}]
```

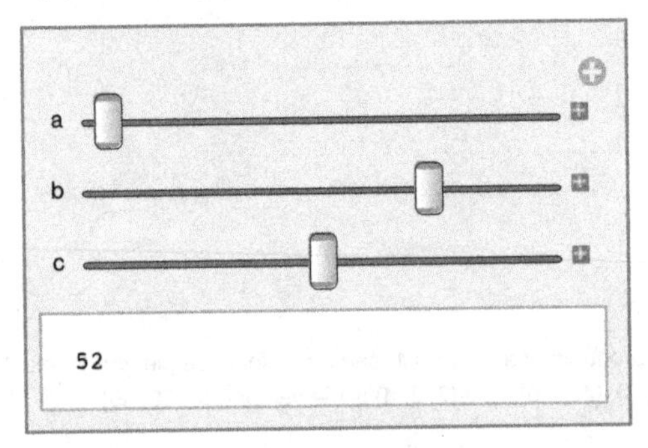

We use **Manipulate**, **Abs**, and **Det** to explore the volume of parallelepipeds. If we let a = - 2, b = 1, and c = 0, for example, the volume of the parallelepiped determined by the vectors {1, 2, -6}, {3, 4, 5}, and {2, 0, 8} is 52 cubic units.

W

Well-conditioned matrix

A square matrix is *well-conditioned* if its condition number is only slightly above 1. The assessment of whether a matrix is or is not well-conditioned is context-dependent.

Illustration

- A well-conditioned 2-by-2 matrix

```
MatrixForm[A = {{1, 0}, {0, 1.1}}]
```

$$\begin{pmatrix} 1 & 0 \\ 0 & 1.1 \end{pmatrix}$$

```
s = SingularValueList[A]
```

```
{1.1, 1.}
```

```
conditionnumberA = s[[1]] / s[[2]]
```

```
1.1
```

- A well-conditioned 3-by-3 matrix

```
MatrixForm[A = DiagonalMatrix[{1, 1.01, 1}]]
```

$$\begin{pmatrix} 1. & 0. & 0. \\ 0. & 1.01 & 0. \\ 0. & 0. & 1. \end{pmatrix}$$

```
s = SingularValueList[A]
```

```
{1.01, 1., 1.}
```

```
conditionnumber = 1.01 / 1.
```

```
1.01
```

Wronskian

Wronskians are arrays of derivatives of differentiable functions in determinant notation. They are used to study differential equations and, for example, to show that a set of solutions is linearly independent.

In *Mathematica*, Wronskians can be computed easily by using the built-in **Wronskian** function.

Illustration

- A Wronskian of two functions

```
Wronskian[{Exp[x], Exp[2 x]}, x]
```

e^{3x}

We can express the Wronskian as the determinant of the functions and their first derivatives:

```
MatrixForm[A = {{Exp[x], Exp[2 x]}, {D[Exp[x], x], D[Exp[2 x], x]}}]
```

$$\begin{pmatrix} e^x & e^{2x} \\ e^x & 2e^{2x} \end{pmatrix}$$

```
Det[A]
```

e^{3x}

- A Wronskian of three functions

```
Expand[Wronskian[{x^5, Exp[x], Exp[2 x]}, x]]
```

$20 e^{3x} x^3 - 15 e^{3x} x^4 + 2 e^{3x} x^5$

Again, we can obtain the Wronskian as the determinant of the functions and their first and second derivatives.

```
MatrixForm[A = {{x^5, Exp[x], Exp[2 x]}, {D[x^5, x], D[Exp[x], x], D[Exp[2 x], x]},
   {D[x^5, {x, 2}], D[Exp[x], {x, 2}], D[Exp[2 x], {x, 2}]}}]
```

$$\begin{pmatrix} x^5 & e^x & e^{2x} \\ 5x^4 & e^x & 2e^{2x} \\ 20x^3 & e^x & 4e^{2x} \end{pmatrix}$$

```
Det[A]
```

$20 e^{3x} x^3 - 15 e^{3x} x^4 + 2 e^{3x} x^5$

If $x = 1$, for example, then the Wronskian is not equal to zero:

```
Det[A] /. {x → 1}
```

$7 e^3$

Therefore the functions x^5, e^x, and e^{2x} are linearly independent for $x = 1$.

Manipulation

- Exploring Wronskian determinants

```
Manipulate[Expand[Wronskian[{x^n, Exp[m x]}, x]], {n, 1, 5, 1}, {m, 1, 5, 1}]
```

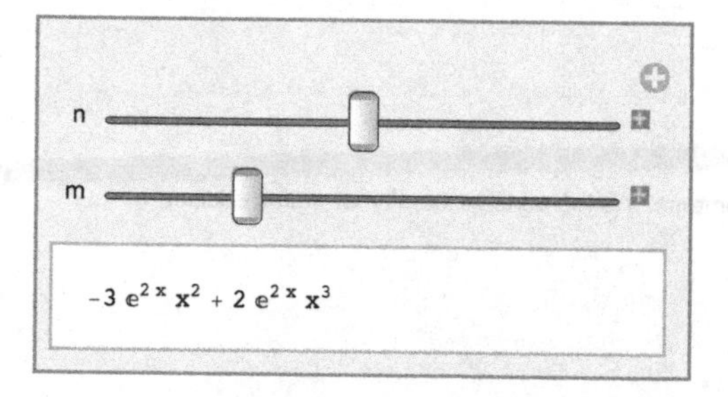

We use **Manipulate**, **Expand**, and **Wronskian** to explore the Wronskian determinant of two differentiable functions. If we let n = 3 and m = 2, for example, the manipulation produces the Wronskian determinant

$$-3\, e^{2x} x^2 + 2\, e^{2x} x^3$$

of x^3 and e^{2x}.

Z

Zero matrix

A *zero matrix* is a matrix made up entirely of zero elements. It is the additive identity for matrix addition.

Illustration

- A zero matrix as an additive identity

```
MatrixForm[Z = {{0, 0, 0}, {0, 0, 0}, {0, 0, 0}}]
```

$$\begin{pmatrix} 0 & 0 & 0 \\ 0 & 0 & 0 \\ 0 & 0 & 0 \end{pmatrix}$$

```
MatrixForm[A = RandomInteger[{0, 9}, {3, 3}]]
```

$$\begin{pmatrix} 3 & 3 & 6 \\ 0 & 7 & 4 \\ 1 & 7 & 5 \end{pmatrix}$$

```
A + Z == A
```

```
True
```

- A 2-by-5 zero matrix

```
MatrixForm[ConstantArray[0, {2, 5}]]
```

$$\begin{pmatrix} 0 & 0 & 0 & 0 & 0 \\ 0 & 0 & 0 & 0 & 0 \end{pmatrix}$$

- A 2-by-2 zero matrix

```
MatrixForm[Array[0 &, {2, 2}]]
```

$$\begin{pmatrix} 0 & 0 \\ 0 & 0 \end{pmatrix}$$

- A 3-by-4 zero matrix

```
MatrixForm[Normal[SparseArray[{i_, j_} → 0, {3, 4}]]]
```

$$\begin{pmatrix} 0 & 0 & 0 & 0 \\ 0 & 0 & 0 & 0 \\ 0 & 0 & 0 & 0 \end{pmatrix}$$

- Converting a nonzero matrix to a zero matrix

```
A = RandomInteger[{1, 5}, {4, 5}];

MatrixForm[A = {{2, 4, 4, 3, 3}, {2, 4, 5, 2, 5}, {3, 2, 1, 5, 3}, {3, 5, 2, 1, 4}}]
```

$$\begin{pmatrix} 2 & 4 & 4 & 3 & 3 \\ 2 & 4 & 5 & 2 & 5 \\ 3 & 2 & 1 & 5 & 3 \\ 3 & 5 & 2 & 1 & 4 \end{pmatrix}$$

```
S = SparseArray[{}, {4, 5}];

MatrixForm[A S]
```

$$\begin{pmatrix} 0 & 0 & 0 & 0 & 0 \\ 0 & 0 & 0 & 0 & 0 \\ 0 & 0 & 0 & 0 & 0 \\ 0 & 0 & 0 & 0 & 0 \end{pmatrix}$$

- Creating a 2-by-4 zero matrix using scalar multiplication

```
A = RandomInteger[{0, 9}, {2, 4}];
```

$$A = \begin{pmatrix} 0 & 7 & 6 & 9 \\ 2 & 8 & 5 & 0 \end{pmatrix};$$

```
MatrixForm[Z = 0 A]
```

$$\begin{pmatrix} 0 & 0 & 0 & 0 \\ 0 & 0 & 0 & 0 \end{pmatrix}$$

Zero space

A *zero space* is a vector space whose only vector is a zero vector. All vector spaces have a zero-dimensional subspace whose only vector is the zero vector of the space. It is convenient to consider the empty set { } to be the basis of the zero subspace. All subspaces of a given vector space have the zero vector in common. If this is the only common vector, the subspaces are said to be *disjoint*.

Illustration

- The zero subspace of $\mathbb{R}$ is the space $Z_1 = \{0\}$.
- The zero subspace of $\mathbb{R}^2$ is the space $Z_2 = \{\{0, 0\}\}$.
- The zero subspace of $\mathbb{R}^3$ is the space $Z_3 = \{\{0, 0, 0\}\}$.
- The zero subspace of $\mathbb{R}^{2\times3}$ is the space $Z_{2\times3} = \left\{\begin{pmatrix} 0 & 0 & 0 \\ 0 & 0 & 0 \end{pmatrix}\right\}$.
- The zero subspace of $\mathbb{R}[t]$ is the space $Z_0 = \{\mathbf{0}\}$, where $\mathbf{0}$ is the zero polynomial.

Zero vector

The zero vector of a vector space *V* is the vector **0** with the property that **v** + **0** = **v** for all vectors **v** in *V*.

Illustration

- The zero vector of $\mathbb{R}^5$

```
zero = {0, 0, 0, 0, 0};
```

```
{a, b, c, d, e} + zero == {a, b, c, d, e}
```

```
True
```

- The zero vector in the polynomial space $\mathbb{R}$[t,3]

```
zero = 0 + 0 t + 0 t^2 + 0 t^3;
```

```
a + b t + c t^2 + d t^3 + zero == a + b t + c t^2 + d t^3
```

```
True
```

- The zero vector in the matrix space $\mathbb{R}^{2 \times 3}$

```
MatrixForm[zero = {{0, 0, 0}, {0, 0, 0}}]
```

$$\begin{pmatrix} 0 & 0 & 0 \\ 0 & 0 & 0 \end{pmatrix}$$

$$\begin{pmatrix} a & b & c \\ d & e & f \end{pmatrix} + \text{zero} == \begin{pmatrix} a & b & c \\ d & e & f \end{pmatrix}$$

```
True
```

Index

H

I

J

K

L

Printed in the United States
By Bookmasters